MACHINES FOR MAKING GODS

Machines for Making Gods

MORMONISM, TRANSHUMANISM, AND WORLDS WITHOUT END

Jon Bialecki

FORDHAM UNIVERSITY PRESS   NEW YORK 2022

Fordham University Press has no responsibility for the persistence or
accuracy of URLs for external or third-party Internet websites referred
to in this publication and does not guarantee that any content on such
websites is, or will remain, accurate or appropriate.

Fordham University Press also publishes its books in a variety of electronic
formats. Some content that appears in print may not be available in
electronic books.

Visit us online at www.fordhampress.com.

Library of Congress Cataloging-in-Publication Data available online
at https://catalog.loc.gov.

Printed in the United States of America
24 23 22   5 4 3 2 1

First edition

For Judy: the most human person I know,
and yet already far more than merely human

Contents

Preface

§i

This is a book about religion and science, secularism and belief, about founding histories and imagined futures. It is about two movements that in important and often unsuspected ways rhyme with one another: The Church of Jesus Christ of Latter-day Saints and secular transhumanism. It is about a specific set of people who have taken hold of these two movements and brought them together as a way to reimagine their faith, to bring an ethics and an eschatology to emerging and anticipated technologies. It is about the play of universalistic ambitions and clannish guardedness. But most of all, this is a book about speculation, and the role speculation plays in religiosity.

§ii

This book harkens back to one of the earliest milestones in my intellectual life. My first introduction to something like the idea of religious transhumanism occurred at the end of second grade. My parents had been worried that I was underperforming in school, and they were advised that I needed more challenging reading material. My mother, knowing about my fondness for science fiction (there was a lot of running around in the backyard holding plastic phasers and wearing cardboard Spock ears), decided to take me to the public library and check out, on my behalf, a few works in that genre. She picked out two books based on their titles alone, titles associated with known science fiction movie properties. These would be things that I had not seen so that they would be fresh to me, but because she was unlikely to watch this material on her own, and certainly hadn't seen these particular properties, she couldn't know their specific contents either. And so, after checking them out, she handed me those tomes, telling me that she and my father thought it was time that I start reading the kind of books adults read. The two exemplars of adult knowledge that served as my initiation were *2001: A Space Odyssey,* and *Chariots of the Gods.* (The latter book technically being 'non-fiction,' but still somehow cataloged and shelved with the science fiction titles, which was perhaps a bit of editorial comment from some librarian as to the plausibility of the book's claims). They were both difficult reads, but there was one shared element in those books that I could glean: ignorant of the details of these two books' contents, my mother had picked two books that each claim in their own ways that God is actually a space alien.

This disturbing revelation no doubt played a part in my adult decision to become an anthropologist of religion.

But this childhood literary incident also left another mark: a visceral conviction, based upon nothing more than on a single idiosyncratic experience, that there was an irreconcilable, natural enmity between scientific speculation and religious faith. Despite my shaky evidential basis for thinking this, I was not alone in this position. Whether or not it was always the case historically—and putting to the side the fact that even the most conservative religious believers in America still acknowledge that science has a special capacity to make certain kinds of factual claims—there is at least the common perception that religion and science stand in tension with one another, constituting differing and irreconcilable ways of understanding the world.[1] Full-throated religious advocacy for science, and especially for the sorts of science that either promise or threaten to play God, seems difficult to imagine. Perhaps some liberal Mainline Protestant denomination, or some sect-like Quakers or Unitarian-Universalists, see themselves as pro-science. And the contemporary Catholic Church also sees no tension between religion and science, even having a Vatican Observatory. But there is no such science-affirming gesture from the sort of politically conservative and theologically traditional forms of Christianity that have dominated American conversations about the authority and social role of religion during the late twentieth and early twenty-first centuries.

Though, as with all totalizing claims, there are often surprising exceptions.

§iii

I first heard of the Mormon Transhumanist Association in 2012, eleven years after the date set by Clark and Kubrick for humanity's first manned expedition to Jupiter. While teaching at the University of California, San Diego (UCSD),

I found myself writing a literature review article on the topic of the anthropology of religion. I thought that, as part of that article, it would be clever to juxtapose two ongoing anthropological conversations. One was a discussion of the Neolithic settlement of Çatalhöyük, which revolved around whether this site, filled with evidence of sacrifice, ritual burial, and Venus-like figurines, should be understood as religious.[2] (Those familiar with nominalist trends in contemporary anthropology will not be surprised to hear that the consensus answer was *no*.) The other was *the singularity*, a moment in the near future when some computer scientists and speculative thinkers believed that (as I put it at the time) "exponential increases in technology would lead to human-crafted artificial intelligences that would surpass their makers in speed and capacities."[3] While some futurists feared this, others saw it as a moment in which (to quote the poet Richard Brautigan) "machines of loving grace" would use their superhuman computational powers to end human want, solving all our problems, including the problem of death.[4] Rather than being human, we would transform through technology or merge with machines, to become transhuman. Sometimes this moment was also disparagingly referred to as "the Rapture of the Nerds," a phrase that likened this sudden transformation of the relation between humanity and advancing technology to the eschatological hope many conservative Protestants have of being suddenly whisked up into heaven on the eve of the apocalypse.[5] The purpose of my juxtaposition was to make the arch comment that while literally Stone Age ritual practices were read as a kind of secularism before the fact, potential near-term advances in computing capacities were not seen as secular, but as religion thriving in somewhat slipshod and unconvincing disguise.

I mentioned this still in-progress essay during a hallway encounter with a department colleague who happened to have some family ties to the Mormons (or the Church of Jesus Christ of Latter-day Saints). We chitchatted about the piece for a while, and then he said, offhandedly, "you know, there are Mormon Transhumanists." At that time, I had little interest in, and no knowledge of, Mormonism, even though I was an anthropologist of religion who wrote on American Christianity. I had always assumed Mormonism was a monoculture, incapable of producing surprise (a notion of which I have since been quite disabused). So, I was incredulous to hear of such a group of religious transhumanists—and of *Mormon* religious transhumanists at that. My colleague walked me to my office, flipped open my laptop for me, and after a quick search, showed me something that would end up profoundly impacting my life. On my computer screen was the online edition of an influential Mormon periodical, *Sunstone*, a magazine equivalent to other respectable, somewhat literary religious magazines, such as *Tikkun* or *Sojourners* or *First Things*. Going through the pertinent back issue, a few pages after an essay by Harold Bloom, there was a rather lengthy article about Mormon Transhumanism. The article was not a report or a critique from some outside commentator on the subject but rather a manifesto collectively credited to something called the "Mormon Transhumanist Association." It put forth an argument for a parallelism between Mormon doctrine and near-future, humanity-transcending technology; it was illustrated primarily with reproductions of nineteenth-century engravings, images of winged chariot wheels and antiquated, unidentifiable mechanical devices.[6] The details of the argument were beyond me, and from my vantage point, the essay quickly went into the weeds, especially when it came to Mormonism (though the transhumanism was a hard follow as well). But despite my illiteracy in these topics, I was fascinated.

The fascination stuck. I began to follow these Mormon Transhumanists. First, I looked through their information-packed websites. Then I started looking at the Association's official Facebook page. A cluster of Twitter-follows on my part soon followed, as did adding myself to an email list. And I found myself musing about them, not as an ethnographic project, but more as a puzzle. (I was not the only one to do so; the Mormon Transhumanist Association has received journalistic attention from multiple venues, including *The New Yorker, Slate, The Washington Times,* and *Ozy*).[7] How could there be such a thing, and how would it work? After relocating to the University of Edinburgh, I found myself thinking of the MTA even more. And then I broke. Via Twitter, one day, I asked if I could attend their upcoming annual conference; I was welcomed with virtual open arms. And so, when flying out from Edinburgh to visit my wife and daughter (who were still living in San Diego), I set up a convoluted route that allowed for a very short stopover in Salt Lake City. By the end of the day there, I was determined that this would be my next ethnographic project. Within a few weeks, I asked the board of the Mormon Transhumanist Association if I could begin ethnographic work with them, and I received their blessing.

§iv

How to carry out this investigation, though, was at that point still a puzzle. Where 'was' the Mormon Transhumanist Association? And who constituted it? With some initial help from the association, I would go on to interview members, from some of the group's founders to some of its most marginally committed participants. (I had one person share with me that he was not into all that "*Star Trek* stuff" that came up in the MTA, which surprised me, because from my vantage point at the time it was nothing but "*Star Trek* stuff.")

I became an even more devout follower of their media production, and I engaged with them where they interacted most—the internet. I attended their annual in-person conferences, and when I could, their monthly Provo, Utah, meetups. I also threw myself into the larger online media ecology that the Mormon Transhumanist Association was a part of. More social media, many more web pages, and podcasts—endless podcasts. (For reasons having to do with founder effects, but also with Mormon facility with the casual-yet-confessional speech genre that makes up much of the sacrament meetings in the Church of Jesus Christ of Latter-day Saints, Mormons excel at making podcasts, with *excelling* referring not just to quality but also number. At one point I found myself listening to over three to four hours of Mormon-oriented podcasts a day, and still could not keep up with the continuing flow.) I attended local sacrament meetings as well, though here my presence and the depth of my engagement was limited by the fact that I never received approval for extended ward-focused fieldwork from any of the local Church authorities. I could not, as I would have liked, continuously be in Utah over the several years of research I devoted to Mormon Transhumanism. Even after being granted research leave from the University of Edinburgh, and later changing my status with the university to allow myself to live permanently in the US, I still had obligations to my long-suffering wife and daughter, who had endured years of a trans-Atlantic commute. But I visited Utah as often as I could and stayed for as long as I could, and during each trip, people were generous with their time and their thoughts. I even decided, as a participant-observer, to speak at the MTA's annual conference instead of merely attend the gathering. My first talk was a report on the interim findings from my fieldwork, and the second was my attempt at giving a 'transhumanist' paper. (The first one was received well, the second politely.)

The learning curve for this project was steeper than anything I had taken on before. It turned out that more than a decade's worth of studying Charismatic and Pentecostal Christianity gave me little traction with Mormon history, doctrine, and sensibilities; and the scientific, mathematical, and engineering aspects of transhumanism were far from my comfort zone (though my amateur interest in mathematics gave me a leg up from time to time). Other parts of the problem came together quickly. I found that there were Mormon Transhumanists in part because Mormonism already looks much like transhumanism. Mormonism is deeply invested in a materialism that while spiritually charged, is a materialism all the same. Mormonism views miracles as being, in essence, not a suspension of nature but rather an exercising of divine technology. And the Mormon doctrine of *theosis*, or becoming god-like, already paved the way for thinking in religious terms through the technical-scientific overcoming of human limits. These parallelisms were no great discovery on my part; they were something Mormon Transhumanists consciously knew and had clearly elaborated multiple times, in multiple formats. A narrative for the social aspect of Mormon Transhumanism was more hard-won, though. Part of the answer to why there is such a thing as Mormon Transhumanism is that Mormons are, on average, highly educated; and another piece is the booming technology industry, not just in Utah, but throughout the West. (It's no accident that outside of Provo, which houses much of Utah's tech corridor, the two other concentrations of Mormon Transhumanists are in the Bay Area and Seattle.) But I found there was more to the puzzle. Many of the adherents I observed indicated that for them, Mormon Transhumanism offers a new way of being Mormon, which allows them to navigate thorny moments of Church history and problematic aspects of Church culture without forcing a break from either a Mormon identity or a wider circle of Mormon kin and compatriots. This transformed Mormonism has helped many with this problem of belonging, but it does not help all. I also spoke with Mormon Transhumanists who felt that they had to break with the Church. This break was often motivated by intellectual doubt, but also by traumatic experiences with the Church, and often specifically by traumatic experiences regarding the Church's very con-servative vision of what proper sexuality and family look like. But even those who no longer saw themselves as Mormon—or alternately saw themselves as looking for new ways to "Morm" (as they jokingly said) outside of the institu-tional Church—still found in the MTA ways to relate to their Mormon past and their cultural inheritance. And due to this discovery of novel ways to relate to the faith, some dissident Mormons who most likely would have had little interest in transhumanism absent the MTA found in the association a friendly, open-minded climate for those who saw themselves as a part of the loyal op-position. The MTA, which was originally a home for white, male tech sector workers educated at Brigham Young University (BYU), slowly became more diverse, even if the breadth of that diversity was limited by Utah Mormonism's somewhat narrow demographics.

This particular solution to the problem of Church membership that the Mormon Transhumanist Association facilitated is unique to that association. But as I conducted my research, I found that the underlying problem was not specific to the MTA at all. After a century of growth, and after what felt like a wave of wider interest in Mormonism from the larger gentile world, it seemed that in the eyes of some of the faithful, Mormonism was beginning to lose its way. The same internet technology that gave rise to the Mormon Transhumanist Association also allowed a multiplicity of avenues for the dissemination of information about difficult Church history and hurtful Church policy. Many

Mormons were angry with the Church, sometimes in open conflict with the Church, and often ended up leaving the Church. The bulk of those leaving the Church were not angry intellectuals but simply young adults who wished to quietly slip away from the Church's restrictions, but due to the prominence of the Mormon Internet, it felt like there was also a middle-class, middle-aged, liberal exodus. And the institutional Church, which already had a defensive cast to it due to its rough and bloody nineteenth century, and a conservative streak thanks to its governing gerontocracy, seemed to be acting in ways that aggravated, rather than mitigated, this crisis.

Mormon Transhumanists, though, were used to paranoia and defensiveness, having been the recipients of it from another quarter. Transhumanism, which was born thanks to the advent of secularism and secularity, had an atheist cast to it, and often a New Atheist cast. From early on, transhumanists commonly saw religion as (at best) superstition, an outmoded attempt to solve technical problems through magical thinking. They often saw religion as not just a relic or a hindrance but as an active threat to transhumanist aspirations. Many transhumanists understood organized religion as being able to exist primarily because it trucked in access to some paradise located on the far side of death. And so they reasoned that once religion was faced with the possibility of denying death, of realizing a paradise in this world instead of some heavenly or future one, then all of the institutional might that religion had to offer would be used to crush transhumanism. Religion was the enemy, and religious transhumanism was an oxymoron, or alternately a treacherous fifth column. Thus, while some leaders of the secular transhumanist community accepted Mormon Transhumanism, seeing the movement as a means to expand the possible audience of transhumanism to an unreached vast number of faithful and religious, other transhumanists greeted the idea of Mormon Transhumanism with derision and sometimes hostility.

Mormonism and transhumanism, different as they are, therefore often rhyme with one another. These two objects mirror each other not just in their transhuman aspirations, or in the way that their hopeful universalisms run against their suspicion of outside threats. They also share a kind of Americanness. There are non-American transhumanisms, such as Indian transhumanism. Russian transhumanism is also vibrant thanks to a particularly rich heritage in the nineteenth-century school of proto-transhuman mysticism-cum-philosophy called Cosmism.[8] And Mormonism is truly a world religion, the majority of its members now located outside of the United States. But both Mormonism and transhumanism are deeply infused with the settler-culture romance of the frontier, whether that frontier is geographical or metaphorical. Both Mormonism and Transhumanism are also shaped by strong elements of speculative reason. While there have been moments in the twentieth and twenty-first centuries in Mormonism when the speculative impulse has gone a little dormant, Mormon speculation has never been absent in the Mormon Transhumanist Association, where those early forms of Mormon thought were revived and reconfigured. Transhumanist speculation was no stranger to association members, either. So, unsurprisingly, in Mormon Transhumanism, these two speculative streams flowed together to make a torrent. This torrent of speculation has provided for many a way out of a more conventional and narrowly drawn Mormonism. And even as this speculation may at times rush ahead to the furthest horizon, imagining cosmological pasts and resurrected futures, it still finds ways of bringing these horizons to here and now, opening up other sometimes surprising ways of living. As such, speculation pushes against the certitude of belief and the regimentation of discipline. Sometimes such pushing is necessary. Sometimes such pushing is a matter of life and death, whether metaphorical or real.[9]

§v

This work is a qualitative, social ethnographic portrait of Mormon Transhumanism and Mormon Transhumanists. But it is also an attempt to chart the emergent structure of a collective Mormon Transhumanist imagination, often by setting it alongside analogous and sometimes even cognate Mormon and transhumanist exercises in speculation. As such, it is full of folds, twists, and inversions as I trace the anxieties and hopes that run through all three of these ways of thinking. Because of this, while not taking up the full theoretical mechanisms of formal structuralist theory, this book at times has a structuralist flavor, a kind of "structuralism without structure," in the words of anthropologist Eduardo Viveiros de Castro.[10] But there is another reason to turn to structuralist thought here, beyond its capacity to map the topology of the imagination. Anthropological structuralism has been used to parse wildly different domains, from American sumptuary rules to Japanese architecture.[11] But above all else, structuralism achieved its most mature form, and shows its greatest utility, in the analysis of myths. And many of the people in the MTA who produce speculative material consider their visions to be myths, though myths both of and for the future. Because of the futurial nature of these myths, these narratives and visions are often virtual as well, both as in the kind of virtual reality associated with computers and artificial worlds and as in virtuality as a never quite closing imaginative potentiality.

It is because of this speculative aspect that this is a much different work than my previous ethnography on Charismatic Christianity. For that book, many of my informants were intellectuals, organic and otherwise. But the Pentecostal-Charismatic miracle, which was the focus of that volume, was an emergent and often embodied event, one that challenged and changed the will, but which was ultimately not understood as being controlled by those subject to it. The miracle was a way of finding sometimes life-changing divine surprise in churches, prayer groups, and day-to-day life, and the evidence I drew on when writing that earlier text was often what I had seen with my eyes, or what I felt in my body, as I opened myself to miracles by praying alongside believers. For this book, the importance of speculation warranted a focus on a purposefully intellectual activity that was vital to many of my informants: the production, the circulation, and most of all the knitting together of texts (where *texts* is given a rather expansive meaning). Attention is given to how the creation and mediation of these texts shaped cosmological scientific/religious myths. But even more attention is paid to readings of these texts. This is not a reading in search of some esoteric origin, an exercise of what is sometimes called a herme-neutics of suspicion. Rather, this is a bluntly exoteric reading, attending to the contours and geography of what we might call the *surface* of these speculative works. Given the often-heady nature of their content, such a surface reading is often challenge enough.

If I were forced to give a commonality between my last book and this one, it would be that they both are concerned with the production and the social life of novelty. I created an account of novelty in the prior work through chart-ing the way that the same series of abstract relations that constituted the miracle could be expressed in numerous modes in the lives of believers, and in many forms of a miraculously inflected Protestantized and post-protestant Christi-anity. In this book, by contrast, the focus is not novelty changing the will through surprise, but novelty as willed into existence, as people transform their present by imagining their future. The philosopher Gilles Deleuze wrote that "[a] book of philosophy should be in part a very particular species of detective novel, in part a kind of science fiction."[12] While neither of my two monographs are works of philosophy, each adheres to this principle in its own way. My previous book was a kind detective story, a search for a secret commonality in an ocean of Christian difference, and then after identifying a suspect—the miracle—an investigation into all the covert, individuated forms of work that this secret commonality has wrought. This book, and particularly the final third of this book, is a kind of science fiction.[13] But though I am the author of this book, I am not the author of this science fiction. It was authored by my interlocutors in this project, people who were sometimes Mormon, sometimes transhumanists, and often both. Their work, as presented here, is science fiction in the sense that it envisions ancient pasts and far futures, but it is also science fiction in the sense that it is materialist, agentive, rationalist, and a form of knowledge—all terms loosely glossed as ways of interpreting the word *science*. My interlocutors' work also represents carefully, purposefully crafted inventions or fabulations, and hence fictions as well in the etymological sense of "that which, or something that, is imaginatively invented" or "that which is fashioned or framed."[14] Whether this particular stream of science fiction presages the actual future, in the way that other science fictions have done before, is something we cannot know. But regardless of whether this science fiction does or does not shape or predict the future, this speculative thought as religious science fiction is certainly working at making the present. And to be honest, the present, which also needs to be made, is often science fiction enough.

§vi

One of the shared tenets of Mormonism, transhumanism, and Mormon Transhumanism is that not only does the future come fast, it is coming faster all the time. Whether it's due to the apocalyptic press of the latter days, or caused by the exponential rise in machine intelligence thanks to Moore's law, there will be no slackening in the pace of things. My experience writing this book bears out the truth of this claim. During the process, several well-written books germane to my subjects came out; while I did try to integrate this new material as much as I could, unfortunately, given the staggered pace of publication, revising the whole of the book after each new publication would mean that my own work would never see press. I do wish to take the time to briefly acknowledge these works, though.

Important to this work is a crisis among Mormon intellectuals, an internet-facilitated explosion in the access Saints have to knowledge about aspects of the Church's history and contemporary structure and practices that, in prior decades, would have been hard to come by. One of the chief predictors of whether a Church member will fall into this situation is, ironically, how seriously they take the Church. The more committed one is, the greater the chance that this commitment will be shaken. E. Marshall Brooks's *Disenchanted Lives: Apostasy and Ex-Mormonism Among the Latter-Day Saints* is a sensitive and moving documentation of the pain that this turning away brings, and how it builds a community of self-avowed "Ex-Mormons" who help each other work through alienation, psychic and sexual language, and yearning for aspects of a Church they rejected.[15] It is also a critical reflection on secularism, pointing out that it is not necessarily corrosive doubt from outside religion, but dedication to the truth-claims made by those inside religion, that leads to unbelief.

Ayala Fader's *Hidden Heretics: Jewish Doubt in the Digital Age* tells a similar tale, against quite a different backdrop.[16] It documents Hasidic ultra-Orthodox Jews who lead a "double life," presenting themselves as faithful members of a tight-knit, all-encompassing religious sect, while at the same time accessing questionable online sources of information and forbidden forms of internet-conveyed critique. Fader traces how her interlocuters often at first enter these computer-virtual realms trying to quell uncertainties but are then shaken by different visions of what religious and patriarchal authority could be, exacerbating their doubts, sometimes to the point of crisis, thus endangering the stability of their families and sense of place in the community. It is also a portrait of rabbis, educators, activists, and even Jewish life coaches charged with guarding against this injury from internet-mediated contact with the secular world, and with repairing the wounds when that contact occurs. Like Brooks's book, Fader's is a sometimes wrenching portrait of people trying to live with an incredible gulf between what they believe and what they appear to practice, and how this gap can lead to a loss of faith.

Both these books involve attempts by the religious to navigate the social chasm that secularism brings. Secularism is a theme in this book not just because it stands as religion's other in the folk social taxonomy of most Americans, but also because secularism has an intimate and formative relationship with transhumanism. Abou Farman's *On Not Dying: Secular Immortality in the Age of Technoscience* commandingly lays out this genealogy, showing how secularism is one of the core possibility conditions for something like transhumanism, but also how transhumanism breaches many of the secular/non-secular divides that helped generate it. It also presents a trenchant critique of transhumanism, identifying biases in what transhumanism imagines to be existential threats to the species, what are considered mere unfortunate deaths, and the way that the technoliberal order that some transhumanists imagine hides forms of violence that fall short of killing while still doing economic, physical, and psychic harm.

Transhumanism is not the only force to kick back against secularism. Peter Coviello's *Make Yourself Gods: Mormons and the Unfinished Business of American Secularism* presents nineteenth-century Mormonism as a resistance movement to the secular consensus of the day.[17] Understanding secularism as the controlling, embodied code of the modern age, early polygamous Mormonism is seen by Coviello as a communal rejection of capitalism, and as a politico-religious movement that queers gender and problematizes race. Twenty-first-century Mormonism is not nineteenth-century Mormonism, but there are still some resonances. Mormon Transhumanists often look to the nineteenth century's speculative energies as a model for thinking and being Mormon in the twenty-first, albeit making allowances for changes in the ethics and politics of gender and race. Some would strongly agree with Coviello's understanding of Mormonism as queer: there is a movement with some loose ties to Mormon Transhumanism that argues for a literal queer polygamy: a form of sacralized, non-heteronormative, feminist non-monogamy. For some Mormons, the nineteenth century is closer than it appears, even as they stand in the twenty-first, facing centuries and millennia to come.

There is a temptation to think of transhumanism's unmarked form—American secular transhumanism—as the original and paradigmatic form of transhumanism. Anya Bernstein's *The Future of Immortality; Remaking Life and Death in Contemporary Russia*, pushed against that tendency by presenting a history and an ethnographic portrait of Russian Transhumanism.[18] This is a vision of transhumanism not as an American import, but as an autochthonous movement, informed by nineteenth-century Orthodox mysticism, twentieth-century Soviet ambitions, and twenty-first-century Russian capitalism. On top of providing an artful sketch of a fast-changing movement with a surprisingly long history, this text is a reminder that there is no single ur-transhumanism, and even that which we take as the standard or default is just one varietal.

These works are sometimes parallel, sometimes orthogonal, and sometimes they trace the same arc as this book, even as they work toward different arguments or different ends. Taken together with this book, though, they suggest tension between the open society and the closed society—or rather, the dilations and the constrictions of the social—as well as the sometimes injurious, sometimes intoxicated, sometimes world-making properties of speculative thought, whether in its religious or secular guise. They also highlight that there can be no final measure of this faculty: not only is it always in development and flux, but, even if one were able to accomplish the impossible and present a single synchronic slice of it for investigation or review, the variations contained in that sliver of the collective and individual imagination would border on the infinite.

§vii

I close with a note to the reader. The introduction, which is called *Series Zero*, for reasons that are explained in that section, does double duty. It is an initial description of the three groups ethnographically discussed here, offering a handle on who and what they are. But it is also an introduction to the *anthropological* issues at stake. As such, some of its stretches are the most theoretical aspects of the book, dealing with issues of science and myth, the role of imagination and speculation, the place of religion in a secular society, and the complex relationship between the forces that work for an open society, and those that work for a closed one. I have worked to write this in an approachable manner, but for those uninterested in anthropology, it might be a bit of hard sledding. Those who would rather not drink from that theoretical cup, and would rather get to the description of Mormonism, transhumanism, and Mormon Transhumanism, should consider reading sections 1 through 5 of Series Zero, and then jumping ahead to section 14. While there will be theoretical stakes in the discussions after that, and moments where theory is used to open up descriptions of these groups, all that will occur in the context of a thick ethnographic description, making it easier to follow, and hopefully spending less time lost in the conceptual weeds.

A Note on Names and Terms

This book is about organizations and individuals with high internet profiles. This subject presents a particular challenge to the anthropologist's obligation to guard confidentiality. To protect the identity of conversation partners, this book avoids proper names and pseudonyms. Further, certain acts, events, or traits have been either disaggregated, just as some events or people have at times been conflated. All this has been done to further hamper any attempt at identifying those people who were willing to share with me. The only exceptions to this are individuals who have attached their names to widely circulated printed or online material that is relevant to this book. Given the importance of online communication in this ethnography, and specifically of communication that occurred in open forums, this may still not be enough to prevent the truly dogged from acquiring names of people who are a part of this presentation. To anyone so interested in doing so, I urge restraint; privacy is a scant resource these days, and preserving it is a good in and of itself.

Individual names are not the only problem; the use of the proper name for a major religious organization that has an outsized presence in this book also needs to be addressed. While I do call this institution *The Church of Jesus Christ of Latter-day Saints,* and at other times refer to it simply as the *Church* (note the use of capital letter), I also at times refer to this organization as the *Mormon Church*, or the *Mormon faith*, or as *Mormonism*. I further use *Mormon* as an adjective for projects, objects, and institutions that trace themselves back to Joseph Smith. This is not an uncontroversial choice. In the August 2018 General Conference, President Russell M. Nelson announced that members should only use the Church's full name. He also rejected the use of the term *Mormon* to refer to the institution, its members, or the associated culture.

(The Church's official statement to this effect was first posted online at a website then named, ironically enough, the "Mormon Newsroom.") Nelson also requested that journalists, scholars, and other writers respect this naming convention.

I respectfully refrain from following President Nelson's admonitions here for two reasons. The first is that much of the research on this project was done before this announcement. Hence my engaging in such a change in nomenclature would be ahistorical at best, revisionary at worse. The second reason is that the Mormon Transhumanist Association, the group I worked with the most while researching this book, has declined to change its name. Part of the reason why the Mormon Transhumanist Association took this position has to do with its formal charter: as a nonprofit corporation, this change would require amending its constitution. But the chief reason they give for retaining their name is that the term *Mormon*, though originally derogatory, was embraced by Joseph Smith, the founder of the Church of Jesus Christ of Latter-day Saints. Further, a large part of the nineteenth-century religious speculative movement that the MTA draws inspiration from understood itself to be engaged in thought and debate about Mormonism as well. The MTA also notes that several religious movements also trace their origins to Joseph Smith and include the Book of Mormon in their canon. It is my hope that more orthodox readers belonging to the Church of Jesus Christ of Latter-day Saints will understand the logic behind my choice; if not, I am deeply sorry for the offense, and I hope that it is clear that this choice does not come from a place of disrespect.

There are also moments in this book where I discuss temple ordinances (*ordinances* being a word that in a non-Mormon context, could be glossed as sacraments or rituals). Only Mormons in good standing with the Church, who have been vetted by Church authorities and issued temple recommends, can participate in temple ordinances. And they are further sworn to secrecy about what occurs in the temple, and particularly about the "signs and tokens" revealed during what is called the *Endowment*.[19] These are things and events that are to be kept "secret from the world." As a non-member in the Church, I have never entered a consecrated Temple; my information comes either from public documents, or from conversations with Church members who shared with me material that they felt did not violate their oath. Like many members of the Church of Jesus Christ of Latter-day Saints, my interlocutors felt that while there were elements of what occurs in the temple that should be kept secret, refusing to discuss non-secret aspects of the temple only breaks down dialogue and fuels unwarranted suspicion. I have certainly steered far clear of writing about or even asking about the signs and tokens that stand at the center of the endowment ceremony, a task made easier by my complete lack of knowledge of that specific material. Again, for those offended, I beg forgiveness and hope that it is clear that this choice also does not come from a place of disrespect.

Finally, the word *virtual* is important to this book. This term, first appearing in English in the twelfth century, and borrowed from an older Latin term, is polysemous. Among the meanings are the potentiality of an object—contrasting what is actualized with what is supposed or imagined—and, more recently, a computerized or digitalized simulation of something. Both definitions will be important to this discussion. When it is not clear from the context, the term *computer-virtual* will be used for the latter, even though the latter definition's strict meaning is technically more expansive.

A book . . . should . . . in part be a kind of science fiction. . . . Science fiction in yet another sense, one in which the weakness becomes manifest. How else can one write but of those things which one doesn't know, or knows badly? It is precisely there that we imagine having something to say. We write only at the frontiers of our knowledge, at the border which separates our knowledge from ignorance and transforms one into the other. Only in this manner are we resolved to write. . . . We are well aware that we have spoken about science in a manner which was not scientific.

—GILLES DELEUZE

Series Zero: "Children of God *would* try to play God"

§1

It is a clear day, not cold, but crisp; we are up early this morning, but then a lot of Utah starts early in the morning. The sky is a blue vault, broken only by the line of craggy mountains to the east, sharp peaks that dominate the horizon. These mountains are patchy with snow; in the morning light, they show slate gray underneath the white blanket. Running in ridges to both the east and the west of the city, the mountains echo each other, their rising and falling making a steady visual rhythm.

City blocks are long in downtown Salt Lake. The streets are broad, the sidewalks empty. Together, these blocks make an expansive grid pattern; traveling on foot, everything has a sparse but ordered feel. We are currently breaking with that order. We are rushing east, in a hurry, cutting diagonally through block-sized, tree-bordered parking lots. Close to our destination, we cross a park. Where the paths cross in the center of the park, there is an ornate, late-nineteenth-century tan stone building that rises for five stories, its clock tower rising even higher: the Salt Lake City and County Building. We don't know this at the time, but this striking building was originally intended by anti-Mormon partisans to stand as a rival to the Salt Lake City Temple, with its clock tower spire an echo of the temple's towers. But having city hall serve as a rival to the temple never quite took.

We cross the street. Glancing down at a GPS map, we can see that we're almost at our destination: The Salt Lake City Public Library. It's five stories of modernist curving glass and metal, looking out through the park's trees at the decidedly not modernist City and County building across the street. It's eight

1

o'clock, opening time for the library, and there's a fair-sized crowd out front. Some of the patrons are the homeless, dependent on places like the public library for shelter from the elements. But there are others, a lot of others for a weekday opening time at a public library, all in button-down shirts and clean, well-pressed chinos. The doors open, we enter. Heading up to the conference center on the third floor, we pass by a sizable graphic novels collection (the largest library holding of this genre in the United States, apparently). We also walk by a large glass case holding an 1860 reissue of John James Audubon's *The Birds of America*. The book is enormous, printed on what the book trade are called *double elephant folios*.[1] The book is open, showing on one side the title page, and on the other, Audubon's depictions of the pinnated grouse.

We walk past the case and toward the conference room. This space is busy. We see that the majority of the crowd has gathered here. The age of the people varies; some are quite old, a few are teenagers (or even younger), but the bulk of people present are men somewhere between the ages of twenty and forty. For the most part, the people present look like the Mormon stereotype: white, short hair, clean-shaven faces, wearing what could pass for office-casual clothing.[2] There are exceptions: a few men with beards and goatees to complement their otherwise conservative appearance, and a handful with a kind of messy long hair more often associated with computer programmers than office managers. There is a lot of catching up going on; most, but not all, of the people here seem to already know one another. The fact that this is *catching up* suggests that there are extended periods in which they do not see one another, at least in person. It also suggests that despite possible distances in space and time, they remain deeply connected. They ask each other about new children, changing careers, aging family members. Awkwardly, knowing no-one, we shuffle through the crowd and then take a seat.

At the front of the room, there is the usual conference setup: a podium and a PowerPoint screen. There are also the requisite large audio speakers to carry the voices of anyone addressing the spacious room. What flanks those speakers, though, stands out. On either side of the podium are black vertical banners put up this morning by the group. Each banner has the same single illustration, placed above the group's name: "Mormon Transhumanist Association" (the text used to spell out that name is a minimalist, sans-serif font). Both the letters and the accompanying image are white for maximum contrast against the banner's black background. The illustration on the banner is an apparent engraving of a wooden cartwheel, with extended wings jutting out from the wheel's hubs. The effect it gives is of something culled from a nineteenth-century treatise, an illustration taken from some yellow-paged guide to arcane religious symbols. Years down the line, I would learn that the image was taken from some clip art; later, the group ended up "hiring someone to convert the bitmap to a vector graphic so that [they] could blow it up to an arbitrary size for signs and stuff." The same piece of clip art was also used for the cover art of a science-fiction paperback otherwise unconnected to Mormon Transhumanism; among the few people in the group who have read it (there are a lot of science-fiction readers in this group), there are mixed opinions regarding whether that novel was any good.

Things begin in earnest. A man comes to the podium, leans forward slightly to speak into the mic. He has very tight, short hair and a robin-egg-blue dress shirt that's open at the neck. His eyes are quick, intelligent, intense. He starts by welcoming us to the conference, and he encourages us to get to know one another over the course of the day, saying "you may not know this yet, but there are incredibly interesting, and incredibly good, people sitting next to you." He has an even, calm cadence to his voice, and is comfortable with public speaking. (Many of the speakers today will be at ease talking to the public—an effect, perhaps, of the Mormon tradition of planned sacrament meeting talks and impromptu presentations during fast and testimony meetings.) He gives us his name and mentions that he is the president of the association, "appointed to that position by the board of directors." There is an election for new board members right now, he mentions, encouraging people to join, run, and vote; "the association is very democratic," he adds, by way of encouragement.

He shifts from pleasantries and introductions to his substantive material with hardly any transition at all. Suddenly officially inaugurating the conference, he adds to that speech act the observation that these are "challenging times," and he references communicative technology affecting shifting social norms. He mentions nihilistic religious fundamentalists, but also "anti-religious zealots . . . making a fetish of small-minded ridicule that would shame the pure in heart away from our greatest hopes." Mormonism, too, is in crisis, he tells us. He mentions debates about the historicity of the Book of Mormon, the attrition of young Mormons due to controversial Church political stances, and a series of excommunications of members associated with advocating for feminism and gay rights. But progressive Mormonism is no better, he says. It has an "all too predictable politics of its own" and "functions more as a secularizing anemia than as a force for renewed strength." Given all this, we might, he says, feel an "increasing discomfort. We might sense real risk for our people, for our culture, [our] heritage, our family. We might feel our grip less trustworthy, or our footing not so sure. Yet our love, or at least our self-preservation, brought us together today." It is clear what the proper referent for "our people" is.

He mentions transhumanism, "a new way to think about the future of humanity." Transhumanism is important, he tells us, because human nature is not static. There is biological evolution, but there is also cultural and technological evolution, he states, and says the common expectation is that while biological evolution may be stalled, humans will continue to evolve. He acknowledges that there are risks to transhumanism, that it is a dangerous idea, perhaps the most dangerous idea: "We might agree with our critics that we are trying to play God." But, he adds, "for us, who know we are not dead, why not believe that most dangerous idea? . . . We have heard this story before. It's our calling, and our choice hasn't changed. Children of God *would* try to play God. We'd learn how to be God."

He states that while most transhumanism is secular, it has religious roots; he recalls how early Christian authors were influenced by Neoplatonism, a philosophical movement that he describes as "the popular science of their day." He mentions the role that Christianity played in bringing Aristotle ("again, popular science") to the West and also recounts nineteenth- and twentieth-century visionaries who tried to meld science and religion. Mormonism itself, of course, is a religious transhumanism, he says. Mormons are like Christ, working to redeem and atone the world; being like Christ includes fulfilling this task "through suffering and death, if needed." There is a small pause after he says that last sentence.

By now, there is something in his tone and cadence that, for all his careful self-constraint, seems positively mantic. "Mormon transhumanists have many myths and visions," he says. These are based on "an abiding love of Mormonism," on "a deep hope for an ecumenical Christianity," on "substantial research on emerging technology and its trends," on "just some plain imagination on how it all might fit together." Some of the narratives "may be shocking," he warns. But that is for the good, he says, as it is the shocking elements contained in these myths that work to spur our thoughts and imaginations.

He then starts speaking of Gods, who create worlds without end. He mentions space being "formless and empty," and recounts a slightly abbreviated version of the opening passages of Genesis chapter one, remarkable only in the fact that the chief actor in those passages is referred to in the plural. (We do not know it at the time, but the language used here is an allusion to one of the sacred texts particular to Mormons and also to parts of the script of Mormon initiation in the Temple, which is a center of Mormon ritual activity.) From creation, he works forward in time. He references a war in heaven over whether humans could act as creators. He sketches the founding of Judaism and Christianity, stressing both as syncretic religions that brought together disparate preexisting strands to make something new. His narrative leaps over the ages, jumping from the foundation of the Christian religion to that religion spreading to America, and then jumping again to the growth of America as a world power. He also mentions the birth of a new religion in this new world, an event not unlike the birth of Judaism and Christianity in the old. "Its adherents combined Christian doctrine" (the word *Christian* is emphasized as he says it) "with mythology about native Americans to make scripture. They pioneered from Illinois, established a colony in Deseret, and began to build a temple."

There is another pause, then: "Today, we live in telestial glory, an adolescent civilization.[3] In the fullness of times, filled as if by an unstoppable rolling river pouring from heaven, our knowledge becomes unprecedented. Nothing is withheld, whether the laws of the earth or the bounds of the heavens, whether there be one God or many Gods. Everything becomes manifest." He sees humanity "emulating the work of Christ" by using technology to heal the lame, to let the blind see, to clothe the naked. As he speaks, despite a purposeful clarity in his talk, tenses start to become strange, and the sense of a clear-cut chronological sequence begins to become fuzzy. While he uses the present tense, it is not clear if he is recalling the past, analyzing the present, or anticipating the future. This is not the result of confusion on his part. It is purposeful. It stems from a desire to laminate past, present, and future into a single extended moment. He wishes to unmoor us from the here-and-now—or rather, make every moment past and future a fractal aspect of the here-and-now.

The telling of the myth, or perhaps the sharing of the prophecy, continues. Biotechnology will bring back extinct species, he says, and possibly "reprogram" ecologies. "Some recall prophecies about the renewal of our world," he adds. "Hearts turning to our ancestors, we remember them, and machine learning algorithms begin to process massive family history databases, perhaps to redeem our dead." This is just the beginning of a full catalog of miracles. Personalized medicine, along with advances in geriatric sciences and technology, will become sharp enough to prevent death, and while some worry about death's necessity, "new voices repeat old stories about those who were more blessed for their desire to avoid death altogether." He foresees reproductive technology that will help the infertile, gay couples, and "individuals and groups" to conceive their own genetic children, noting that "some recoil from threats to traditions, others celebrate gifts to new families." Revelations and wonders continue: solar energy, distributed information networks, nanotechnology, cellular robotics, the evolution of the internet into a "distributed reputation network," and also the internet's transmutation "into a composite of virtual and natural reality." This changes missionary work, he notes in passing. He observes that this explosion of knowledge will allow the wealthy to finally conquer mortality, but he also foresees the resulting "stunning socioeconomic disparities." He sees robotic moon bases, space colonies, and the return of the pioneer spirits. Morphological possibilities expand, allowing humans to remake themselves; this causes some to warn about "desecrating the image of God," while others "recall prophecies about the ordinance of transfiguration."[4]

His attention returns to the problem of the dead (perhaps he never left them) as he imagines reviving the cryonically preserved while also envisioning computers simulating "our dead ancestors individually." He is still speaking calmly, but as his voice remains carefully controlled, narratively, he is now rising to a crescendo: "Some say the possibility was ordained before the world was, to allow us to redeem our dead, perhaps to perform the ordinance of resurrection." He talks about artificial and enhanced intelligences with intellects that dwarf our own. And with a pause, he says, "then something special happens. We encounter each other, and the personification of our world, instrumented to embody a vast mind, with an intimacy we could never before have imagined. In that day, we will live in terrestrial glory." Later, we will learn that these invocations of telestial and terrestrial glory are references to the first and second of three Mormon Kingdoms of Heaven.

The prediction of terrestrial glory is spoken as if it is a climax, but the speaker is not yet finished. As the millennium rolls on, he tells us, religion and technology will both become more than can be conceived of now (he uses the word *evolve*), and more than could even be understood, "except, perhaps, loosely by analogy." And then there is what he says next, which may be an instance of that loose analogical thought, though it could perhaps be taken very literally as well: "We will see, and feel, and know the Messiah, the return of Christ, in the embodied personification of the light and the life of our world, with and in and whom we will be one, in a world beyond present notions of enmity, beyond present notions of poverty, suffering, and death, the living transfigured, and the dead resurrected to immortality. We will fulfill prophecies, and we will repeat others, forthtelling and provoking ourselves to greater challenges, to celestial glory and beyond, in higher orders of worlds without end. As we share these religious narratives, expertly or not, we are engaging the function of prophecy—not in any institutional sense, that would usurp some other's authority, but rather in the broad sense to which Moses aspired when wishing that all of us were prophets." We will have then entered the celestial, the third, and the highest, degree of Mormon heaven.

Prophecy, he tells us then, is about potential, connected to living engagement. Maybe, he concludes, Mormonism will be little more than a footnote in religious history; it may also be the case that our labors do not matter, as it will be God who does all the work, vitiating all our human efforts. "But it could yet turn out that the grace of God is best expressed in all the means at hand, from prophecy to technology. It could turn out that it's up to us to learn how to become Gods ourselves, the same as all other Gods have done before. That, to me, is worth trusting in, worth working for."

He is finished speaking, but all this was just prologue. Later that day, there will be presentations on the use of personalized software to streamline government, on using existing brain-scanning technology to investigate qualia, on social media and empathy. But there will also be talks on subjects such as the "fractal lineage of gods," a slideshow presentation which at one point quotes a hymn entitled "If You Could Hie to Kolob" (*Kolob* being the name for the star or planet that, in some expressions of the Mormon imagination, is either nearest to God, or which God calls his home). This reference will not be the only time that hymns come up in the speeches given that day. At one point, there is a keynote presentation intended to help think out loud as to why we might want to "be smarter, live longer." This presentation addresses that topic by reflecting on affect, literature, and music. When the musical notation for a hymn appears on a PowerPoint slide show, the speaker spontaneously asks the audience to sing. "Who's got perfect pitch? Can you give me an F?" she inquires. No one claims perfect pitch, but it turns out that someone *does* have a pitch pipe on them. As she makes choral director motions, almost everyone in the conference joins her in singing the Charles Wesley hymn "Love Divine, All Loves Excelling." As she sings at the podium, in the window behind her, standing out against the vault-blue sky, towering over the copse of tree in the park, is the temple-like clock tower of the Salt Lake City and County Building. For a few moments, the conference feels like a sacrament meeting.[5]

Later, when dusk falls, and the sky sinks from an azure blue to a deep purple, the meeting moves five long Salt Lake City blocks up north to a region of the city known as Temple Square. That sprawling ten-acre space is filled with visitor centers, conference halls, Church office buildings, and the titular Salt Lake Temple. On its edge lies the Joseph Smith Memorial Building, a ten-story *fin-de-siècle* luxury hotel converted into an auxiliary LDS Church building. That building is our destination. When we step inside, we are greeted by a large greying man wearing a white dress shirt and an American flag tie. The interior of the building is all marble columns, cream-colored paint with gold trim, ornate crown molding, and large potted ferns. A statue of Joseph Smith dominates the center of the chamber. In statue form, Smith is holding a book, assessing his audience, and looking as if he is pausing reflectively in the middle of a sermon. A massive chandelier hangs over the center of the two-story lobby. Through a side door on the second story of that foyer, in a reception hall rented for the event, conference attendees meet for a catered dinner (there is a vegetarian option for those who care for it). The room is decorated in roughly the same aesthetics as the foyer, but in this case, there are also rows upon rows of dour oil portraits of black-suited nineteenth- and twentieth-century Mormon "prophets, seers, and revelators," depictions of the highest echelon of leadership in a Church where almost every male believer holds the priesthood.

Back inside the banquet room, the conversations make quite a din. People talk, taking up in real time and space online discussions and arguments that have been stretching (in some cases) for years. After a spell, a barbershop quartet (the night's entertainment) comes up to sing, and then everything comes to a close, though people will linger for much longer, carrying on their conversations. (It will turn out that barbershop quartets are a frequent feature of these meetings; often the quartets will sing standards, but with the lyrics changed to reflect transhumanist themes.) This gathering is not the first such meeting, and will not be the last. In spaces as varied as universities, libraries, and convention centers, this growing cohort of people has been having conversations like these since 2006. This meeting—the 2015 Annual Conference for Mormon Transhumanist Association—was just one more in a series of assemblies that will stretch out, the group hopes, into a literally infinite future.

§2

Mormonism is the nineteenth-century religious movement founded by Joseph Smith, sometimes called "the American prophet." Smith saw the faith he founded as a restoration of the forgotten doctrines and practices of an original primitive Christian Church, which had lost its way in a "great apostasy" that was supposed to have occurred in the first century or so after Jesus ascended. Before he was lynched in Carthage, Illinois, Smith introduced a number of startling religious innovations, including the discovery and translation of what he claimed was another gospel based on events in the Americas; the institution of temples and the creation of rituals to be carried out in them; the salvaging and radical expansion of the theological doctrine of theosis (that is, the divination of the human); the initiation of a new cosmological imagination in which gods had been raising men up into newer gods for all eternity; the anticipation of a looming apocalypse; and perhaps most famously (or infamously), the doctrine of celestial marriage, which was originally understood as a form of sacred polygamy integral to attaining true godhood. During the nineteenth and twentieth centuries, Mormonism would undergo many changes, the most significant being the abandonment (by the vast majority of Mormons) of polygamy as a this-worldly practice, with a concomitant shift to interpreting celestial marriages as monogamous unions that continued on into the afterlife. The outside world's understanding of the Saints shifted, too. Mormons came to be viewed less as sexually perverse, racially "questionable" frontier cultists who were irrevocably antagonistic to the United States government, and more as deeply conservative, family-oriented patriots. During this long transition, some expressions of Mormonism would break away from the Salt Lake City–headed Church of Jesus Christ of Latter-day Saints. But despite these defections, the Utah-centered institution continued to be the demographically and culturally dominant form of this mode of religion. And during all these transformations, the stamp left on the Church by Joseph Smith meant that in many ways, Mormonism would stand apart as something unlike any other form of Christianity.

When measured against Mormonism, the transhumanist movement is arguably of much more recent vintage. Some academic histories of transhumanism would differ with this assertion, tracing the transhumanist impulse to the Enlightenment-era desire to channel natural forces to human ends. Some would backdate it to Eastern Orthodox mysticism, early Christian strivings for eternal life, Greek legends of mechanical or otherwise artificial forms of life, or even the immortalist aspirations of the titular character in the nearly four-thousand-year-old *Epic of Gilgamesh*.[6] Whatever its provenance, though, contemporary transhumanism is as distinctive as Mormonism. Transhumanism is the aspiration that, and the agitation for, anticipated forms of new technology that would so change the human condition that the beneficiaries of that technology would in a very real way no longer *be* human. Relieved of the burden of mortality, our intellects no longer hobbled by the limits of our biology, our society no longer trapped in an economy of scarcity, our civilization gifted with endless energy and unlimited computing capacity, humankind will have transcended the human state. (Note that there are some sharp barbs on exactly who is a part of transhumanism's first-person-inclusive plural 'we'). The immediate background and social context of transhumanism are varied. They are fed by diverse movements ranging from blue-sky military research initiatives to hobbyists concerned with means of life extension through nutritional supplements, to the cryonic preservation of corpses for reanimation in later ages. The technology entrepreneurs of Silicon Valley and high-tech initiatives elsewhere are closely associated with and influenced by transhumanism as well.[7]

At the surface level, Mormon faith and transhumanist aspirations may seem to be so different that it is hard to imagine how these two concerns could exist alongside one another, let alone serve together as the joint impetus for a social movement. And while this book will argue that these two fields are far more simpatico than they may first appear, the tension between them as a way of understanding and speaking is quite real. As this book progresses, we will tend more to the sort of logic and imagination that characterize Mormonism (in the form of the Church of Jesus Christ of Latter-day Saints), secular transhumanism, and the Mormon Transhumanist Association itself. We will see how these various modes of thought echo each other, steal from each other, oppose each other. And we will see how this speculative play opens up new horizons of envisioned potentiality, and different dreams of how the universe may be configured. But to understand the differences and the similarities, to get a better feeling for the grain of these two discourses, and to see what is facilitated, it helps to attend to how Mormonism and Transhumanism form different styles of speech within the MTA.

§3

To a considerable degree, language is one of the chief products of the MTA, and the circulation of differing discourses is the association's chief labor. While the Mormon Transhumanist Association has many other facets, this is a group that is constituted in large part by formal speech and casual conversation. But the language of the conference talks, as well as of the table talk and online back-and-forth, can sound strange. The same can be said of the exchanges that occur during the group's various in-person meetups and of the back-and-forth found in the blogs, podcasts, and social media feeds that are an important space for this group. But then, the content and style of the speech used in the MTA would also sound strange to those conversant in the language and performances associated with either mainstream Mormonism or secular transhumanism. This sense of strangeness comes from the fact that Mormon Transhumanist presentations and conversations are not being carried out in just one language. Rather, these religious transhumanists are speaking and writing in an agglomeration of two specialized vocabularies. This is a doubled cant, one language suggesting the religious, and another language redolent of radical science and speculative engineering. The references and intended effects are harder to track when listening to these presentations, therefore, because the listener is dealing with multiple codes. Further, these constituent languages are not just drawing on different vocabularies, they also convey and affiliate with different pictures of the world. That is to say, these two vocabularies are themselves statements that express a meaning that exceeds what is being said in a literal sense.[8] The use of these languages is intended to tell the listener what kind of person the speaker is.

But these languages do more than that. These are languages in different stages of development, addressing worlds that are working at different speeds. The language closely associated with speculative science and an engineering imaginary points to ideas that are only now coming into focus as possible or newly established practical projects: machine learning algorithms, reprogramming ecologies, distributed information networks, nanotechnology, cellular robotics, distributed reputation networks. Other parts of that same language come across instead as instantiations of science fiction dreams—sometimes quite old science fiction dreams—that seem to have taken on a new veneer of plausibility: space colonies, planetary exploration, artificial intelligence. Still other aspects of their speech are couched in a language of philosophical-scientific speculation, though some of this old philosophical language is given life through their ability to be plumbed by new concepts (such as qualia), or by relying on source domains such as contemporary mathematics and aesthetics (recall the reference to the fractal lineage of gods). Altogether, this is an amalgam of languages that points to a kind of historically grounded technical and investigatory rationality, which is oriented toward expanding scientific futures and bleeding-edge technologies.

The other vocabulary, the vocabulary of religious particularity, can also seem strange to the ears of those not familiar with it. The terms are odd to those outside of Mormonism: Melchizedek and Aaronic; Kolob and Liahona; ordinances and degrees of glory. Not all the terms are entirely alien, but even much of the language recognizable to the ears of the non-Mormons works to new effect in this context. Christ and God are recognizably Christian terms. But in this context, the words seem to be orienting themselves with a compass other than that American Evangelicals or Catholics use to theologically position themselves. As employed here, Christ is as much a template for human action as he is a soteriological figure. He is, in short, not just a savior but a brother and, perhaps most importantly, a way of working in and redeeming the world. Other terms seem to have not only used different sensibilities but perhaps different referents. *Gods*, in the plural inflection, seems hard to reconcile with the way that the singular form of the word appears in mainline Protestant churches, Evangelical meetings, and Catholic masses. And some of this language, such as the self-conscious and marked use of terms like *myth* or *mythology*, seems more suited for the analysis or disparagement of someone else's alien religion than for how it is used here: presenting visions of the speaker's own faith. Finally, there are aspects of this vocabulary, such as *prophet*, *seer*, and *revelator*, that seem taken from Old Testament or early Church arcana, while there are others, like *Kolob*, which sound like loan words from some half-forgotten or half-dreamed Semitic or Egyptian tongue.

So, this is a split tongue, an undissolved admixture of science and religion. Yet there are other parts of this bifurcated language that sound very ready to hand. The language of post-secularism, of religious discord, and moral orders (whether under siege or experiencing transformation), is in no way particular to the people in that library hall; these are terms and concepts on the lips of many people of faith in contemporary America. Nor do references to contestation over the meaning and moral valence of terms like *gay* or *feminist* sound alien, even in this framework, even though that specific attitude taken to these terms may not match those of right-leaning American religious cultural warriors. In the same spirit, something similar can be said about the references made to both tensions and anxieties regarding technological advancement and to the concomitant social mutations these now-incipient technologies will catalyze. These are familiar fault lines, too, sites of sometimes bitter contestation about what makes us human, and how we can hold onto it.

To be fair, stressing the alien sensibilities of some of the language of antic-ipated technology may be taken too far. The language of radical futurism that presently sounds distant to many seems increasingly less distant all the time as that future seems to approach. And this *is* a language of a radical future that suggests something is rushing toward us, a shifting proximity conveyed in a way not unlike the rise in pitch and decrease in wavelength as a siren approaches. This language of futurist technology feels, for all its wildness, to belong to a future so close it is difficult to distinguish from the present; this is not the language of some temporal horizon set at a fixed distance, a moment which is always ahead of us and which we will never inhabit. Likewise, the language of the Latter-day Saints seems increasingly familiar. References to Mormonism no longer conjure up visions of polygamist Sherlock Holmes villains, rather they are now associated with senators and presidential candi-dates, or in other settings to a major touring Broadway show. The Mormon is increasingly familiar, even if still not increasingly understood. For all the strangeness, there is now something homey, too, about these Latter-day man-ners of speech.

It would be a mistake to think of these languages or vocabularies as entirely independent, autonomous objects. When laid alongside each other, these two languages seem to lose their distinct borders as their inevitable interactions rework their associations. Social-media- or software-abetted smart government sounds different when discussed as part of the larger religious telos to build Zion. And speech about enabling the blind to see, or the lame to walk, or even about the resurrection of the dead, has a different flavor when presented as much as technological achievements as a religious miracle. This phenomenon of mutual transformation of parallel language can even be seen in the term *Mormon Transhumanism*, which may be used to indicate a particular transhu-manist strain of Mormonism, or a Mormon strain of transhumanism, but also appears to indicate a single object, a collectivity, a mode of thought, with an identity all to itself. The phrase seems to point to something separate from and irreducible to either Mormonism or transhumanism alone.

And so, robbed of more context, the term *Mormon Transhumanism*, and the hybrid language associated with that term, sits: both distant and familiar to mainstream American ears. With dependent and combinatory aspects that are yet autonomous in their structure, this linguistic material acts as a bridge between science and religious tradition, as scientific or science-fiction conjecture, as spiritual reflection, and finally as an abstract social or cognitive space of its own. What more context could be added to make the words *Mormon Transhumanism* more than a schematic operation? And more importantly, what does the psychic and social space created by these words enable those who inhabit that space to do? And for those outside of the MTA who wish to engage with what that space entails, what thoughts does the Mormon Transhumanist Association facilitate? And finally, how far could that Mormon Transhumanist horizon of thought extend? Does it show us things only visible from that one vantage point? Or does it open other vistas to us if we attend to it carefully enough?

§4

We could go too far by framing this question solely through reference to language and concepts. That is because Mormon Transhumanism is not just a thought or a mode of speech. It also has a sociological face. It is an institution in the literal sense of the word, a social form and a 501(c)(3) organization. But it is also an institution in the Durkheimian sense, in that it is a group of people caught up in an enthusiasm, forming an enduring community. The sociological edge of transhumanism is something that often is ill-attended to in discussions of transhumanism, but it is crucial. And it is crucial because one of the values of transhumanist thought is the way that it denatures social problems that are seemingly unconnected to it. Not all transhumanist believers are originally driven by transhumanist problems.

A natural question is how large a community Mormon Transhumanism is, and how enduring it will be. This is not an easy question to answer, not because the information is lacking, but because it is not quite clear on its face what these simple questions are asking: both size and age are relative, situated judgments. By one measure, when we talk about the Mormon Transhumanist Association, we are talking about a small, recent agglomeration of people. By this standard, we are also discussing a not-particularly-outsized set of in-person and electronically mediated forums those people create. They meet in halls, not stadiums; they gather in living rooms, not meeting houses; they converse together on the internet, but not in such numbers that their gravity bends the shape of the online virtual spaces they inhabit. Together, the people and the social spaces that these people have carved out collectively form what anthropologists sometimes refer to as a *counterpublic*, a haven from a broader community of speech and media ecology that may not share the MTA's sentiments or mission. It is not *the* public but, even though it is always mindful of its minoritarian status, it is still *a* public of its own.[9]

Size and age, though, also must be weighed against the rate of growth, as well as against the other communities that serve as both rivals and interlocutors. The MTA had only fourteen members when it was founded in 2006; it grew to roughly a thousand members by 2020. This gives it something like a mind-blowing 6,150 percent rate of growth from its founding. This is a growth rate on the order of the logarithmic, like a hockey stick jutting up from linear to exponential growth six years in, starting in 2012. Age and size also must be considered by acknowledging the proper *scale* of comparison. The Mormon Transhumanist Association is both the largest and the longest-running religious transhumanist movement. And it has lasted far longer than most secular transhumanist groups, whose lifespan tends to be mayfly-like. These groups often no sooner agglomerate than they are ripped apart by dissension, or decohere as interests and debates within them shifts or wanes. (There are a few transhumanist listserves or even ancient online bulletin boards that have a much longer history, but almost without exception these decades-old sites are moribund). The short lives of the vast majority of other groups becomes more explicable, and the Mormon Transhumanist Association's longevity becomes more of an outlier, when one considers that for all the in-person meetings, these transhumanist collectivities are chiefly computer-virtual groups, quickened in the distributed and accelerated space of the internet. The internet is itself a system shot through by exponential and near-exponential growth, in both the speed and in the volume of communication, in the complexity of its infrastructure, and in the number of both human, and now also inhuman, participants. Being lost in the vast proliferation of groups and discussions, being rent into pieces by dissent, collapsing under the weight of too much membership gain, or being outcompeted by some other entity suddenly appearing from beyond in the virtual horizon, all this is the order of the day for large swaths of the internet, and especially for the transhumanist internet. This quicksilver constitution of most groups gives special weight to Mormon Transhumanism's longevity, and thus, perhaps, to its importance as well.

§5

The value of an idea is not just in how common it is, but also where it sits in relation to other ideas, and to what conceptual spaces that idea opens up. Likewise, in anthropology, the value of a group is not its size, but rather the unique structures of thought it traces out, the *sui generis* way of life the group encapsulates, and the light the group reflects on still other human forms of living and practice. It turns out that the light that shines from Mormon Transhumanism illuminates some rather distant objects, objects that are both demographically and conceptually several orders of magnitude larger.

For example, the Mormon Transhumanist Association tells us something about the broader category of religious transhumanism. Religious transhumanism, after all, is more than the Mormon Transhumanist Association. There are Buddhist transhumanists, Christian transhumanists, Muslim transhumanists, and (unsurprisingly) Wiccan transhumanists. (There are even a few transhumanists who, although avowedly secular, would unironically invoke the name of the devil himself.) Further, religious transhumanism does not always come in the form of transhuman instantiations of some preexisting historically human religiosities. There are religious forms that are particular to transhumanism alone. These are religious transhumanist groups with no non-transhumanist analog, meeting in converted churches to sing hymns about cryonics, or ritually recording what they call *mind-files* that they broadcast into space to create cosmic immortality.[10]

These fusings of religion and transhumanism are *seemingly* hybrid forms that can teach us about sociality on the internet and also about what effect new forms of mediation and new technical imaginaries are having on the category of religion. These religious transhumanist groups are diverse collectivities, with their own histories, as well as with their own anticipations of the future; therefore no one group can stand for religious transhumanism writ large. But because of Mormon Transhumanism's longevity, and thanks to its habit of mentoring different religious transhumanists who identity with other traditions, the MTA has made both common cause with, and considerable impression upon, many other religious transhumanist individuals and groups. To a large degree, the history of contemporary religious transhumanism is the history of Mormon Transhumanism. It is true that each of these groups, sharing an interest in working through the combined problem of religion and technology, would equally allow insight into the sort of variation that these combined issues give birth to. But, for the historical reasons laid out, it is possible to argue that among these equals, Mormon Transhumanism is more equal than others.

The previous paragraph used the word *seemingly* as a modifier for the hybridity encompassed by Mormon Transhumanism. The reason for that word choice has to do with both Mormonism and with transhumanism in its non-Mormon form. In some ways Mormon Transhumanism is definitely a compound, as we saw in the discussion of language, and as we will see in many other ways. But in other ways, like some hypostatic theological accounting of Christ, the MTA both is fully transhumanist and fully Mormon. As will be shown in the closing moments of this book, Mormon Transhumanism imagines itself as the proper telos of both movements. This means that it is a complete form of Mormonism, a complete form of transhumanism. And therefore, in theory, at least, it could faithfully continue the lineage of all other Mormonisms and all other transhumanisms if those were to exhaust themselves or fade away. This is not to say that the Mormonism or the transhumanism of the MTA is the same as other forms of Mormonism or transhumanism; as a Mormonism, it has aspects peculiar to it, and as a transhumanism, it has features that are not found in other modes of transhumanism. But then, neither Mormonism nor transhumanism is a self-identical movement. Both Mormonism and transhumanism are internally var-iegated, and each can be conceived of as multiple expressions of the same concept, different variations on a common theme or problem. There are many Mormon-isms, but they are all Mormon, and there are many transhumanisms, but they are all transhumanist.

§6

Therefore, Mormon Transhumanism, in all its internal differences, and in all the ways that it differs from the two traditions it stems from, tells us something about both Mormonism and transhumanism. And similarly, Mormonism and transhumanism teach us something about the MTA. This is helpful, as much has gone unsaid about these three collectivities both generally and as anthro-pological objects. It turns out that for different reasons, all three groups are surprisingly anthropologically underserved. It is excusable that Mormon Trans-humanism has not been attended to yet, given its relatively small size and recent emergence. But it is more concerning that neither Mormonism nor transhu-manism has been given the attention that is its due. That does not mean that there has been no attention paid at all. Both groups have had their dedicated and able ethnographers. But there have not been *enough* of them, and the con-versation between them has not congealed into shared academic discussions or debates. The reason for this, though, differs for each movement.

Let's start with transhumanism. Transhumanism is in part underserved because of the novelty of the movement. We have seen that there are transhu-manists who claim that there have been transhumanist strains in Western thought from its founding moments. But as a self-conscious movement, modern transhumanism is usually understood as coming into being in the 1980s.[11] As stated earlier, it was put together from diverse but already existing groups, such as amateur cryonicists, home computer builders, space enthusiasts, antireligious libertarians, and some scientists who may not have been fringe, but certainly were not mainstream. Still, none of these groups were that sizable or stable on their own. Striking as the idea of transhumanism was, it was still statistically unlikely to steal the attention of more than a few ethnographers. Therefore, the movement became lost in the weeds, obscured by all the other competing social forms that might catch the anthropologist's eye.

There are profound methodological challenges to studying transhumanism as well. Where, for instance, is transhumanism located? Physically, there are particular nodes, such as cryonics centers, anti-aging laboratories, Silicon Valley startups, periodic transhumanist conferences, and the like. But as the use of the word *node* suggests, these sites are dispersed, and each only conveys a small sense of the movement, a grasp on only a bit of its contours. Furthermore, there are also a significant number of dispersed transhumanists with no strong ties to any of these nodes. These dispersed transhumanists form an amateur futurist social penumbra that even as it is free of direct involvement with any of these nodes, is both affected by these nodes and, through distance-mediating capillary action, affects them. By resorting to conceptual frameworks such as multi-sited ethnography, anthropology has become better at piecing together such distributed social networks. Still, to anthropology as an academic discipline, these sorts of delocalized bodies still present serious theoretical and methodological challenges.[12]

But transhumanism may be even more distributed than this list of scattered transhumanist locales suggests. In addition to these nodes, there are other computer-virtual spaces where bodies, networks, and objects become attached to or facilitate the social.[13] It could be said that more than anything else, transhumanism's location is in a Republic of Emails, a web of electronically mediated forms of communication which historically have ranged from ARPNET to Reddit. Through conventions, laboratories, and tech-startup offices, this movement is increasingly present in what some transhumanists might call *meat space*. But it is in virtual spaces that transhumanism can be said to be most present. Hence the difficulty. Mediated (and at least initially) almost anonymous, the internet is a challenging ethnographic environment. Now, as with multi-sited ethnography, over the last few decades, anthropology has developed tools to handle computer-virtual ethnography as well.[14] But again, the shale-brittle and quick-moving nature of these virtual institutions complicates a certain kind of easy ethnographic engagement or naive methodological technique.

Despite these conceptual and methodological difficulties, there is reason to hope that the relative lack of anthropological attention paid to transhumanism is a passing situation. As transhumanism as a phenomenon becomes more pressing, so has the anthropological attention devoted to it, bringing more academic hands to this common (ethnographic) task. Anthropologists have addressed topics that we will rehearse later in this text, as well as other important questions such as the variation that can be seen in transhumanism between and within national borders, the way that transhumanism reconfigures the line between life and death, and how transhumanism changes both the nature and scale of time. And perhaps most centrally, the complex connection between secularism (whether thought of as a condition or as an ideology) and the genesis of transhumanism has been and will continue to be addressed academically both within and beyond the anthropological literature on the movement.[15]

The relatively sparse condition of the anthropology on Mormonism, by contrast, seems to have no moment of imminent rectification. This is due to several factors. The first, and probably most crippling, limitation has to do with a widespread (though rarely explicit) anthropological disdain for Mormonism. This disdain, it should be understood, is just a small portion of a much broader problem. For various reasons, for a considerable time Christianity, as a comparative anthropological category, was an unacceptable ethnographic object, a repugnant other that was at once disturbingly close to anthropological practices and too alien to engage with as anything other than a rival.[16] Many anthropologists thought that they already understood Christianity from their time at home; many had gone into the field in the first place to study something quite different than the default domestic religion of the Euro-American West. However, due to the blunt demographic force of Christian growth in the global South, and also because of the intensifying political power of conservative Christianity in American politics, in time, regardless of how they may have felt personally, anthropologists had to make their peace with Christianity. They were forced to acknowledge that, like all other social forms, Christianity was at least *potentially* relevant and worth their attention, and at times the ethnographic study of self-identified Christian cultures can be helpful.[17] Indeed, there eventually was a moment when anthropologists started not just studying Christian cultures on the sly but making up for lost time and purposefully turning their attention to Christian forms.

At first, it was the Christianities that were most frequently met in the field that received the most attention. Which is to say, various fast-moving forms of Pentecostalism that were sweeping some regions of the global South were documented by the anthropologists who worked in those spaces. And fundamentalism and evangelicalism were the focus for those anthropologists who did their work at home. In time, other forms of Christianity also became objects of ethnographic attention, and religious movements such as Calvinism, Eastern

Orthodoxy, and Roman Catholicism found their Malinowskis, those ethnographers willing to take on underserved subjects.[18] But for some reason, unlike other forms of Christianities, Mormonism was always just an intellectual step too far. To quote one of the few ethnographers who attempted to give the Church of Jesus Christ of Latter-day Saints some time, the anthropological attitude toward Mormonism was that anthropology considered that faith "at once unworthy of serious interest and as a scandalous threat."[19]

Why should this be? It has been argued that despite anthropology's long antipathy to Christianity, the discipline had been unknowingly poisoned by some overly Protestant sensibilities that only came to the surface when objects like Mormonism were encountered. The argument was that certain North American/European sensibilities about what religion was really like silently fueled a dormant anti-Mormon animus. But like some recessive allele, this hostility and disdain could only express itself under the right circumstances.[20] No doubt, this argument has some merit. I remember being a young anthropologist who was part of the efflorescence of the ethnographic study of various Christianities. At the time, I had the sense that even as groups like Evangelicals or Pentecostals were interesting, there was something stultifying about Mormonism. I assumed (quite incorrectly) that the strict dress code and dietary taboos of the group meant that the religion was crushingly conformist, and hence not worthy of my time. In retrospect, that I could ever have associated Mormonism with conformity suggests the effectiveness of long LDS media campaigns to paint Mormons as family-friendly pseudoprotestants, and not as persecuted, ex-polygamous prophets scouring the desert as they built Zion and worked to become Gods.

This anthropological neglect should not be taken as a claim that there has been no work on Mormonism *tout court*. Mormon Fundamentalists have been given extended attention in a series of books. Mormon kinship has been carefully charted (with kinship being, admittedly, perhaps the most exquisitely anthropological topic). There have been accounts of Mormon eschatology (or as Mormons refer to it, the *plan of salvation*) that were deeply informed by anthropological theory. And various Mormonisms abroad, from the United Kingdom to Bolivia, have been faithfully reported on.[21] And there is a younger generation of scholars who have written sharp dissertations on some central Mormon issues, such as secrecy around temple practice and changes in the Mormon conceptions of what constitutes ethical and efficacious language.[22] And of course, there is also a long-running, vibrant Mormon sociology—a disciplinary cousin to an anthropology of Mormonism that can also be used to chart these dynamics. So perhaps there is reason to hope on this front after all.[23] But the amount of anthropological attention directly addressing Mormonism still seems strangely lacking when compared to what is given to other modes of religiosity.

§7

An aside: is the anthropology of Christianity relevant when discussing the various expressions of Mormonism? It is conceivable, after all, that Mormonism is distinct enough that the ethnographic record concerning self-avowed Christian communities might not shed any light on the problem. This is to ask, in other words, whether we should consider Mormons to be Christian, at least from an anthropological point of view.

This is no easy question to answer. Or rather, this is a question that defies producing any *single* answer, as all the various takes are conditioned by the positions of the people answering the question. Ideas about the proper status of Mormonism among Protestants, for instance, varies in both how negative the judgment is, and whether there is much hope for a Mormon return to what is understood as orthodox religion. For the most part, Protestants do not see Mormonism as a legitimate Christian form.[24] One reason for this is that when this question falls to Protestants, it is often informed by a normative logic. What ends up being asked is whether Mormons by-and-large adhere to what the inquiring group of Protestants holds as definitional of Christianity as a belief, practice, or authority. Therefore, Protestant judgments about the Christianness of Mormons informs us of the structure of the imaginary and social relations of the particular set of Christians we are canvassing at the moment. But they do not inform us of whether there are some shared generative processes common to members of some set that can be defensibly classified as Christians.

As to what the various Mormon communities (including here not just the Church of Jesus Christ of Latter-day Saints, but also the various other branches and break-away groups such as fundamentalist Mormons) believe about this question, opinions differ widely. Many, particularly among the largest, Utah-based group, see the issue of Mormon Christianity as a fatuous question. These Saints answer with an unhesitating *yes*, pointing to evidence such as the presence of Jesus's name in the official title of the Salt Lake Church, the Church's articles of faith, numerous statements from Church leadership, and the very Christo-centric reasoning that informs the *Book of Mormon*. Other believers, both within and outside of the institutional Salt Lake Church, understand things differently. They see themselves as having the same relation to Christianity as Christianity has to Judaism, or Islam to Christianity: as a successor that at once builds on and breaks with the prior tradition. Such debates are themselves informative from an ethnographic perspective, as some anthropologists have suggested that these sorts of conversations are not just instructive about the views of various sets of people but could serve as the crux for the anthropological studies of Christian populations everywhere.[25]

One way of answering this problem without falling into normative and particularly Christian logic is by asking if Mormonism is engaging with what might be called a Christian problem. Asked this way, Christianity is a function neither of creed nor Church. Instead, it assumes an understanding of Christianity as a form that is defined not by the substantive ideational or social content, but by engagement with certain generative Christian problematics.[26] Therefore, rather than measure a group by the answers it gives, we can measure a group by the concerns, anxieties, and hopes that structure its choices. The virtue of thinking of Christianity in this way is that it allows for a nearly infinite number of simulacra without privileging any as paradigmatic or controlling. Nothing is designated as *ideal*, and instead of there being any kind of *original*, there are only different forms that come earlier or later chronologically. As long as what is produced is seen as an expression of the defining problematic, we are dealing with an instance of that problematic.[27]

This still leaves us with the problems of 1) identifying what particular problems Mormons are dealing with, 2) asking if this problem is a concern that is present and determining (either actively or latently) in Christian populations more generally, and 3) assuming the previous question is affirmed, identifying what it is about the particular stance toward this problem that results in Mormon religiosity as a distinct form. Issues regarding the importance, meaning, and effect of ritual are one axis here in the complex of problems informing Mormon religiosity: Mormonism's mix of temple and meetinghouse allow for a simultaneous endorsement of high and low Church rare among Western Christianities. (We should note, though, that issue of adherence to ritual form versus practice of a religious sincerity is in no way particular to Christianity).[28] And the importance of the temple, along with the concept of priesthood and the institution of the prophet, points to a very particular answer to another Christian problematic: the degree to which the appearance of Jesus Christ marks some kind of ontological, cosmic break that is mirrored by local historical and social ruptures.[29] In the Mormon case, rather than rupture, a sort of ontological continuum is imagined, and the historical break (which is also importantly reconfigured as a *return*) is instead located in the early-nineteenth-century American frontier. And it is certain that we can also see that much of Mormon theology is driven by an assumption that the difference between the human and the divine (a common Christian problematic) is seen as less of a gulf and more of a vast but traversable continuum. Again, while this is not a problem unique to Christianity, it is one that is recurrent enough to be seen as part of a larger Christian complex of generative issues.

But it is iteration and repetition as a Christian problem that stands out as having the greatest effect on shaping Mormonism.[30] Christianity is constantly faced with questions as to how novel certain moments, events, and figures are. An example of this can be seen in the typological biblical hermeneutics found (among other Christian groups) in the American literalist tradition. In this hermeneutic, characters and events from the Hebrew Bible are seen as presaging New Testament figures; likewise, New Testament and post-biblical figures, including contemporary actors, can be understood as echoing events and actors from the Hebrew Bible or the New Testament. Jerry Falwell, as part of a fund-raising effort, praying and driving for six days in a row around the mountain that would be the future site of Liberty University is not *just* Jerry Falwell. He is also a shadow or a return of Joshua, circling the walled city of Jericho for six days in preparation for God striking down those walls on the seventh.[31] The tradition opened by Joseph Smith took this repetition/iteration problematic and allowed it to dilate fully, paving the way for numerous, and sometimes seemingly infinite, subsequent and parallel instances of some prior grounding moment. Consider how *The Book of Mormon* is at times presented as *another* Testament of Jesus Christ, running contemporaneously with the periods covered by the Hebrew Bible, the New Testament, and the early Church. Recall also that the Church has found other additional texts that it presents as scriptures in the *Book of Abraham* and the *Book of Moses*. The singular Solomonic Temple in Jerusalem becomes a network of over one hundred and fifty temples (more than two hundred if planned temples are included). Prophets come in enchained succession, instead of arriving in singular, punctual appearances. And Gods beget worlds and humans, who in turn become world- and human-creating Gods.

This recursive, plural streak to the Mormon imaginary is one of the factors (though far from the only one) that makes Mormonism capable of being articulated in a way that can resonate so profoundly with transhumanism. But this recursive streak also helps us with a very particular problem. If we take it as given that in anthropology, the analytic classifications used for groups should, whenever possible, closely parallel the understanding that groups have of themselves, then the lack of unanimity among Mormons on the question of whether they are Christian should preclude our being able to situate Mormonism.[32] Like the cat in Schrödinger's box, Mormonism, as presented by its many adherents, seems to be simultaneously Christian and not Christian, given that different people give different answers to that question. We can see Mormonism as both Christian and not Christian at the same moment thanks to repetition. Repetition, when it is allowed to open up and create infinities (whether real or merely as a kind of tangible but not yet actualized potentiality), has the potential to create new problems.[33] And these new problems so change the structured relations of problems associated with Christianity that Mormonism becomes something different. Consider what would happen if prophets or sacred texts proliferated to such an extent that the sources and character of truth escaped the landscape of problems involving singular authorities, whether they be human or written. But these kinds of complex-breaking arrangement of new problems only occur at the moments when infinities are brought into the here-and-now in a way that demands a reaction to them. The Church of Jesus Christ of Latter-day Saints, while opening up and sustaining repetition, works to not let these repetitions approach an immanent infinite. Even as there is an open multiplicity of sacred texts, they are kept to a manageable set. Even as there are continually multiple prophets, the number of prophets at any one time is capped (fifteen, if you're curious). And then there is the additional restraint that comes from how some believers use this religious material. The focus for most Mormons is not the kind of cosmological and ontological horizons that threaten to metastasize with the introduction of present infinities; most are instead concerned with the ethical use of the religion, and with the maintenance of family in the here-and-now. And it is the creation of certainties, rather than the proliferation of authorities and possibilities, that these more grounded believers prize. Hence, for them, the infinite is kept under check, and they do not find the warp and woof of day-to-day religion to be so impressed by problems of proliferation that it risks mutating into something beyond Christianity.

But also remember this. For many of our Mormon Transhumanists, what are viewed as conventional morality and traditional families are important parts of their lives, but this is not the case for all of them. And for both those who do prize ethics and family in the same ways as their co-religionists, and for those who do not, it is the infinite in Mormon Transhumanism that captures their attention. . . .

§8

So, religious transhumanism, Mormonism, and secular transhumanism have not had the sustained ethnographic attention that perhaps they could have received. By itself, though, this is no warrant for this project. Nor is it enough for that matter to say that Mormon Transhumanism itself deserves ethnographic attention now because it has not enjoyed it previously. Such consideration may be a boon to those who are curious for whatever personal or prurient reasons about any of these three movements. And such an ethnography would also have value as a document attesting to the contours of these movements at a particular time and place. It would be a contribution to an archive that future scholars could access should these groups be of interest years down the line. But unless anthropology is going to be some kind of societal butterfly-collecting, something more is needed.

There *is* something more. A book about Mormon Transhumanism, which by its nature must also be about religious transhumanism, secular transhumanism, and Mormonism, does things that a work on any one of these objects singularly could not accomplish. Such a book perforce addresses four crucial issues. First, it takes up both the current shape and possible future instantiations of the relationship between science and religion. Second, it allows us to think about the way that time, both as deep history and far futures, but also as a press of current events, is created. Third, because both transhumanism and Mormonism are effectively universalist movements that also see themselves as fiercely embattled by ideological foes, this is a meditation on the generative tensions between the open and closed society. Even as these groups are open to all, they are accessible only on their own terms and in their own way of being. And fourth, addressing the MTA provides a path to an entirely different horizon of thought, and to a framework which is novel, and yet which is in its way a return to some of the most influential anthropological writing not only within the discipline but outside of it as well. It is a way to think in the context of religion about the social work done by speculation.

But before we get to that last point, let us first discuss those other issues, each of which contains dualities of their own: science and religion; the power of speculation; mytho-cosmic and quotidian time; open and closed society.

§9

In at least one stream of thought, the history of modernity is an account of the West carving out culture and nature into two separate domains. And the way-marks and strata in that history are not just the development of ideas such as *culture* or *society*. These turning points are also the successive production of novel concepts about the natural world that underscore both its subservience to, and yet also its autonomy from, the human. And simultaneous with the operation sundering nature and culture was a parallel shift making God transcendent. This transvaluation of divinity was not carried out for the ends of glorifying God, but rather to transform him from a live presence to a mere matter of metaphysics, and thus eliminate him as a (necessary) causal force in either society or nature.[34] It should be noted that this is not the only narrative account that can be given for the emergence of the modern age. The current era can also be imagined as being brought about by the process of religious purification gone mad. In this framing, secularism was an accidental byproduct of attempts to rid religion of areligious superstition and questionable enchantment. Secularism was also an unforeseen side effect of projects working to inculcate a sense of individual ethical responsibility and produce tools for the management of a moral, individuated self. Rather than provide faithful, reformed subjects, though, these projects overshot the mark as far as individuated responsibility goes. They created a sense of detachment so strong that humans can be thought of as authentic, autonomous (or buffered) persons no longer necessarily dependent upon the divine. This same shift also allowed for a breaking of religious consensus. It made religion, which in previous eras seemed foundational, singular, universal, and compulsory, appear instead to be contingent, multiple, and voluntarist.[35]

Whether or not one accepts either or both accounts of modernity, there is one thing that can be stated definitively about where any history of modernity leaves religion in the West: the European medieval Church, the center of 'religion,' turned into the contemporary, multiple 'religion' such as Protestantism or Catholicism. This meant a concomitant loss of unity, of authority. Part of this process involved the various Christian collectivities losing police powers to both compel action and create institutional authority; these were capacities that were transferred into the hands of the state. This is not to say that the state continued to retain a monopoly on them; often these powers were transferred to other social organs, such as high culture or the academy, and as those powers are lost by religion, and as acceptably religious forms become pluralized, religion itself becomes mere personal belief. The fact that "religion has a universal function in belief is one indication of how marginal religion has become in modern industrial society as the site for producing disciplined knowledge and personal discipline."[36] Now unable to compel assent, religion could only ask for faith.

Further, the various expressions of religion became ranked in this new dispensation, sorted into *good* religions that did not challenge the secular consensus, and *bad* religions that ran athwart of it.[37] (Incidentally, nineteenth-century Mormonism, a theocracy-leaning religion breaking with default theology and worse yet, monogamy, was the very image of bad religion.)[38] Religion was not left entirely bereft by this development. Ever since its creation as an institutional and governing faculty, religion had the power to address ultimate issues, making authoritative claims about eschatology and soteriology. It could speak of the cosmos, and of the transcendent, in ways that other social organs could not. And even after the advent of secular society, religion continued to hold that power. What had changed, though, is that religion no longer held that ability exclusively. It now had an other of sorts in science.

Science was not always a competitor to religion; or rather, in previous epochs, what became science was not a competitor to what religion itself would become. This suggests that neither science nor religion is intrinsically the other's foe. In prior dispensations, *Scientia* and *Religio* were respectively defined as first a disciplined capacity for inferences, and second as a properly directed piousness. These were both understood more as complementary virtues than as abstract entities or separate modes of reasoning and understanding, or alternately as the external, concretized institutions that religion and science would later become. As later forms of *Scientia* began to be externalized into first modes of thought and then set bodies of knowledge, these expressions retained a sort of friendly familiarity with religion. There was no understanding, for example, of natural philosophy as being corrosive of either the Church's claims about this world or the next, or of the Church's institutional authority.[39] And many of the technological achievements of the late medieval and early modern eras that would be retrospectively understood as science were at the time thought of as religious. Rather than serving to impeach religion, these innovations were seen as part of a salvific process of regaining lost Adamic powers, and hence as foretastes of the millennium.[40]

But as science solidified into an autonomous set of interlocking disciplines, the relationship between science and religion broke down. Theories such as Copernicus's account of a heliocentric universe, or Darwin's account of the origin of species through natural selection (to choose two accounts that have aged well), or Freud's speculation regarding the breadth and depth of unconscious processes in human behavior (to choose one that has fared less well over time) can be visualized as inflection points, moments where the curve of history sharply changes. These are the instances where prior narcissistic conceits about the human must be dropped as untenable in light of then-novel cognitive tools. These new hypotheses gave us greater power to articulate the connections in a sea of otherwise incompatible phenomena, but at the cost of our losing the sense of our conscious selves having a central role in the universe. And twentieth- and twenty-first-century science were no kinder. In that later dispensation, disciplines from genomics to cosmology appear to open up vistas that dwarf the human, destroy our species' uniqueness, dissolve the subject, and place into question both the rationality of our decision-making and the freedom of our choices. Further, science has offered a variety of its own end-times scenarios which do not just eat at the centrality and purposefulness of the human, but which also suggest that our extinction would be as pointless as our birth and life. In a real way, science has heightened humanity's knowledge and prowess, but at the cost of laying siege to the concept and importance of the human itself.

These scientific claims that reconfigured what it meant to be human, while foreclosing humanity any greater importance, are instances of what anthropologist Abou Farman has called "Cosmological Collapse." In Farman's words, as science developed,

> it subsequently generated a series of its own end-time scenarios, from the inevitable heat-death of the universe to the end of the species to the end of the planet via climate change or an impending asteroid collision or. . . . Thus, the effects of disenchantment are heightened through the twined movement of scientific rationality reaching all the way from atoms up to multiverses, whilst falling categorically short of the metaphysical extremities [of religion].[41]

These are eschatological statements and disenchanting claims that, despite whatever the actual intention of the scientists making them were, in practice seem to mock the type of end-times hopes conventionally ascribed to religion, while simultaneously not producing any new sense of meaning or purpose to take the place of what has been undone.

This account should not be understood as a totalizing claim. There are many deeply religiously committed scientists who not only receive the sort of institutional support and education used to vet proper science, but who also become in their own right successful producers of scientific concepts, instruments, and institutions. Perhaps even more important, many of these believing scientists continue to see their work as serving a religious purpose, be that one of personal self-work that inculcates religious sensibilities or of the production knowledge that celebrates the divine.[42] Furthermore, it is a mistake to think of science as a homogenous entity, either in how it has acted over time, or in the numerous ways that it is constituted now. Throughout its history, the reasoning used by various scientific disciplines is subject to disjunctive fits. These were moments where knowledge was not accumulated through repeated applications of a standard set of research protocols that were associated with a stable set of epistemological and ontological assumptions. Instead, proliferating and incompatible pictures of reality and knowledge clashed with one another. This internally differentiated conflict should be no surprise. The complexity of the social negotiation and production of scientific thought—the act of corralling a host of institutions, workers, and instruments—means that there is a heterogenous aspect to science at whatever scale one is investigating.[43] Thus any claim made about the relation that science has with religion can at best only be about preponderance and particular communities, and certainly not absolute statements about science as a whole.

It is therefore incorrect to see science as engaged solely in the work of producing more arguments resulting in cosmological collapse, even if it has a marked tendency to dissolve humanity as both author and chief subject of history. There are some scientists, science popularizers, and lay enthusiasts who feel that current technical projects will not work to further erode the human's importance on the cosmological stage. Instead, they anticipate that such products will only heighten it. These advocates believe that these projects will reverse cosmological collapse by giving humanity (or what humanity ends up becoming or creating) computational, engineering, and biological capacities heretofore unimagined. Invested with these powers, humans or post-humans will be able to operate not just on a planetary scale but an interplanetary one, and in time perhaps even a cosmological one, harnessing the powers of stars, and then of galaxies, and spreading intelligence throughout the universe by converting the dead material that constitutes most of the universe into smart matter.[44] This would reinstall the centrality of humanity to the cosmos, either as the principal actor or as the progenitor of the principal actor. This might also mean the salvation of a sort not only for humanity as a species but for individual (post)humans as well. The computational power that would reanimate humanity's importance might, among other things, also be used to resurrect the dead—if not bodily, then at least as individuals existing in entirely simulated worlds spread through this computational universe. In this scenario, instead of rendering humanity purposeless, scientific thought would be in a position to offer a new teleology to the cosmos, and a new shared project for the species.[45]

This is not an entirely unprecedented development, and ideas of science undoing rather than accelerating cosmological collapse are not the exclusive product of contemporary science. Similar dreams have informed the work of some important Russian intellectuals and scientists since the nineteenth century, and some Anglo-American science fiction authors since the twentieth.[46] And while these kinds of imaginings are, without doubt, the most significant driving factor in the establishment of contemporary transhumanism, as a community transhumanism has grown beyond the social and institutional confines of science. This may be a dream about what is achievable by a regimen of scientific investigation and technological innovation, but it is not science's dream alone.

The touchier question is whether these technological aspirations should be considered religious either in their genealogy or their essence. This is the way that many contemporary theologians, philosophers of religion, and other fellow travelers have understood the transhumanist impulse.[47] Dreams of immortality, of a perfected social order, and of an AI apocalypse, it is argued, are merely the return of a familiar transcendent impulse.[48] (There are also some transhumanists who also would agree with purposefully frame transhumanist aspiration as religion, it should be noted, albeit for purely instrumental reasons. It has been suggested that absent a "firm, transcendent rudder-like religion," posthuman societies would inevitably collapse. Or even worse, it is feared that the social will for progress would become weak, and we would not even begin to engage in the research that would allow humanity to become a post-human species that could populate the galaxy.)[49] Instrumentalism aside, some theologians argue that, whether or not it is consciously acknowledged as such by transhumanists, due to its eschatological and soteriology nature, this newly envisioned post-human telos must be understood as religious, and transhumanism itself classified as crypto-religion. To these observers, transhumanism is little different than saucer-cults, or similar faux science faiths and techno-science spiritualities. And complaints made by transhumanists about being atheist, being hostile to faith, or otherwise not engaging in the work of religion, are at best false consciousness.[50]

There are strong arguments that go against reading transhuman as irredeemably religious, however. First, this idea is too Procrustean, cutting everything up according to preexisting and perhaps exhausted categorical borders.[51] There is, after all, something slightly tiresome about the claim that all transhumanism is religion. The claim that secular transhumanism *is* religion simply because secular transhumanism shares either a genealogical link or an intellectual kinship with religion can be seen as rhyming with certain theological conceits stating that because of its original grounding in religious reform, secularism is actually religious, and therefore is hypocritical and biased in its denial of other religiosities.[52] Both statements assume that there can never be something new under the sun and that we are doomed to having our thought eternally imprisoned in a limited set of categories and themes. It is a sort of theological imperialism, and a questionable claim to know those religion-denying transhumanists better than they know themselves.[53]

But second, when weighing whether transhumanism is religion, there is the more important fact that at the contemporary moment, there are few discursive and institutional commonalities shared by scientific-transhumanist and religious visions. Recall that following the secular settlement, each of these visions is now coming from a different space in the social landscape of knowledge and authority. They are valued for different things by most civil societies. They are predicated on different institutions (compare the laboratory to the Church, or the university to the seminary), they make their imaginative inferences based on different evidence (scientific publications and instrumental data for one, scripture, revelation, and doctrines for the other). And perhaps most importantly, these are messages articulated in entirely different languages. It seems an error to easily conflate the vast, and almost incommensurable, gulfs in ontology, epistemology, authority, and practice that lie between contemporary expressions of science and religion.

§10

This is not to say that all of those who make claims equating transhumanism and religion are necessarily mendacious. For all the differences between these two platforms, there is something about these two ways of speaking regarding ultimate matters that makes them rhyme with one another. First, there is again the simple fact that they *are* both addressing ultimate matters—issues which, by definition, tax reason and outrun experience. They are events and ways of being that are situated on the far side of an "[absolute] split between the historical worlds-to-follow" and the worlds of the here-and-now.[54] And this brings us to a second similarity: when these two modes of thought and action are addressing the topic of ultimate matters, both science (including here scientific-informed transhumanism) and religion trade at times in something close to spiritualism. The sort of spiritualisms that are engendered here are each grounded in different things. There is the classic religious spiritualism that stems from the sort of openness to phenomenological encounters that Max Weber dubbed religious athleticism. Then there is the aspirational spiritualism of transhumanism, which feels transcendent because it aspires to overcoming our evolutionary and technological limits, and to having us operate at the spaces where our knowledge and imagination break down.[55] Thus, despite profound differences, both these modes of thought are capable of producing affectively charged moments that push the imaginations far past their quotidian boundaries.

And it is this working at the limits of the everyday which gives us what is the third and perhaps most substantial commonality between these two forms of speculation. In addressing events, beings, and modes of existence situated beyond our furthest horizon, these statements partake in a certain freedom, and can be seen as in essence creative.[56] And as such, there is little constraint about what sort of musings they could engage in. This is not unfettered freedom. The constituent ideas incorporated or referenced in the exercise of this creative work must be recognizable to whatever community those ideas are addressing. This means that much of this speculative work results in material that is not only cast in the recognizable language of a particular intellectual enterprise but is still beholden to the identifiable logic of that particular enterprise as well. And they are of course conditioned by the circumstances under which they were generated, which includes the self-reflexive cultural understanding of the nature of speculation and imagining.[57] Then there is the fact that their circulation is shaped and limited by their means of expression or transmission.[58] But still, as imaginative work, they are relatively untaxed by the material affordances that constrain most human activity.[59]

It is, therefore, possible to consider the formal aspects of different exercises of this speculative faculty, and map different routes that speculation might take. Much of this speculative work takes the freedom given to it to explore the full run of the conceptual space produced by core concepts. This is seen most commonly in moments where one takes a specific idea and extrapolates—what happens if computing power continues to become less expensive and more powerful at the same rate as it has over the last half-century? If life spans continue to increase, if nanotechnology continues to improve, if human or human-artificial hybrid intelligence continues to advance? Similar exploration occurs in religious thought, with speculation regarding the meaning of scriptures, the nature of ritual, the sort of character involved in being Christlike on this earth. In both cases, one is interpreting according to established rules, utilizing well-understood assumptions to trace out the arc from a known present for an imagined further moment.

Speculation's subjunctive freedom, though, allows for more than the exploration of the possibility-space created by a concept. It allows for combinatorial moments, where possibilities are created by mapping one concept onto another. As an example, mapping the evolutionary process through which intelligence was created biologically onto the means through which, say, artificial intelligence is engineered allows for expansions and transformations of the possibility-space associated with both endeavors. In this scenario, evolutionary algorithms for the design of computing chips become imaginable, which allows for the creation of forms that would most likely escape any top-down conscious and intentional design attempt.[60] But it is also possible to map the understandings of often quite different domains onto one another. Post-Newtonian notions of materiality and post-Copernican notions of the cosmos, mapped onto Christian pneumatology and creation narratives, produce the embodied Mormon God dwelling among a plurality of worlds. It is here we see speculation not as the exploration of a domain, but as a form of virtual crossbreeding of ideas, where the combinatory is not the mere exhaustion of a space, but the formation of new possibility-spaces through a sort of intellectual hybridization.

This latter form of combinatory practice—speculative production of hopeful monsters—is also the moment of intellectual danger. The concepts forged in this way may not take. There is no guarantee that merely mapping one domain onto another will be productive in any meaningful way. In this book, we will see examples of this in some attempts to map different funereal practices onto one another, where incomensurabilities between the two approaches destroy the project before it can take hold. On top of the danger that new concepts may be stillborn, there is the risk that new hybridized concepts could create intellectual spaces equivalent to a wasteland or desert. Certain forms of apocalyptic thought addressed in this book, for instance, may have a subjective allure as a fantasy space, and a political importance as a way of marking one's relationship with authority. But when mapped onto other concepts, they are sterile. Absent any millennial model of a positive project, the nihilism of the apocalypse makes it empty and unproductive.

Further, even if a novel and productive intellectual conceptual space is formed, there is no guarantee that it can have any coherence apart from the parasitic one that comes from its dependence on two already existent, already intelligible concepts. While this is not a fatal flaw, this will often (though not always) mark the newly created concept as derivative and secondary, rather than as being an independent, fully explorable space that can be used as the grounds for more speculative activity. It will never be seen as foundational, only as contingent.

Finally, the ideas produced through this combinatory mechanism can be dangerous in some environments. As will be shown in this ethnography, issues of gender are quite salient for large numbers of contemporary Saints. And as we will see, mapping new conceptions of gender onto existing Mormon understandings of human nature and the plan of salvation is understood as a threat both by many orthodox believers and by the institutional Church. This mapping produces conceptual spaces that are too open while simultaneously exacerbating rejected practices and still painful psychic wounds. Not all of these novel ideas regarding gender are necessarily dangerous, and even ideas that are understood as dangerous could be assimilated through further speculative labor, but engaging in and promoting more speculative activity along those lines could ultimately demand that the Church reimagine what counts as proper forms of sociability and belonging. There is always an authoritarian temptation, therefore, to constrict these potential conceptual transformations by way of habituation, belief, and exercises of juridical authority through discipline, doxa, and governance. At times, this temptation is inescapable. While speculation can create difference, it is habituation, belief, and authority that create a narrative coherence and an identity. Without eventually returning to coherence and identity, combinatory speculation is just an intellectual witch's flight cutting across a landscape of undifferentiated chaos without ordering that landscape in any way. These constraining and identity-generating forces are thus necessary to create the sort of territorialization that gives speculation not just a raw power but a capacity to have and create meaning. But these are also conservative forces that dampen the freedom of speculation, and that always challenge any speculative activity that escapes the mere exploration of the possibility-space appurtenant to an already acceptable concept. Worse yet, these limitations can freeze institutions at the joints in times when they may most need speculation as an engine of transformation.

Speculation can, however, slip even these constraints if the situation allows. That is because speculation always has the possibility of being not just exploratory or combinatory but transformational. Exploration can push past the limits of a conceptual space, and combinations can result in forms that bear little resemblance to whatever ideas were juxtaposed or exacerbated to create them. This is the hardest form of speculative creativity to map because it does more than distend, reorder, and hybridize existing conceptual spaces. It instead makes concepts into entirely new forms, thereby erasing current boundaries and producing potentialities heretofore beyond any grasp. It does not shuffle or exhaust the *content* of thought but reorders the *grammar* of thought. Limits are transformed into starting lines, new conjunctions are made possible, and new practices of discernment and discrimination are created as old ones are erased.

These transformational shifts are on the same order as the changes in episteme in the Foucauldian sense of the term (that is, to the unconscious cognitive grammar shared by a collectivity at a given time).[61] Perhaps unsurprisingly, these transformational changes are often the ones most difficult to see at the time of their occurrence. The fact that the same objects can still be seen on the horizon distracts from the fact that the very ground we stand on while gazing at those objects is shifting. Therefore, it is difficult to confidently say that the sort of speculation encountered when considering various religious or secular transhumanisms is of that transformational kind and magnitude. But it might be safe to say that some of the speculation here indexes that kind of transformation. Much of it points to a moment when the imaginative plasticity of religion is matched by the creative plasticity of technical achievement, thereby collapsing any attempt to maintain a boundary between the two. As such levels of prowess are approached (or, alternately, are imagined to be approached), the boundaries between science and religion will have to be renegotiated, or perhaps removed. So, it is perhaps better not to ask whether transhumanism is a form of religion. We should instead ask: what are the material and ideational conditions under which the two disciplines may themselves commingle and fuse, becoming something, or somethings, else?

§11

Another aside: conceived of as done here, speculation can be understood as either a subspecies of a greater imaginative faculty, or as imagination itself operating under a different name. Perhaps. It is undoubtedly true that the imaginative faculty, whether thought of in a generic (*imagination*) or determinate, reified manner (*the imaginary*), is a category that has a well-established career in philosophy. But it is also turned to increasingly often in anthropology, albeit in different ways.[62] Classical Kantian visions of the imagination as working syntheses of sense impressions have long had their effect in the discipline. And continental psychoanalytic understandings of the imaginary as a register of unconscious, imagistic, and aggressive fantasy have been part of the anthropological tool kit as well.[63] But as of late, anthropology has been dominated by the work of philosophers and social theorists who view the imagination and the imaginary not as registers and faculties that operate at the individual level, but rather as the collective possession or even engine of societies. The imagination is seen as arising from the circulation of print media in a shared language, giving birth to an *imagined community* that takes the form of the nation.[64] Or *the imaginary* is alternately seen as a collectively shared, thoroughly social supplement to structured systems of meaning, a creative ethos particular to a society that governs how signification is expressed and how institutions function.[65] Or it is conceived of as a *social imaginary*, a mixture of norms and expectations that individuals within a society implicitly rely on to reflexively navigate, anticipate, and express that society itself.[66] These different senses of the imaginary are not always carefully differentiated. At times anthropologists freely use them in ways that allow these vying definitions to (sometimes quite productively) bleed into each other.[67] But whether handled parsimoniously or promiscuously, there has been a definite shift in anthropology toward viewing imagination and the imaginary as a shared social organ.

Some anthropologists are skeptical of such use, though. They view this expanded, collective articulation of the imaginary as just *culture* given a different name.[68] To them, this is a totalizing account of the imagination. These critics do not wish to isolate the imagination from the social, though. They reject any cordoning off of the imagination from other forms of intellectual and material production, saying that sort of partitioning returns us to a romantic conception where the imagination is an exercise of individual inspiration and genius.[69] But they do wish to make the imagination more granular, allowing for socially produced but still individuated moments of imagination that are tied to concrete practices instead of to a nebulous, expansive society. It is interesting to note that in many of these accounts, technology is the particular mode through which the imagination is produced. In various ethnographic locales, objects such as electric lighting, chat rooms, and open-source software are seen as engines used in the production of the imagination.[70] This approach, though, draws on an understanding of technology as something broadly conceived. For instance, in some accounts, numbers systems are also seen as examples of the sorts of technologies that can facilitate the imagination.[71]

To this way of thinking, it is the exteriority of these technologies that is important. This exteriority is found in their shared nature as forms of communicative mediation, and as affordances created by their mechanic nature when speaking of technology in a more traditional sense of the term. In this way of seeing, imagination and creativity are caused by the use of a preexisting technology in a different milieu, where this technology then works to novel effect. Now resituated in some new territory or circumstance, these prior skills and tools open up new possibilities. Alternately, this way of thought sees the imagination as an internalized rehearsal of the use of technologies, where what is being rehearsed, and how the rehearsal is done, are both functions of the features of the technology as object and activity located in the world.[72] For others, what is essential is that these technologies have a generative capacity that shapes the way the imagination is expressed, even if this shaping is indeterminate since technology does not completely control such expression. The limit and powers of a given technology lead to many possible imaginative forms, but only a few of them will come to pass. There will always be different, unactualized imaginative expressions that exist only as unrealized potentialities.[73]

In this ethnography, we can afford to be agnostic about the exact relationship between technology and imagination. We can stay uncommitted in part because here we are not dealing with the imagination in the broadest sense of the term. Our focus is on speculation, and only as it is found in a limited number of collectivities; this means that we can demure from producing an overarching theory of the imagination. We can also allow ourselves some indeterminacy in that technology here is multiple and exists as both a means and as an object. In this account, there is a concatenation of technologies. We will see myriad interactions between technological processes, ranging from conversation to formal presentations, printing presses, emails, chat rooms, message boards, social media, and virtual reality. And all these communicative and cognitive processes will be deployed for the discussion and imagination of a host of still other now purely virtual post-human technologies, which will range from sure-to-occur innovations to far-future vaporware. The use of particular technologies will give each expression of religious and transhumanist imaginations their shape, as well as open up certain dangers.

But the continual jumping from technology to technology also offers opportunities for these imaginative acts to be reconfigured, and sometimes expanded.

There is another way in which this study differs from much of the other anthropological thought on the imaginary and the imagination. This difference arises from the fact that the imagination does not merely face toward the future. As one commentator has noted, in

> the Western philosophical discussion of imagination . . . 'imagination' can refer both to the recall of things that exist and have been perceptually encountered before but are now absent (as when you conjure in imagination a scene from your childhood) and to the creation of images or ideas of things that do not yet, and may never, exist (as when you innovate a new image, maybe even one that has features you think could never be realised in the world beyond your imagination).[74]

But this also means that the imagination and death have a particularly grim affinity, even as death challenges the imagination.

> [W]e might argue that death is the ultimate challenge to the imagination in this sense, since death requires the imagination to recall something from the past that will never be present in the same way again. More than that, death may create a hinge between these two kinds of imagination; by tasking the imagination with bringing back, in images, people who will never come back in quite the form they formerly took, it insists that the imagination recall them in innovative ways: as ancestors inhabiting poles, as memories attached to mementos, as souls in heaven, or as songs, dances, and designs given by ancestors as gifts. It is thus in reckoning with death that imagination as recollection and as creation come together.[75]

Death as a challenge, and as demand for the imagination, particularly resonates with the points made in this book. That is because death is that which is hopefully to be overcome. This conquest of the grave is imagined to entail more than the acquisition of personal immortality. In some of the accounts we will look at, it also includes the resurrection of those long passed. This is a heady concept in itself. But this is also a thought that can scale upwards in vertiginous ways. As we will see, in Mormon Transhumanism, to imagine the way the future might be built is often to simultaneously imagine how this world came about. In this space, we are dreaming not just of lost people, but of lost worlds, and perhaps lost universes.

§12

The opening monologue that day in the Salt Lake Public Library stated that "Mormon transhumanists have many myths. . . ."

What if we took this claim literally?

Recall the earlier discussion of the structure of speculation *qua* speculation. There is a structuralist sensibility in that discussion's language of exploring all the permutations extant in a conceptual space, of mapping different problematics onto one another, of transformations in the generative grammar of thought that makes concepts possible. And while this book will not spend its time overtly leaning into the mechanics, a particular mixing of identity and opposition, the way key aspects are inverted when (conceptual) borders are crossed—the operations that were so important to Claude Lévi-Strauss's work on myth—informs much of the analysis here.[76] The idea of doubled transformations is particularly important, as the shift from transhumanism to religion and back again consists of something other than a monotonous alteration between immanence and transcendence. Instead, the difference between these imaginaries is situated not at the largest, most abstract horizon but in negations and transvaluations of more proximate and concrete thoughts, actors, and phenomena. It is these specific variations, rather than broad categorical judgments predicated on the religious or areligious mode of thought, that hint at what is at stake in the difference between supernatural and natural actors, and between passive, innocent acceptance and ethically weighty agency.

Two possible objections, however, come to mind. One is about the plasticity of the kind of speculative activity being discussed. The second concerns the relevance of an analytic, originally used for cold (that is, ahistorical) societies, to the burningly hot history-driven societies that transhumanists and millenarian Mormon believers live in now and anticipate coming into creation later. Plasticity first. Thinking through the transformations and structural exhaustion identifiable in speculative activity may necessitate a language close to classical structuralism. But at the same time, it also necessitates a language of strange becomings, which are classically understood as resistant to any structuralist logic predicated on static relations of mutual negative reciprocal determination.[77]

Some would state that Lévi-Strauss's thought has problems that exceed the issue of plasticity. Structuralism, and particularly Lévi-Strauss's structuralism, has a bad name these days. It is seen as denying human agency; as indifferent to any form of experience outside of the sensual contact with the natural world; as too entranced with a romantic primitivism; as insensitive to the forces of both realpolitik and the global macroeconomic peregrinations of capital; as being shot through with too much math envy; as a mechanistic, hidebound, and Procrustean system that wants to grind away complexity and leave only bare recursively nested binaries. There still are those who either see themselves as continuing Lévi-Strauss's project or who recognize in the contours of their thought the influence of Lévi-Strauss. But that does not change the fact that for many in the discipline of anthropology, Lévi-Strauss's structuralism is little more than an obsessive-compulsive French-theoretical curio.

For those not inclined to cherish Lévi-Strauss's legacy, one of the chief objects in evidence is his four-volume *Mythologiques*. For its critics, it is an anthropological shaggy-dog story. It starts with a schematic presentation of a single Brazilian Bororo Indian myth (a tale of a revenge-minded incestuous boy who is mistaken for dead after a series of purposeful cruel challenges administered by his father). This myth is then contrasted with differing variants of that myth, with the scope of comparison eventually covering all of indigenous North and South America by the time that the final book in the series reaches its end. It is exhaustive but exhausting. Lévi-Strauss documents every substitution, inversion, or transposition of the elements of these myths, as he employs structuralist principles to chart what he purports to be a continent-spanning implicit indigenous cognitive system. This multivolume tome can be read as a modern-day version of *The Key to All Mythologies*, the syncretic, rambling, unfinished tome written by Reverend Casaubon in George Elliot's *Middlemarch*. For others, *Mythologiques* is a paranoid set of books. It is a power-mad attempt to make two whole continents' worth of indigenous creativity submit to an intellectual machine constructed on the second floor of an index-card-strewn Parisian university office.

But true to the structuralist play of difference, there are other possible readings of Lévi-Strauss's magnum opus. And in these other readings, the values associated with structuralism are themselves inverted and repositioned. The Brazilian anthropologist Eduardo Viveiros de Castro sees a mutation in Lévi-Strauss's thought occurring during the writing of *Mythologiques*. More precisely, Viveiros de Castro sees something that had always been latent in Lévi-Strauss finally coming to the fore. In this reading, Lévi-Strauss's primary debt is not being toward the linguistic structuralism of Ferdinand de Saussure.[78] Instead, this understanding sees a Lévi-Strauss wounded by the accusations by his fellow anthropologists that he "us[ed] exclusively binary patterns." Here is a Lévi-Strauss who sees himself as relying on an analogy-centered reading of myths influenced by the mathematician and biologist d'Arcy Wentworth Thompson.[79]

Thompson's classic *On Growth and Form* is itself a strange text. It is concerned with the way that variation in features such as rate of development or overall magnitudes found in biological specimens can effectuate torsions and involutions of forms. This allows what is in effect the same topological shape to be topographically expressed in wildly different ways. And this topological logic is then used to forge an account of the variation between not just different biological individuals, but between different members of species of the same genera.[80] As Lévi-Strauss told it later in life, this idea of transformation, which he first encountered during his exile in the United States, is not incidental to his idea of structure. Instead, Lévi-Strauss states that the very notion of such transformations is "inherent in structural analysis":

> I would even say that all the errors, all the abuses committed through the notion of structure are a result of the fact that their authors have not understood that is impossible to conceive of structure separate from the notion of transformation. Structure is not reducible to a single system: a group composed of elements and the relations that unite them. In order to be able to speak of structure, it is necessary for there to be invariant relationships between elements and relations among several sets, so that one can move from one set to another by means of a transformation.[81]

For the idea of structure to make sense, alternative formulations must be not just technically conceivable but actually possible. And for such transformations to occur in the real world, it must be possible to stress, distend, or twist the relations between elements, all in real-time.[82]

Transformation allows us to attend to individual variation, to shifts in emphasis and saliency. But it also allows us to note the thresholds where the force of the torque exercised on local instantiations creates the kind of double twists in mythic imaginaries that (at least in Lévi-Strauss's analysis) mark the thresholds that identify boundaries. Such transformations, double twists, and mutational inversions were always arguably present in his writing, as one would expect if Saussure and Thompson had both entered Lévi-Strauss's mental toolkit at the same time. But as we work through the *Mythologiques* and other subsequent works, these transformational operations do become central to his analysis. Acknowledging the increasing importance of Thompson would make Lévi-Strauss's mode of operation a moving target, growing "closer to dynamic fluxes than algebraic permutations" as his thinking on myth matured.[83]

This shift away from closed structural constellations to topological deformations not only allows for shifts, gradations, and folds in myth. It does more than moving Lévi-Strauss toward the direction of intensities and away from arrays of fixed positions. It opens up the process, allowing for a continual movement, the constant creation of new forms, in short, for a "structuralism without structure."[84] No longer rigid and therefore brittle, this is a system that is always in play, and a style of thought where each movement or actualization is also a moment working toward some other expression. And it also makes *Mythologiques* (and the books by Lévi-Strauss that came after those four volumes reached their conclusion) useful in mapping how variation in thought in other realms could be traversed.

Even if this argument is persuasive, it still needs to be shown that the speculative work discussed here is an apt object for an analytic made for the study of myth.[85] Myth, after all, seems inapposite for future imaginings. We have seen the mythic already. It is no accident, though, that in that convocation of a Mormon Transhumanist Association conference, visions of technological futures were purposefully cast as myths, and not as techno-rational futurist predictions or common-sense extrapolations from the present moment. This juxtaposition of temporal frames is one that will be seen yet in these ethnographic depictions of transhumanism, Mormonism, and Mormon Transhumanism. These are all spaces where the fast-acting, human-scale time of technological innovation and change in social mores unfold at their own rapid pace while also simultaneously creating and transforming mythic time scales of cosmic redemption.

There are good reasons to lean into both the mythic and into cosmological time when struggling with the sort of ethnographic objects we have here. Part of the warrant to speak in terms of quotidian and mythic time has to do with the fact that at least some of the movements we will be talking about understand themselves as religious, or at least can be given that reading by others. There is a relation between myth and religion, though not a correspondence. Lévi-Strauss suggests that much of ritual (one of the most common elements of religion) exists as an inversion of myth. Rites blur together various elements, and thereby leaning into metonymy enables a sliding between those elements that comprise a set, instead of tracing structural oppositions constituted by myth. But still, unlike the difficulty of speaking about religion in spaces such as the indigenous Americas where religion does not exist as an internal category or a word, the purposeful enchainment of myth and religion seems in this instance like a reasonable proposition.

By way of contrast, though, the enchainment of myth and science seems like something else. It is not that myth and science do not have a relationship at all, but rather that one element in this relationship always had the promise of eclipsing the other. Again, Lévi-Strauss notes that in studies of myth, writing marks a threshold: "It only seems to me that in societies without writing, positive knowledge fell well short of the power of the imagination, and it was the task of myths to fill this gap."[86] Myth was then the supplement of the sort of naturalistic knowledge that would later be understood as science. But as a supplement, myth was always at risk as naturalist knowledge became encyclopedic in its scope due to the prolonged artificial memory created by writing. Like a medieval palimpsest, where Greek literature was occluded through inscription by latter Christians texts, myth is the ur-text on the manuscript page that science over-writes. As understood by Lévi-Strauss, such an expansion of positive knowledge through writing would end up exhausting these sorts of issues, rendering myth without a space in the human intellect to express itself.

But the relation between myth and science is more insistent than might be implied by that dynamic. At a certain tipping point, rather than winnow away myth, science starts to instead catalyze myth's growth. Scientific knowledge, expansive to the point of exceeding any one person's mastery, and complex to the point of escaping most people's understanding, no longer colonizes the psychic space that was myth's domain. Rather, myth becomes the only way that science can be presented outside of the intellectual circles of knowledge specialists. As Lévi-Strauss put it,

> With us, positive knowledge so greatly overflows our imaginative powers that our imagination, unable to apprehend the world that is revealed to it, has no other alternative than to turn to myth again. In other words, between the scientist who through calculations gains access to a reality that is unimaginable and the public eager to know something of this reality, which is evidenced mathematically but entirely against the grain of sensible intuitive perception, mythical thought again becomes a mediator, the only means for physicists to communicate with nonphysicists.[87]

Outside of small cadres of experts and professionals who were working with a technical knowledge of phenomena, concepts such as the Big Bang, DNA and RNA, and quantum physics all become mythic narratives, working to document moments and entities that have origins so beyond our lay ken that they seem to have supernatural, rather than naturalistic, powers. And this mythicization of science happens at the same time that much of this science impinges on our lives, becoming real-world phenomena. Through mechanisms such as the recombinant powers of CRISPR, the quantum phenomena of lasers, or the relativistic logic behind GPS satellites, mythic shadows of concrete technical entities live among us again.[88] The question is what work, if any, is accomplished by this speculative thought?

One last point: there are some who might object that myth is an inappropriate framework because the material discussed herein is so flat when compared to the wild Amerindian narratives that constitute *Mythologiques*. To this, I would simply state that the imagined worlds charted in this book may be far

wilder than one might initially suspect; we are trading the zoological concrete-
ness of indigenous American myth for the technological specificity of transhumanist speculation. Still, if, after exploring these stories, the reader finds
them relatively *manqué*, one can't be too surprised if the decades-old settler-
colony myths discussed are so much barer than the millennia-old myths of the
indigenes that the settler-colony uprooted. If these myths strike you as thin and
flat, remember that the recently coined, denuded, sterile myths of the invaders
are still myths, all the same. Also, recall one other significant difference. The
final few pages of *Mythologiques* famously end with a note of cosmic pessimism,
meditating on the epiphenomenal and ultimately meaningless nature of the
human species, which is, like all life, eventually fated to become extinct.[89] This
book takes no stance on the ephemeral nature of life. But unlike the author of
Mythologiques, the *subjects* of the book are infused with an optimistic vision
of (post)human life thriving for eternities.

§13

One of the functions of speculative thought is to dilate the constraints of society, and to rush ahead of already-extant society to produce new social forms.

    The title of this book, *Machines for Making Gods*, is a fragment taken from
the longer phrase, "realizing the essential function of the universe, which is a
machine for making gods." This quote, in turn, is just a segment from the
English language translation of the sentence "*À Elle de demander ensuite si
elle veut vivre seulement, ou fournir en outre l'effort nécessaire port que s'accom-
plisse, jusque sur notre planète réfractaire, la fonction essentielle de l'univers,
qui est une machine à faire des dieux.*"[90] In English, the full sentence reads
something along these lines: "Theirs is the responsibility, then, for deciding if
they want merely to live, or intend to make just the extra effort required for
fulfilling, even on their refractory planet, the essential function of the universe,
which is a machine for the making of gods."[91] Read *gods* as *myth*. The "theirs"
here is us, our own species, seen for just a moment from an alienated and inhuman perspective. This is a statement that is more spoken *about* humanity
than it is spoken *from* a standpoint internal to the species. But that degree of
forced alienation is perhaps necessary to make a statement of this sort and scope.

    This quote is the last sentence of the last book written by the Anglo-French
philosopher Henri Bergson, who wrote during the late nineteenth and early
twentieth centuries; the title of the book (as translated in English) is *The Two
Sources of Morality and Religion*.[92] In some ways, this sentence is a curious way
to close an oeuvre by a writer with Bergson's sensibilities and concerns. This
is especially the case since one of the hallmarks of Bergson's career was a

long-running critique of what might be called mechanical thinking. This objection to machine-like conceptions of the human ranged from his dismissal
of the "cinematographic illusion" for capturing movement as a series of static
pictures instead of as a true unfolding and becoming, to his claim that the
comic is a derisive reaction to something mechanical found in a living being.[93]
The closed, autonomous functioning of the machine was the opposite of the
flowing, intermeshed phenomenon of life. For him to have the capstone of his
writing career be an embrace of the machine as a metaphor for the operation
and telos of the universe, therefore, seems counterintuitive. Perhaps in the end,
for Bergson the gap between the inanimate machine and the living creature
lessened, or possibly even collapsed.

His final book, though, was not so much a break with a particular vision of
the machine, but rather a culmination of Bergson's entire line of thought, a
fitting conclusion to his intellectual career. For all his suspicion of the mechanistic, this work was no mere Luddite rant. It was something far more ambitious. Since its very beginning, Bergson's work was interested in contrasting
forces or instantiations of reality: temporal intensities set against spatially
structured cognition in *Time and Free Will*; virtual memory versus concretized
image in *Matter and Memory*; movement and stillness, auto-poetic conserva-
tiveness and branching diversification, instinct and intelligence in *Creative
Evolution*.[94] In *The Two Sources of Morality and Religion*, Bergson synthesizes
all of these separate earlier oppositions. He does so by at times laminating one
of his prior antinomies onto another later conjoined divide. At other times
this is accomplished by integrating some of these internally opposing dyadic
pairs and juxtaposing them with other of his theoretical bifurcations. This
integrative work was done to fashion an opposition that he saw as having a, and
perhaps *the*, vital role in the governing human sociality: open societies against
closed societies.

Closed societies were grounded in moral obligation and static doctrine, but
this was still a moral obligation to one's fellows, built around static doctrine's
reified image of who it is that one is obligated to, how it is that they should be
identified, and what forms these obligations take. This ethos did not extend
past the group. Open societies, by contrast, were informed by what Bergson
called the "creative emotions." The characteristic operation shared by these
open societies is reaching out across the sort of barriers that mark the limits of
closed groups, the forging of new thought and of new ways of being that overcome difference rather than see difference as the line where sentiment ends
and hostility begins. This final opposition of openness versus closure, while
indicating what Bergson felt were real polarities, was also an ideal type; this
opposition was never total. Neither openness nor closure, creativity nor

cohesion, came in pure, unalloyed forms. In every instance, there was always an admixture of these two potential directions of will and belonging. But in any one of these combinations, there was no law mandating that these two forces should be present in the same ratios, or that they should be expressed with the same strength, insistence, or form.

The final synthesis of his work that produced the entangled opposition of open versus closed societies does not mean that his earlier interests had been completely folded into the project, seamlessly sublimated into this one crowning dichotomy. Bergson's focus on creative emotions shows the continuing importance of affect, but also shows an interest in time and virtuality. Bergson's life-long critique of mechanistic and spatial thinking was based on his belief that these two cognitive frameworks arrested the subjective experience and eclipsed the thought of the way processes unfold over time. Events were not a mere jumping from initial states to conclusions, where both the beginning and the end could be captured in quantitative terms. Instead, the hesitant and quicksilver time between, the flowing moments where events opened up, was just as important. Time was where one could see parallel series of unfoldings synergistically affecting one another in a way that was so subtle, so much in flux that it could best be thought of only *qualitatively*. Bergson often referred to this framing, with its privileged viscous time (instead of snapshot-like spatialized states of affairs) as *duration*, or, since there is a tendency for English translators to use the original French term, as *duree*.

Time was not Bergson's only concern; or rather, time was only one face of what held his attention. He was also interested in the virtual. This use of *virtual* is not the computer-virtual in the sense that our transhumanists might use the term (as a shorthand for a computer-generated immersive fictive reality that one might enter and exit). *Virtual* for Bergson was not about what computers did, it was about what could be done through any process. Virtuality was a way of talking about potentiality as opposed to the merely possible. The *possible* is a framing that suggests a state of affairs lacking only one element—its existence. As Bergson described it, the possible is like a fully formed slice of reality sitting on a shelf, waiting to be taken down and used as-is. Essentially, the possible was not unlike the gridded, quantified view of space as extension that Bergson thought had too much sway on our thinking.[95] But by way of contrast, potentiality instead is hesitant and open. It is a direction one could imagine taking things, even as the details as to what taking that direction would mean are always larval, its results and success promissory and speculative rather than sure. Virtuality was a way of talking about potentiality as not actual, but possible, ways forward—which were still very much real. Specifically, virtuality was used to talk about numerous large-scale transformations, such as the process of biological evolution. But for Bergson, it is clear that bodily movement, thought, and memory were virtual processes as well. The opening of memory or thought might at first suggest a hazy welter of sensations. But this open virtuality would concretize as a subject focused on a specific memory, or even on a particular aspect of a memory, or on a particularized notion as it hardens into a concept. A bodily motion may start with a certain ambiguity. But the direction and force could concretize as that motion continues to be carried out. This, though, was a two-way operation. At times, the memory or concept or body would have to relax, to fall back into a wider mnemonic, cognitive, or physiological unfixedness as one searched for new aspects to conjoin it with, or as one must lose focus on one of its edges in the search for a different edge to latch onto. Again, displaying the specific style of thinking that can be seen in so many of his other intellectual endeavors, for Bergson, this play of concrete memory and hazy recollection was not a special case, but an example of a much wider play of affairs. For Bergson, *everything*, and not just psychic processes, floated between pure actuality and pure virtuality. Because of its relationship with potential new paths, and also because of a certain degree of uncertainty, in Bergson's thought, the virtual and creativity were different threads, different dimensions, of the same process. *Duree* as temporality and virtuality as qualitative state were so mutually implicated that they were inseparable.

Thought this way, openness, creativity, and a continually moving uncertainty come together. This surely describes the groups that we are looking at here. Bergson-derived sensibilities offer ways of talking about the sort of different expressions of each group discussed earlier. This gives us a language for the kind of inevitable self-differentiation that occurs between each concrete instantiation of the groups, both from one another and within each group itself, all due to the internal incommensurabilities that constituted the groups in the first place. The diverse origins of a group, and the tensions that exist within it, mean that each time the group is realized in some manner, be it in different spaces or different times, the group is not a pure repetition. And each repetition of the group also changes with time. This dynamic is why we can compare different strands of Mormonism, different flavors of transhumanism, without privileging any one expression. Bergson's work eases the way.

But Bergson's thought and technical vocabulary regarding openness and closure have a particular aptness for the case(s) at hand. Mormon visions of a restoration of the gospel that opens the way for radical theosis, and transhumanist visions of overcoming the human, both promise new vistas, new ways of socializing and belonging, even if their exact contours are unfixed and underdetermined. But both of these groups also have elements of paranoia. There is often a fixation on enemies (real or imagined) who are seen as not merely different but antithetical. This openness and this closing-in sometimes work in conjunction. Transhumanism and Mormonism are both open to everyone, but only provided that everyone is willing to leave their own specificities, aesthetics, and aspirations aside. And at times, this play of openness and closure is a cross-cutting force which serves not to dilate or constrict a relation to the outside but rather to eat at the fabric of the group itself. This angry turning-inward creates both dissidents and dissension as individuals and groups sometimes struggle to reform a collective, or as they alternately struggle not to be ejected from these broader collectivities of belonging as the morality of the closed society begins to hold sway.

Aspirations for openness set against spastic fits of closure can be a recipe for a group pulling itself apart. So, too, with the MTA. Mormon Transhumanism's laundering of Mormonism into transhumanism, and of transhumanism into Mormonism, in an attempt to cure the ills of each group by viewing them through the prism of the other. And this speculative energy is not limited to the conceptual and the mythic, though this is where it is the most Promethean. While creating new forms of speculative thought, Mormon Transhumanism also creates an openness to a different space. The MTA can be a way to access a secular world, where one can transform the perceived weirdness of Mormonism into a more digestible technological enthusiasm. Conversely, it can also serve to take some of the triumphalist visions and libertarian moral equivocation that characterize much of secular transhumanism and introject a more socially comprehensible, collective veneer of faith and a more carefully articulated religious ethic. But this ideological alchemy does not mean that the problem of openness and closure is cured in that space, either. For all of its desire to be accepting, the inherent religiosity of the MTA closes it off from large segments of transhumanism. There is an apparent symmetrical reaction among the more conservative Mormon believers as well. The air of a technological and nonreligious reimagining of core Mormon tenets makes Mormon Transhumanism seem like a secular, progressive, and perhaps even heretical Mormon Trojan horse to some of their more dogmatically orthodox co-religionist critics.

Social scientists and historians once commonly believed that under the light and heat of modernizing rationality, religion in democratic, technologically sophisticated states would wither away. This has turned out not to be so.[96] The collapse of the secularization hypothesis makes it clear that no matter how fast the *nones* (those who profess no religion) grow in American culture, religion is in no danger of going away. But at the same time, both technological changes and the adaptive economic reconfigurations that occur in response to such change make it appear that transhumanism is here to stay for a while, both in its soft form of near-futurist projections, and in its hard form of cosmological speculation. And in each group, we can see Bergsonian dilation and constriction. If Bergson is right, the tensions between an open society and a closed one are more than just a moment of particular conflict we are enduring for an extended but still delimitable period; rather, these tensions are endemic to our species. Perhaps it will also be endemic to whatever our species becomes (assuming that there is a future for either humanity or whatever it is that humanity will end up fostering). But just because tensions between openness and closure are inescapable, it does not follow that all expressions of that tension have to take the same shape or carry themselves to the same logical or historical conclusions. Bergsonian attention to shifts, changes, and difference points to that truth as well. We will see both synchronic and diachronic variation in Mormonism and transhumanism. We will see the same thing in Mormon Transhumanism as well, and in even sharper relief, as we can look at it in a finer-grained and ethnographic, molecular way, as opposed to the broader molar standpoint required when studying nation- or world-straddling movements like Mormonism or transhumanism. And viewing how these various groups, as well as the elements that comprise them, take on the problem of openness and closure may tell us something about the human capacity to build the Bergsonian open society in the present day, and perhaps a little something about the human (and post-human) species' capacity to build open societies in general.

One last thought, to bring our discussions of Bergson and Lévi-Strauss together: the mythic on one hand, and the dynamic of openness and closure on the other hand, may be more intimately related than they appear. Lévi-Strauss saw "curious analogies" between Bergson's philosophy and the understandings that often characterized the indigenous peoples that Lévi-Strauss's own thought focused on; in Lévi-Strauss's account, both were characterized by a tendency toward holistic thought, an interest in the play between the continuous and the discontinuous, and a sense that these two poles were "complementary perspectives" on the same "truth."[97]

A different aspect of Bergson's thought can be found in the mythic speculation looked at in this book. Lévi-Strauss held that in addition to their apparent or patent message, myths also had a latent message, an aspect of the myth that "is not delivered explicitly, but could be understood implicitly" by that myth's original audience; this is an aspect of the myth "open to elucidation by structural analysis."[98] In the *Mythologiques* series, Lévi-Strauss ultimately held that while the latent content of the American myths he looked at contained all sorts of cognitive aporias arising from indigenous theories of zoology, plant biology, meteorology, geology, and astronomy, the ultimate latent content was a meditation on the theme of the conquest of fire, and how that conquest created a divide between nature and culture.[99] In this book, the myths of the future discussed will contain and convey multiple moments of latent content, delivered in the form of ultimately undecidable problems that generate mutual exclusive concretizations: preservation/destruction; authoritarianism/egalitarianism; divinely fashioned/human-fashioned; fixed gendered and sexual belonging/post-gendered sexual indifference or indistinction; future as difference/future as return. But it might fairly be put forward that ultimately, the latent content covered in this text, the analog to *Mythologiques'* fundamental problem of fire and the nature/culture divide, is the irresolvable tension between the open and the closed society.

§14

Synthesizing these various strands, what do we have? In Mormonism, transhumanism, and Mormon Transhumanism, we have three collectivities, each capable of self-differentiating through numerous distinctive expressions. These three collectivities may or may not be united by something along the lines of religion (though, as seen, there is reason to be skeptical of this claim). But wherever one lands on the question of religion, there is no doubt that even though they are entangled with numerous other issues, for the case at hand, all three collectivities have enchained their core problematics to yet another problematic. (With problematic understood here as internal incommensurabilites that make them distinctive, cast here as virtual questions that are answered through the various ways these collectivities are actualized in specific, concrete moments.) Specifically, they are together all entangled in the messy, open project of not just fulfilling their humanity but overcoming it. These three different modes of enmeshment with overcoming humanity open up issues of the current arrangement between and structure of science/scientism and religion. They create space for both hopeful universalisms and paranoid anxieties about alterity. And they cleave, in both of the antipodal meanings of the word, at once joining and splitting, a quotidian, secular, and rapidly accelerating time of the now with a cosmic-mythic time as expansive as the furthest horizon.

It may be possible to try to present this situation through an attempt to disarticulate these three groups. Such a depiction would work to disentangle the issues of scientism and religion, of openness and closure, of mythic time and quotidian time. One could attempt to address these issues in serial chapters. Or possibly the project could be taken on through some nine-fold combinatory matrix. (I could imagine some multidimensional chart showing the intersections of Mormonism and science/religion, transhumanism and openness/closure, and so forth, going through every permissible permutation in this imaginary grid). And indeed, the first three extended discussions in this book will be shaped by a set of partitions, where Mormonism, transhumanism, and Mormon Transhumanism are hived off from one another so that the relevant specificities of each of these three collectivities can be traced. But once these three objects are laid alongside each other, the ways that they catalyze and impede one another quickly proliferate. That does not mean that there are not shared patterns. There are moments these three objects mirror each other, moments where they appear to be inverted images of each other, moments where they exacerbate, curtail, or conflict with each other's expressions.

But these patterns would be obscured if represented through the disarticulation of this complexity into individual elements. Even if we could make issues of religion/scientism, openness/closure, and secular/mythic temporality separate, autonomous, and orthogonal axes, we could not chart the mutual orbits of our three bodies in this conceptual space with the clarity of a single equation. (And even if we could attempt such, our efforts would be undone because these three objects do not constitute a closed system.) The only way that these patterns can be made visible is to watch them unfold and interact in the spaces they occupy, like various sets of ripples in a pond crossing each other and either intensifying or nullifying sections of the expanding wave pattern, or like viewing the churning of a kaleidoscope, where various patterns come into being as the reflected pieces tumble about.

It may be possible to force the patterns made by these three bodies into the conventional ethnographic genre; I have certainly tried. But even while trying to engage in that work, I have been forced to experiment with the form. Hence the structure of the book: a line of folios, or in this case *series*, comprised of thematically and topically linked moments and passages designed to show the specific features of the particular entities depicted, while facilitating a comparative understanding of the members of the *clade*, as found in one (expansive) territory. Each section or scene is a thought, an observation, an event, a snatch of conversation, or a bit of found text, which are then threaded together like beads on a string. Here, each chapter is comprised of a series of these scenes, strung together by transitions operating as hard-cuts, jump-cuts, cross-cuts, and match-cuts, depicting ethnographic moments, passages of texts, and concepts that are taken from various expressions of Mormonism, transhumanism, and Mormon Transhumanism. Taken synoptically, the scenes shaped and joined by these cuts allow the foldings, inversions, and twists in religion/science, openness/closure, and cosmic/secular temporalities to play out in a way that makes the patterns visible. These patterns play out in various conceptual spaces, terrains of possibility where the contours are shaped by shared problems that are answered in manifold ways. In the first three series, we see the shifting contours and fault lines first of the Church of Jesus Christ of Latter-day Saints, then of secular transhumanism, and finally of Mormon Transhumanism. At this point, we pull back a little, peripherally taking in elements of the Church of Jesus Christ of Latter-day Saints while placing Mormon Transhumanism at the center of our vision. In the fourth series, we watch the play of secrecy and revelation in Mormon religiosity, and in the series immediately following, we turn to discipline, belief, and speculative thought, observing how Mormon Transhumanism is shaped by and reacts to these social facts. We then start traversing all these collectivities, leaving behind the last vestiges of limited focus on groups. Instead, we follow various threads as they run across different communities and different practices. In the sixth series, we follow the paths left by the dead, looking at the resonance patterns found in burials, cryonics, and cremation in transhumanism, Mormon Transhumanism, and Mormonism itself. The seventh section likewise takes off from Mormon apocalyptic and millennial strains, and in the eighth series, the various visions of creations and cosmologies that are either held to or anticipated. The ninth series takes up gender, sexuality, and kinship by touring propagations, procreations, and poly-gamies as they knit together various new forms of family, and open up spaces for worlds without end. This last passage is a moment where speculation, having unchained itself from particular social problems, returns to the concrete social to illustrate the way that speculation loops back at critical moments to recon-figure problems and produce new forms of social life.

This approach—first carving these conceptual spaces and collectivities, these events, texts, and ideas, into scenes, and then stringing them together to form series after series—hints at something subjective, even arbitrary. One could imagine numerous, and perhaps endless, other moments and concepts strung together in different ways, covering other facets of our three collectiv-ities. We can imagine the construction of different series than the ones we use here, strung together from different scenes, that traverse some of these same conceptual spaces, but in different ways. Other paths might cut across the care for the dead on a different trajectory, or explore other sides of the millennial, or the cosmic, or the procreative, the pleasurable, the polymorphic and polyg-amous. It is our strong suspicion that, while some of these other tranches might present these objects in a new light, over time, as the process is repeated, and as slice after slice is cut through these imaginary terrains, we would slowly see the patterns found in our original series repeated. Over time as the visions compounded, collectively, they would begin to mirror each other to such a degree that we could not only see the collectivities in higher resolution but be able to grasp these resonance patterns in the abstract. As the number of series carved out asymptotically approach some conceivable but still imaginary ultimate zenith (that is), we could do something like what is done in stable versions of the three-body problem in physics. We could be able to summarize what we have been looking at and what we have found, not only trace out histories and futures, thereby getting a sense for the trajectories that these collectivities are on. But we would also be able to articulate the periodicities, harmonies, and symmetries in the epistemological frameworks, social constrictions and dilations, and temporal imaginaries we have been patiently mapping.

At the birth of cultural anthropology as an academic discipline, the Victorian-era armchair ethnographer Edward Tylor defined religion as "the belief in Spiritual Beings."[100] For Tyler, a recovering Quaker, religion in the contemporary world was merely what he called a *survival*, an outmoded relic of a time when culture was less advanced, less rational. But at the same time, Tylor also saw animism, which he considered to be the ur-religion and the genealogical precursor of all other modes of religious practice and belief, as a rational, and hence proto-scientific hypothesis, a theory built to parsimoniously explain phenomena as varied as dreams, fugue states, visions, disease, and death. Tyler's ethnocentric account of cultural evolution (a process he imagined as reaching completion with the development of rationalist, European civilization) is rightfully no longer an acceptable proposition in current anthropological thought. His original definition was overly centered on the somewhat parochial Protestant concept of religion as belief. It also leaned too heavily on the particularly narrow and incorrect idea of the universality of *spirit* as a specific category of entities or matter that could be differentiated from the rest of the world. There have been attempts to cure Tylor's definition by replacing the term *spiritual beings* with the more expansive *superhuman beings* or, more capaciously, reworking the definition root and branch by rearticulating it as "relations with the more than human."[101] Thought of that way, what we have here is a moment where scientistic dreams and religious yearnings, long separated in the Euro-American imagination, if not in the actuality of human history, dovetail in a surprising manner. And as they do, they would both seek a very particular relation with the more-than-human. This would not be a relation of supplication, adoration, veneration, or utilitarian implementation, but a relation of a fraught and hopeful *becoming*, which is made actual through a determined practice of *speculation*. Theosis, the idea of the divinization of the human, may have a deep history in Christian theology, reaching all the way to the earliest moments of the pre-Constantinian church.[102] But here, in a world where we can find post-Copernican religion, and post-Turing technological aspirations, we see these various strivings for a divine becoming at once bleeding into and disrupting one another, as different collectivities struggle to build the technical, intellectual, and social mechanisms necessary for the construction of Bergson's machine for making gods.

PART I

Dramatis personae

First Series: Mormonisms

§15

Apostle and Second Counselor to the First Presidency Dieter F. Uchtdorf had two problems.[1] One of them was medical; one of them was not. Which is not to say that the two problems were unrelated.

We do not know what the medical problem was. Uchtdorf chose to keep the particulars of it private, despite the fact that it was being spoken about on one of the largest stages imaginable. The stage was the General Conference for the Church of Jesus Christ of the Latter-day Saints. Those last three words, The Latter-day Saints, is another common name for the Church, though that title is more commonly shortened to LDS, or often by members to Saints; and this is different from the title that the Church was given when it was first established in 1830 as the Church of Christ. Despite this wealth of possible titles, most commonly, the church and its members are known as Mormons. This last moniker is a broader term, including not only the Salt Lake–headed Church but all sorts of fundamentalist groups (many of which still practice the nineteenth-century doctrine of earthly polygamy) who would not be welcome to the General Conference. It is perhaps the capacious nature of the name that led to a push in 2018 by Church leadership to use only the full name of the Church and reject all other appellations. Use of any name other than the Church of Jesus Christ of Latter-day Saints was, in the view of the Church's paramount leader, "a victory for Satan."[2] As in the case of numerous previous pushes to use only the official name, those other names will certainly endure.[3]

President Uchtdorf may have been addressing only believing members of the Church and not any of those various polygamists and sectarians that are seen as an anathema to the organization. But as usual in general conference, no one was particularly concerned with policing who can attend this event. There are a few sex-segregated meetings during the annual conference. There is one priesthood meeting where only men and boys over twelve can attend, and there is also a separate session called the "General Women's Session," which is usually held the week before and intended for a female audience, though open to all.[4] For most sessions, tickets are given away quite freely by LDS leaders, as well as to anyone who writes in for them; there are also standby tickets available. Following the dress code for Church events (suits, ties, and white shirts for men, knee-length or lower dresses or skirts for women) is more important than proving one's religious *bona fides*, at least when it comes to entrance into the conference hall.

The meeting is held in the LDS Conference Center, located in Temple Square, which is the very heart of Salt Lake City. Inspired by the platt for the planned city of Zion sketch in the nineteenth century by Joseph Smith, the founder of Mormonism, Salt Lake City's streets form a grid system. And Temple Square lies right where the guiding x and y coordinates of the grid intersect, in what is called the *origin* in the Cartesian coordinate system of graphing. The Conference Center that stands almost exactly at that nexus is a cavernous, ornate building that can seat more than twenty thousand people; it has been claimed that the Center is the largest theater-style auditorium ever built. And for the annual conference, every seat is full. But the audience is much larger than those twenty thousand people who are present. Ideally, every Church member in good standing should watch, if not attend, the conference. Given the physical impossibility of that obligation for an organization that claims fifteen million members,[5] to further the fulfillment of this imperative, the conference proceedings are broadcast live in some of the other church-owned buildings around Temple Square. General Conference is also shown live on church-owned television and radio stations, as well as the national cable channel owned by Brigham Young University, which is the premier LDS religious school (located in Provo, just forty-five minutes south of Salt Lake City). And if we take broadcast to mean 'cast broadly,' then we should not be surprised if, like in the parable of the sower, some seeds fall where they shouldn't.[6] Sometimes, even women and children will watch the all-male priesthood sessions as they are broadcast. Some members have told me that when, as children, they watched televised priesthood sessions in the company of their mothers and grandmothers, they felt the multitonal thrill of being edgy yet pious at the same time.

But the audience for General Conference is even bigger than this. Via satellite and (more importantly for this conversation) via the internet, General Conference proceedings are distributed all over the world, simultaneously translated live into over seventy languages, including Apache, Hmong, Malagasy, Palauan, Swahili, Urdu, and Yapese. After the fact, conference proceedings are later translated into even more languages, including Igbo, Twi, and Yoruba, and distributed via DVD.

This is an undertaking made possible by a great deal of tireless labor but also by rapidly developing communication technology. The very architecture of the building presumed that this technology would advance at such a rate that it would outpace the life of the building itself. The communications infrastructure of the building was built to be modular so that new and unanticipated ways of digitally transferring and manipulating information could just be snapped into place rather than having to have the technology of the year 2000 CE (the buildings' completion date) hardwired into the structure's skeleton forever. But this focus on the technological possibilities of the hub should not trick us into thinking solely radially, with a center of agents radiating words to a peripheral ring of receptive patients (in the linguistic sense of the term). The in-the-moment communications technology summoned up by a General Conference is not only hierarchical but works laterally as well; upwards of 155,000 tweets might hashtag any typical General Conference meeting during its five-session run.[7] General Conference is when the leadership of the Church speaks. But it is also a time when the members of the Mormon internet speak to themselves.

The sort of communications technology used to simultaneously translate and distribute a General Conference talk in over seventy languages is arguably central to any understanding of the Latter-day Saints as a specific cultural sensibility, as a set of propositions and practices, and as a series of tightly interlocked social institutions. It has become common, in both critical theory and anthropology, to understand mediation as being essential to religious phenomena.[8] Whatever the merits of this argument are for other forms of religiosity, it is undoubtedly true for Mormonism. It was the publication of a book—what has been famously presented as "Another Testament of Jesus Christ" and derisively called by the first wave of critics "Joseph's Golden Bible"—that marked the social genesis of Mormonism as a faith. It was also a series of nineteenth-century newspapers like *The Evening and the Morning Star, Times and Seasons,* and *The Latter-day Saints' Millennial Star,* that promulgated that book and that served as a site for the production and distribution of a theology associated with it. Together, that book, those newspapers, and that theology all knitted together a trans-Atlantic Mormon community and fueled the massive influx of British and Europeans to the rolling intentional community that the Saints called "Zion." This investment in mediated communication continued as the latter days stretched on to include the twentieth century and as outreach branched out into new channels and modes. The Church is responsible for the production of the show *Music & the Spoken Word*, featuring the Mormon Tabernacle Choir (now known as the Tabernacle Choir), which has been broadcasting weekly since July 1929 and may well be the longest-running radio series in the world. The Church also reinvented its public image in the 1970s through an innovative family-centered television advertising campaign. These commercials mostly featured fractured families, earnest children, and harried, callous parents in thirty-second morality plays. It worked to overwrite the un-derstanding of the Church as former polygamists with the image of the Church as a group that cherishes the Nuclear Family. This has long been religion as circulation, at the fastest speed and highest texture technically possible.

This interest in media has only intensified in the era of the internet. The Church has a robust and well-attended web presence, including not just a wealth of online material for believers (at LDS.org),[9] but also an outward-facing webpage that includes the opportunity to chat live with a Mormon missionary. Church leaders are also masters of search engine optimization (or SEO), ensuring that almost all searches involving LDS related topics, and often searches on issues related to Christianity more generally, have LDS-operated websites near the top of the results.[10] There is even an official church statement regarding SEO, where they state that they "view SEO as a method to spread the gospel online and encourage others to treat it as such." (The webpage goes on to present the "standard guidelines that must be followed on all church websites in an effort to effectively spread the gospel online," including techniques for choosing and implementing title tagging and key-words, meta descriptors, image Alt Tags, XML Sitemaps information, URL canonicalization, and information architecture, so as to receive the greatest possible ranking).[11]

Just like the Twitter live-tweet shadow commentary on the centralized General Conference broadcast and live stream, the Mormon web presence is not limited in any way to official websites. There are a host of private websites, podcasts, and influential social media accounts that concern themselves with Mormonism or are written consciously addressing a predominately Mormon audience. Topics vary. Discussions vary. They include such issues as detailed meditations on Mormon history, Mormon theology, brass-knuckled Mormon apologetics, Mormon feminists, Mormon "mommy blogs," social and institu-tional internal critiques of the Church, wry social commentary on larger secular society, Mormons and politics (both left- and right-leaning), Mormon anarchism (yes, it exists), Mormon activists, the works of Mormon artists, Mormon science fiction, the concerns of Mormon academics, and various accounts discussing what it is like to be on the periphery of the Mormon social world in one way or another (that is, to be a convert to Mormonism, to be a Mormon of Color, an indigenous or first-nations Mormon, a Mormon from outside the United States, or an LBGTQ Mormon). And there is a large amount of Mormon-centered internet humor and parody; the seriousness of purpose and wholesome aesthetic of the Church seems to invite satire. These various internet-mediated discussions often link up, playing, debating, or building on one another. While no one is policing membership in this republic of blogs and emails, and the term used for it is somewhat tongue-in-cheek, the loose community formed by this circuit of online discussions about the Mormon faith is sometimes re-ferred to as the Bloggernacle.[12]

The Bloggernacle happens to be President Uchtdorf's other problem.

§16

Standing physically and virtually in front of an audience of millions, Uchtdorf tried to explain his problem with the internet by way of discussing his medical issue. Speaking in his light but noticeable German accent (Uchtdorf was then the only member of the highest two echelons of Mormon authority to be born outside of the United States), he started off by noting the passing of three other Mormon leaders and promising that three replacements for them would be affirmed by the unilateral consent of those present during this very conference. Having touched on the mortality of Church leaders, he shifted topics slightly and shared a little about his own recovery:

After a recent medical procedure, my very capable doctors explained what I needed to do to heal properly. But first I had to relearn some-thing about myself I should have known for a long time: as a patient, I'm not very patient.

Consequently, I decided to expedite the healing process by under-taking my own Internet search. I suppose I expected to discover truth of which my doctors were unaware or had tried to keep from me.

It took me a little while before I realized the irony of what I was doing. Of course, researching things for ourselves is not a bad idea. But I was disregarding truth I could rely on and instead found myself being drawn to the often-outlandish claims of Internet lore. . . . Hope-fully we will learn that when we chase after shadows, we are pursuing matters that have little substance and value.[13]

The language here is that of a parable. The message is unmistakable: the search for truth should not be a search at all but rather a trusted reliance. Do not look outward for the key to faith. Instead, Uchtdorf says, people should rely on the positive experiences that they undoubtedly have as members of the Church of Jesus Christ of the Latter-day Saints. And if their experience is not as positive as Uchtdorf says it should be? Then, he tells us a bit later in the talk, they should consider simplifying their spiritual life and relying on God's grace to overcome their shortfalls. All this is part of the discipleship that makes up being a faithful Mormon.

There is a lot to say about this passage. The turn to an anecdote about health may have been more motivated than it first appears; coming right after discussing the passing of three other church leaders, one could read it as a verbal guarantee that Uchtdorf himself is hale enough. This is important in a movement where the average age of church leadership was eighty years, an average age higher than any time in the history of the Church. But Uchtdorf's discussion did something more than show that he is still kicking. It set up an analogy between Uchtdorf himself not relying on the truth of his doctor and those Saints in the audience who may not be trusting God and are turning someplace else for information.

But other aspects of this analogy are uncertain. Where are these untrusting Saints turning towards to get this questionable information? And who would play the role of the doctor, the dispenser of Truth? Here, possible readings diverge. This could be seen as a simple fable, stressing a kind of reliance on God. There certainly were Mormons who read it this way. But it could also be read in a way where the gap between the source domain and the target domain of the message was much closer. For many Mormons, this parable was understood as a statement warning individuals from relying on internet information rather than on material from church leaders like himself.

It could be read this way for two reasons. One reason is that Uchtdorf and his colleagues in the highest strata of Mormon leadership are understood to be (supposedly) speaking for God. The Church is led by the Quorum of the Twelve and the First Presidency (which is where Uchtdorf had his role as second counselor to the then-president, Thomas Monson). These individuals both individually and collectively bear the title of "Prophet, Seer, and Revelator." This status is more than an honorific. It is held by "all who have received the fullness of the keys of the Melchizedek Priesthood associated with the apostleship," and the origins of this title go back to Joseph Smith himself. For those with this title, when speaking in a prophetic mode, their voice is God's voice, and while these particular humans may be the author and animator of what it said, the principle behind the message is undoubtedly God.[14] This ultimate responsibility and ownership of speech is especially the case with testimony given from the podium during the General Conference.

This status does not entirely immunize the Prophets, Seers, and Revelators. Hubris and error can creep into the works, twisting or eclipsing God's intent. In a previous conference talk, Uchtdorf himself said

> [T]o be perfectly frank, there have been times when members or lead-ers in the Church have simply made mistakes. There may have been things said or done that were not in harmony with our values, princi-ples, or doctrine. I suppose the Church would only be perfect if it were run by perfect beings. God is perfect, and His doctrine is pure. But he works through us, his imperfect children. And imperfect people make mistakes.[15]

Mistakes made by Church leaders or members are not the only cause of doubt. In that 2013 talk, Uchtdorf lays out a typology of uncertainty and suspicion:

> Some struggle with unanswered questions about things that have been done or said in the past. We openly acknowledge that in nearly 200 years of Church history—along with an uninterrupted line of inspired, honorable, and divine events—there have been some things said and done that could cause people to question.
>
> Sometimes questions arise because we simply don't have all the in-formation and we just need a bit more patience. When the entire truth is eventually known, things that didn't make sense to us before will be resolved to our satisfaction.
>
> Sometimes there is a difference of opinion as to what the "facts" really mean. A question that creates doubt in some can, after careful investigation, build faith in others.[16]

Unlike the latter talk, which had a faithless Uchtdorf wandering through cyberspace, this earlier talk didn't mention the internet. It didn't have to. The associations that link doubt, the internet, and apostasy together are common knowledge to his Mormon audience.

There is a reason for this association. The Bloggernacle has functioned to ease the acquisition of information about the Church, information that, in prior eras, would be a struggle to obtain. Rather than explore libraries or brave pur-chasing anti-Mormon tracts from Evangelicals, one can just fumble with a smartphone for fifteen minutes or so and be shown an entirely different vision of Mormon history. As one Mormon academic said during a talk outlining the challenges for present-day Church apologists:

People already know quite a bit about the Church, and with a few
clicks of a mouse they can know a lot more, including negative information that was difficult to come by in the 1970s. The story of the Restoration is less clear-cut than we had imagined, the sorts of evidences
we used to put forward are less persuasive than we had hoped, and
troubling issues in Church history have not faded away but instead
have been magnified.[17]

Much of the troubling history that is referred to involves celestial marriage,
more commonly known as polygamy. Before the internet, the Church did not
bring much attention to the fact that the practice of polygamy began with
Joseph Smith and not his successor, Brigham Young. Likewise, there was little
circulation of the news that many of Joseph Smith's wives were already married
to other men or that one of his plural wives was only fourteen years old when
they were wedded.[18] Before the internet, many Saints were unaware that the
end of polygamy was not the end of polygamy. Even though polygamy was
formally banned when a decree proposed by Church President Willford Woodruff was affirmed at an 1890 General Conference, many Saints, including
leaders in the Twelve Apostles, performed and participated in plural marriage
well into the first decades of the twentieth century. Some of these marriages
were carried out in temples, which are some of the most sacred and policed
spaces in Mormon religious geography.

But celestial marriage is just one aspect of Mormon history that has greater
circulation due to new information technology. Before the internet, many
Mormons were unaware that there were multiple and seemingly incompatible
accounts of the First Vision, where God and Jesus appeared before a fourteen-
year-old Joseph Smith to prepare him for his later status as prophet. Before
the internet, many Saints were unaware that Joseph and his family were sued
for fraud because of their activities as diviners for treasure, using folk magic
to identify supposedly buried gold. Before the internet, it was a common
belief that when Joseph Smith was dictating his translation of the Book of
Mormon from the reformed Egyptian characters, he was gazing directly
at the gold plates that contained the book. (The image of Joseph looking
directly at the plates was the most common way Joseph's transition was por-
trayed in Mormon visual culture.) After the internet, many Saints learned,
to their discomfort, that Smith's translation consisted of him staring at a seer
stone. This method involved Smith sticking his head into a stone-containing
hat to block out light so that "a spiritual light could shine" from the stone,
displaying both the Egyptian and the English translation one character at
a time.[19]

It goes on. There was the Mountain Meadows massacre when gentile settlers passing through Mormon territory were killed by a joint posse of Mormons and native Americans (*gentile* being a term repurposed from Judaism, used for all non-believers in the Mormon faith). There was Brigham Young's doctrine of "blood atonement," which served as a theological justification for these killings and others as well. There was the ban on African Americans becoming priests. The priesthood is a status that (with this one now-revoked racial exception) was almost universally enjoyed by all-male Mormons over age twelve who were in good standing with the Church. But this ban on Black men in the priesthood was only lifted in 1978. Following that, some facts emerged that question not the history of the Church, but the veracity or origins of scripture and ritual. Many of the most sacred and secret rituals carried out in LDS Temples appear to be surprisingly similar to those of various Masonic rites, to the point of plagiarism in the eyes of some critics. (This similarity is a fact commonly explained by Mormon apologists by claiming that both sets of rites date back to the days of King Solomon's Temple; it just happens that, due to divine providence and revelation, the Mormon rites are a more accurate reflection of those ancient practices). There is the general scientific consensus stating a lack of any archeological or population genetic evidence for the presence of any of the large Hebrew New World settlements described in the Book of Mormon. There is the Church's nervous secret purchase of the "Salamander Letter," a forged document that purported to prove that Joseph Smith engaged in dream-communication with angels via a shape-shifting salamander. (The purchase was, at least according to most accounts, an attempt to keep this document from the public eye, though the Church did release the text of the letter to the public in 1985 before its fraudulent nature had been conclusively proven. The forger later engaged in a bombing campaign in an ultimately unsuccessful attempt to further the selling of more fraudulent documents to the Church.)

§17

Again, this information was not completely occluded before wide public access to the internet. There were books and documentaries out there that contained this knowledge. However, when this information came from outside sources, it often had the cast of anti-Mormon propaganda. Part of the reason that, prior to the advent of the internet, this information seemed biased and unreliable was the tenor and origins of these books. Usually written by Evangelicals, they were nominally directed at Mormons. But the anger and pride this proselytizing material contained made it seem much more like evangelical self-congratulation than an honest attempt to educate wayward Mormon brethren. Another reason this information seems new is that the Church also worked hard to make sure that antagonistic understandings of Mormon history would not take. During much of the twentieth century, the Church set up several disciplinary regimes that were designed to obscure these facts. Instead, these exercises in self-reformation worked to mold Latter-day Saints in such a way that they appeared more like respectable American protestant sects.

It is easy to be tempted to judge the Church, either for their underlying information or for their stance towards withholding its promulgation. Perhaps too easy. It should be remembered that what the Church did was carried out as a reaction to the very specific American Mormon experience. For large swaths of the nineteenth century, Mormons were the one group of white settlers who received the kind of genocidal attention from the state and federal governments that was usually reserved for people of color and Native Americans. Mormons were presented as a collection of racial degenerates, the dregs of Europe, and the most backward of Americans congealed together through bizarre sexual rites into a treacherous group that threatened democracy itself. And their faith was presented as savagery rather than religion.[20] More than once, the federal army threatened open war with the Mormon Church, and more than once, state militias attacked Mormon settlements, burning buildings and slaughtering civilians. Even after the Mormon relocation to Utah and the end of the United States military incursion in the Utah War, federal agents would spy on and sometimes raid Mormon communities. Statehood for Utah was supposed to be armor against those sorts of injuries, but even then, it was only flimsy protection. After the federal government reluctantly granted Utah statehood, the first Mormon senator elected from that former territory was denied being seated in the senate for four years. During this time, Congress conducted a series of almost McCarthyite, anti-Mormon hearings to suss out supposedly strange practices and learn who was the real puppet master behind this seemingly innocuous Mormon politician.[21]

Hence the hard twentieth-century drive to reinvent the Mormon Church and to bury their history. The drive has worked—to a degree. This long struggle to present the Mormon Church as a family-friendly, patriotic, anticommunist organization that produced model citizens has been largely successful, but not entirely successful. On the one hand, Mormons have been joined in dialogue with critical contemporary figures such as Richard Mouw, the former head of Fuller Evangelical Seminary (easily one of the most important educational centers, if not the most important educational center, for American Evangelicalism). Writing in *First Things*, one of the premier theologically conservative American journals, Mouw has stated that due to "theological self-criticism and self-correction," Mormons are "approaching orthodoxy" (meaning here *evangelical* orthodoxy) and should be considered suitable theological conversation partners for American Christians.[22] Not all Protestants want to sit down at the same metaphorical table with their Mormon brothers. Many theologically conservative Evangelicals still classify the Church as a cult. Throughout the nineteen eighties and nineties, evangelical churches would often host screenings for anti-Mormon documentaries like *The God Makers*. (And even now, former child star and Fundamentalist Protestant Kirk Cameron has made anti-Mormon documentaries.)[23] But again, this constant ideological hectoring just meant that many Mormons had, to some degree, become immunized to outside attacks.

But, along with the providence of this information changing as its immediate sources went from Evangelical pamphlets to Bloggernacle webpages, the ethical charge of this has changed. Now the problem is, now that this information is widely distributed via the newest generation of digital information technology, these attacks are not from the outside. The Bloggernacle's diversity of opinions means a diversity of attitudes towards the Church and the Church's culture. Some of these internet figures are faithful, true-blue Mormons, but there are also reformists. And then there are those who have ended up outside the church's reach. Many outside critics earned that position over time, starting as reformists who then drift into the opposition, often explaining this drift as being a result of bad experiences with Church leaders and other Mormon apologists.

Many Mormons never search for this information. And many more learn of this more complex and shaded version of the Mormon past and are either not thrown into a crisis of faith, or they find ways of accommodating to this picture of the Church while still retaining some form of belief. Even more pressing, apparent inconsistencies and worrisome secrets have been overcome by believers in other religious traditions. But other Mormons are shaken to the core by these discoveries. It is not uncommon for these Saints to come to feel that their trust in the Church has been betrayed and that they have been purposefully misled.

For many, a sense of betrayal is just the first reason why they are angry.

The revolt against supposedly hidden history often goes hand-in-hand with frustration about the values of the Church. With a universal priesthood that is only given to males, a conscious elaboration of marriage and the family as the chief purpose for women, and a complex eschatology that ties the fate of women in the afterlife to that of their husbands, past polygamy scandals can look like another facet of a contemporary and perhaps unescapable Mormon sexism. The same heteronormativity raises issues for non-heterosexual Mormons, particularly for younger non-heterosexual Mormons. This is a group of individuals who often have trouble feeling validated by the Church and who, at times, are subjected to treatments involving various levels of coercion to *correct* their sexual preferences or gender identity. The Church's strong political opposition to same-sex marriage, such as what has been displayed in the Church-backed push for California Proposition 8, cements this sense of alienation. LGBTQ Mormons and their supporters are not the only alienated constituencies. Due to the historic ban on African American Mormon men being recognized as full priesthood members, many Mormons of color in the United States feel misunderstood and marginalized by both the Church and their fellow believers.

These various concerns may have always been present, but the internet allows for something else. Much like Twitter allows for a virtual, democratized General Conference that shadows the institutional, hierarchical one, the internet also provides paths for Mormons who are disturbed by the church's stand on gender, sexuality, or race to find each other and make communities. Much of the Bloggernacle is oriented toward these concerns. When the term was first invented in 2004, the Bloggernacle was understood as enveloping the entirety of interlinked Mormon online discussion. But the Bloggernacle has come to be seen as a progressive Mormon project, with many more traditional Mormons seeing their own internet blogs, discussions, and podcasts as an entirely different conversation, occurring in furtherance of an altogether different project.

§18

The Bloggernacle is not just for an airing of views. Sometimes it is a platform for taking action. Sometimes it is a target of action.

In October of 2013, a largely internet-organized activist group called Ordain Women requested 150 General Conference tickets for the all-male priesthood session. Rather than watch the conference through any of the other modes through which it was distributed, they desired to be there as equals beside their male Mormon co-religionists. Their request was denied. And when, despite that denial, they showed up at Temple Square to stand in line for same-day tickets to that session, they were turned away. For the April 2014 General Conference, Ordain Women again requested tickets, this time 250 in number, for the General Conference Priesthood session. They were again denied, and when they asked to protest this on site, they were told that they were limited to the free speech zone at the edge of Temple Square. Rather than comply, they took their protest to the Tabernacle, a Temple Square former temple site now used for performances by the eponymous Mormon Tabernacle Choir. It is a space that in many ways is almost as close to the ritual heart of the Church of

Jesus Christ of the Latter-day Saints as one can get outside of the Salt Lake City Temple itself.

The Church rose to this provocation. After Kate Kelly, the founder of Ordain Women, was given a series of warnings from the head of her local Church stake (an organizational unit roughly equivalent to a diocese in size), she was excommunicated *in absentia* that June.[24] Kate Kelly at least has the consolation of not having to be alone. In February of 2015, the Church, again through the office of the local stake, also took the step of excommunicating John Dehlin, host of a popular podcast and known for his advocacy for LGBTQ Mormons.[25] On April 17, 2016, a disciplinary hearing against Jeremy Runnells was only ended by Runnells' resignation from the Church.[26] Runnells was the author of *Letter to a C.E.S. Director*—C.E.S. standing for Continuing Education System, the Church's organ of in-house religious education. The letter was an over eighty-page downloadable document containing a laundry list of supposed plagiarisms, anachronisms, and serious inconsistencies contained in both the Mormon Sacred Scriptures (specifically the Book of Mormon and the Book of Abraham) and in the Church's presentation of early Church history. On September 16, 2018, Sam Young, a businessman and former Bishop for a ward in Texas, announced at a press conference that was purposefully held against the backdrop of The Salt Lake Temple, that he had been excommunicated because of an extended campaign of open protest. This campaign, which in-cluded at one point a three-week hunger strike, was held to change Church policy so that Mormon bishops could no longer ask minors, for whom they had a pastoral duty, about sexual matters in *in camera* interviews.[27] These four figures are not universally loved by all progressive Mormons. But all the same, these highly visible excommunications sent a message.

Actually, these excommunications sent more than a message. They arguably shifted the climate. These are just the excommunications that have been given the most public attention in the past few years. There have been many other excommunications where the subject of the procedure either did not seek out publicity, or where their lack of standing as a public figure meant that Church action against them received little media attention. This is not the first wave of excommunications in Mormon history, it should be noted. Excommunica-tion and other sanctions that fall short of excommunication, such as disfellow-ship, have long been part of the Church's disciplinary regimen. As just one example, in 1993, the Church excommunicated a set of people (including historians, Old Testament scholars, theologians, and feminists) who became known as the September Six.[28] But in this current set of excommunications, the common denominator seemed to be questioning the Church on the internet, and it raised the specter that the Church could hold any member accountable for the online content they create or even for the comments from others that they choose to disseminate via online fora.

If the excommunication campaign was an attempt to calm the waters, it did not succeed. The church faced another major upheaval in the fall of 2015 when portions of the *Church Handbook* (a document usually only available to bishops and stake presidents, both lay positions that have a great deal of disciplinary authority over local members) were leaked. The specific leaked portions stated that people in gay marriages are to be considered apostates and that the children of parents in gay marriages were ineligible to participate in major Church rites such as baptism and entrance into the priesthood. (After age eighteen, they could be given these rites if they disavowed same-sex marriage—and even then, their application to participate in these rites had to be approved by the office of the First Presidency.) This leak led to another wave of protests, protests which included mass resignations from the Church. Similarly, on the final day of the Fall 2016 General Conference, a series of church videos were anonymously leaked.[29] Running more than seven and a half hours altogether, these videos consisted of the highest echelons of Mormon Church leadership being briefed on topics as varied as Chelsea Manning's passing confidential military information to WikiLeaks, how to counter political movements in support of gay marriage, how to reach out to Muslims, the state of the medical marijuana movement, and on the tendency for the current generation of Mormons to wait considerably later to marry. A driving concern about homosexuality is a recurrent theme in these leaked videos, sometimes breaking into conversation as seeming non-sequiturs. At one point, a church leader asks who is behind the gay rights movement. In another video, during (ironically) a briefing for the Church heads about the potential danger of leaks that could damage the Church, the person making the presentation is interrupted in an almost comedic non-sequitur by an Apostle demanding to know if Julian Assange is gay.

These stories of leaks, protests, and excommunications all received attention from print and broadcast media. But the actual material, the leaked information about handbook passages and leadership briefing videos, was first promulgated by way of the internet.

§19

There is another reason why some believers have trouble leaving, even as elements of Mormonism may unsettle or even repulse them. The problem is that if they wish to move on yet remain believers in some ways, there is no other place for them to go. Compare the Latter-day Saints to the Roman Catholic Church. Catholicism is a warren of different liturgical, disciplinary, and territorial authorities; those who wish for a more liberal or a more conservative instantiation of the Catholic faith can affiliate with groups ranging from the Catholic Workers Movement to Opus Dei, depending on their predilections.[30] And Protestantism has mastered the denomination as a technology for differentiating and proliferating forms of Christianity, making it easy for a believer to shift adherence to suit his or her tastes. Even Orthodoxy is filled with multiple possible authorities that one can choose to relate to, even if the generally territorial logic contracts the scope of available options.[31]

But the Church does not afford its adherents such opportunities. There are Fundamentalist Mormons and other autonomous non-Fundamentalist groups that trace their ancestry to both Joseph Smith and the Book of Mormon, but not necessarily through Brigham Young or the Utah Church. But in another way, there is no Church at all. There is nothing at all but a great enchained mesh of individual Latter-day Saints. The reason for this lies in how the priesthood is understood. The Church of Jesus Christ of Latter-day Saints forms an ecclesiastical structure. But it is also a chain of successive transmissions of the priesthood that all lead back to Joseph Smith. As part of what is called the restoration of offices, priestly statuses that were supposedly lost by a quick-to-apostate early Christianity, Joseph Smith was ordained into the Aaronic and the Melchizedek Priesthoods. As memorialized in bronze statues in Temple Square, this ordination was performed by the apostles Peter, James, and John, who had to return to earth to reinstate these lost rites.[32] And the Melchizedek Priesthood comes with *priesthood keys*, capacities to preform ecclesiological rituals and functions. All these priesthoods and priesthood keys are transmitted directly from person to person, by an officeholder laying hands on the recipient. Like that of apostolic succession, this is not just an idea or an ideological justification.[33]

It is important to stress that there is no genealogy of the Restoration Priesthood apart from the one given to Joseph Smith. It is a concrete chain of particular individuals that goes back (via the passingly returned-to-earth apostles and then Joseph Smith) to Jesus Christ. Some Mormon priesthood holders even carry with them documents that trace their "priesthood chain of authority," often handwritten in on a blank, church-provided sheet of paper that looks surprisingly like a tax form. Those who don't know their line of succession can contact the Church, which will tell them if it can be determined from church records. The church provides both a toll-free one-eight-hundred number and an email address (lineofauthority@ldschurch.org) for members to use to request this information.

This chain of succession means that while there is room for faithful Mormons to agree to disagree about topics such as doctrine or proper scriptural interpretation, to set up oneself as a separate group and perform ordinances for someone (ordinance being the term for rites or sacraments) that has already had that same ordinance performed for them is in effect to challenge the ecclesiological authority and the line of succession transmitted within the Church, and possibly of all other lines of succession. After all, if the original ordinance was from someone who could validly exercise it, there would be no need to perform it yet again. This is also why common Protestant and post-Protestant practices, such as re-baptizing, are so inflammatory in a Mormon context. Such an act suggests that instead of the Church controlling who can perform certain ordinances, either anyone has the capacity to perform these ordinances (thereby questioning the unique nature of the priesthood), or the previously conferred LDS baptisms are invalid. The first option makes the Church's authority unremarkable, while the other option makes it corrupt.

This has constrained the capacity of Mormons to peacefully branch out to parallel but non-competing religious groupings. A group is either the real restoration of the gospel brought by Joseph Smith, or it is an apostate fragment in revolt. This structure is why there are no Mormon denominations: the idea of parallel groups with different theological or liturgical practices which still somehow shared some legitimacy is illogical given these presumptions. One could imagine still trying to find a new home in spite of this limitation on imaginable difference. But there again, the choices are limited. The only thing close to a denominational variant of the Church is the Community of Christ, a group with roots to some of the nineteenth-century Mormons who rejected Brigham Young's leadership of the Church and who declined to take the long trek to Utah. While the Community of Christ still recognizes the validity of the Book of Mormon and the Doctrines and Covenants (another sacred text that primarily consists of prophetic revelations made by Joseph Smith), the CoC, as it is sometimes called, has still had problematic aspects in the eyes of many Mormons. Its similarity to mainline Protestant churches, its adherents' interest in social justice, and the movement's comparatively much smaller size keeps it, in the mind of most Saints, from being a real competitor to, or substitute for, the Utah-based Church of Jesus Christ of Latter-day Saints.

But the chief deterrent to leaving is something more than the real lack of alternative Mormon institutions, and more than a perceived or performatively created barrier between Saints and the non-Mormon world. The chief deterrent is that in many ways, leaving Mormonism means leaving the family. As we have noted, the Church institutionally places a strong emphasis on the family; this investment in the family has taken such different forms. Examples of this thematic focus on the family can be seen in family home evenings (setting aside one evening a week where the entire family takes part in a mélange of prayer, religious instruction, and recreational family activities—there is a reason why so many Mormons are so conversant with board games), or in the continuing, institutionally concerted media campaigns lecturing the larger gentile world that family is a good in and of itself. The Church has even released *The Family: A Proclamation to the World*, a document stating the divine nature of the heterosexual family (you can sometimes see framed copies of the proclamation hung inside the houses of particularly devout members). The proclamation does not have the official status of being canonized as Church doctrine. But in effect, it is treated as such. And perhaps most important is the particularity of the Mormon proselytizing pitch: that not only can salvation be achieved, but it is possible to achieve it as a collective, familiar unit, and that one can be with all of one's loved ones in the next life. Given this centrality of family, it is no surprise that one of the leading anthropologists of Mormonism, Fenella Cannel, has repeatedly stressed the importance of kinship and the genealogical imagination to understanding Mormonism. And Bradley Kramer, another anthropologist of Mormonism, has even gone so far to say that Mormonism is not a religion but rather a kinship system with religious elements.[34]

The Church does more than proclaim a model of kinship. It also regulates it. Marriages and adoptions are ceremonies that are not performed in chapels or tabernacles, which are places that are open to everyone regardless of their status (or their lack of status) with the Church. Instead, these rites take place in the inviolate space of temples, closed to non-believers, and are therefore under the complete control of the Church. Both groom and bride must have temple recommends to have access to the temple, a temple recommend being a document signed by a bishop stating that one is a baptized, confirmed member of the Church, in good standing. Good standing means giving the correct answers to a series of formulaic but still probing questions about one's faith: whether one is being chaste, whether one keeps the Word of Wisdom (the prohibition on alcohol, coffee, and tea), and (if one has already received an endowment at the temple) whether one wears temple garments (the garb worn under one's clothes, which is often derided by non-Mormons as the "magic underwear").[35] Good standing also means being able to say that one tithes. Mormons are expected to give ten percent of their earnings to the Church. And finally, good standing means to consent to the following question: "Do you sustain the President of the Church of Jesus Christ of Latter-day Saints as the Prophet, Seer, and Revelator and as the only person on the earth who possesses and is authorized to exercise all priesthood keys? Do you sustain members of the First Presidency and the Quorum of the Twelve Apostles as prophets, seers, and revelators? Do you sustain the other General Authorities and local authorities of the Church?"

Lacking a temple recommend does not just mean that one can't be a part of a temple ordinance. That by itself would be a serious lever through which to police membership. But then many American religious institutions require that you be a member in good standing to receive their sacraments. It goes much further than that. Not being a member in good standing means that one can't even be an *observer* of a temple ordinance, since only those with temple recommends may enter the temple, the site where these rituals are carried out. This rule means that many marriages are conducted with one or two members of the family in a waiting room that is not considered to be part of the consecrated temple structure. Alternately, those without recommends simply stand outside of the temple on the temple grounds. (Often these adult family members without temple recommends end up watching the too-young-for-recommendation children of other family members as a favor for all those who are on the inside.)[36]

This exclusion of non-Mormons, as a church-run website offering advice on the issue puts it, "can be a tender subject."[37] Relatives and friends of converts to the Church are often confused and sometimes angered by not being able to be a part of a marriage. But for those who have grown up in the Church, being outside can be particularly awkward. It is a simultaneous performance of one's fallen nature and one's alienation from the family. Unsurprisingly, people who find themselves questioning or openly doubting aspects of the Church try to find some path through the briars. They find ways to be able to answer the formulaic temple recommend questions in the approved manner while still being honest to themselves, carefully redefining the meaning of common words in their mind so as to be able to assent to them without feeling themselves to be liars. And others just fabricate. Bishops often know that people are not being entirely candid when speaking, and many choose not to press them on the subject. After all, bishops, despite being part of a priesthood of all male believers, are just lay members donating what little free time they have that is not taken up by their jobs or their family. Many have an intimate understanding of the gap between aspiration and actual behavior.

But policing access to temples and the ordinances performed within them is still a tool that can be used to pry apart a non-believer from a church that is laminated to their family. And due to the interwoven nature of family, eschatology, and temple ordinance access, the form of kinship threatened is not just descent, but affinity as well. One marriage partner leaving the Church or losing faith can be the tipping point that instigates a larger family crisis. And it is not uncommon for a spouse who leaves to either be followed by his or her partner or for the marriage to weaken or dissolve under the strain that this shift in religious allegiance brings with it. A wife tells her husband that she doesn't want to hear anything about it when he has a faith crisis; a husband dutifully brings his children along to church with him after the wife has stated that she can no longer go in good faith herself.

This religio-kin laminate also is affixed to a host of other social and professional ties. Utah is the state with the highest rate of affinity fraud, where hucksters prey on the trust placed in them by co-religionists.[38] Slightly less criminal, a surprising number of Mormons are deeply involved in various pyramid-like multi-level marketing schemes. Multi-level marketing (colloquially referred to as MLM) are businesses made of tiered sets of independent partners selling their products, usually various small domestic luxuries and unproven naturalist cures, to their circle of friends. (It is no accident that the two leading therapeutic essential oil corporations—each of which have independently claimed the status of the largest seller of these oils in the world—are run out of Utah.)[39] Affinity fraud and multilevel marketing are just the dark shadow cast by the much more extensive network of Mormons who assist each other in business and educational endeavors.

Affinity fraud and multi-level marketing can function because so much of the larger world of non-fraudulent business deals are catalyzed, or even sometimes just made possible, by shared ties through the faith. For many Mormons, particularly in the "Mormon Corridor" (the band of the American West, including Nevada, Utah, Arizona, and Idaho, which is thick with the descendants of original Mormon pioneers), kinship, religion, and business are perhaps formally separate categories, but in practice they are indivisible. This borderland, a melding of spheres that are typically kept separate in American secular society, may be the closest thing America can have to what anthropologists call the "total social fact," the point where a single phenomenon has effects across multiple domains usually kept separate in modernity, such as kinship, law, and religion, that can no longer be disentangled.[40]

These are all forces and barriers running against each other, creating a social friction slowing the trajectory of those people who are finding it difficult to inhabit Mormonism. We have not exhausted them all. As discussions of genetics, archeology, and the book of Mormon's vision of the pre-Columbian New World suggests, science is another reef that the Church can run aground on. And it is important to stress that anxieties do not always result in people leaving the Church. In addition to those who stay because they are locked into Mormonism as a total social fact, there are also a great many whose faith is shaken, and whose commitment is fractured, but who stay for other reasons. They wish to reform the Church from the inside. They see it is a space for the proper inculcation of morality to their children. They are the descendants of sometimes as much as eight or nine generations of faithful Mormons, and at this point they see their Mormonism as much as an ethnic identity as a specific mode of religious adherence and belonging. These other possibilities do not mean that staying is easier than leaving. It is perhaps best to think of remaining despite doubts as a different kind of painful difficulty rather than as a higher or lower level or degree of suffering.

And it is not all doubt and alienation. There is still a large contingent, quite easily the majority, of what is referred to as TBMs (short for either True Blue Mormons or True-Blooded Mormons, depending on who is asked) who hold fast to the faith without the struggles that some of their co-religionists endure. For these, what is presented as a challenge to leave is seen instead as more positive reasons to stay. For these believers, they are a peculiar people, unlike the gentiles. They are the true Church whose legitimacy is shown by the relative unity when compared to fractious Protestants and Pentecostals. They are a family-facing institution where lines of care and lines of faith mutually strengthen one another. They are a religion which is informed by a timeless sense of gender and which is not confused by a post-modern insistence on dissolving the roles that constitute the most fundamental of human institutions. In the mind of the conservative and orthodox, all this is not a barrier preventing escape but a wall shoring up a garden from a hostile and chaotic world. For them, it is these rules, practices, beliefs, and enmeshed belongings that thankfully demarcate the borders of Zion. And for these believers, outside of what is found in Zion, there can be neither salvation, nor propriety, nor true family.

§20

There is one more thing that works as an incentive for continued adherence, both for the faithful and the skeptical. Many Axial Age religions promise some salvation or reward, usually situated in a future post-mortal scene or state. Mormonism does the same. But the scale and scope of the reward here are arguably far greater than that of most other forms of American religion. The reward for proper discipleship is *vast*. In the same talk where President Uchtdorf spoke of his medical encounter and alluded to the corrosive nature of the internet on Mormon religious life, he closed by stressing Mormons' effective universalism. "Even the lowest kingdom of glory . . . surpasses all understating," he said, reminding believers that, with the exception of a tiny number of explicit apostates who are hurdled into the void of non-being after death, after a cleansing fire in "spirit prison," almost all of humanity will be born again in a perfected body, inhabiting a "telestial" kingdom. This afterlife is a paradise offered to all non-believers, even to "liars, and sorcerers, and adulterers, and whoremongers, and whosoever loves and makes a lie."[41]

But this does not mean that the Church is universalist in the religious sense of the term; or rather, this does not mean that the afterlife is genuinely egalitarian. Something far greater is on offer for those who are faithful in the mortal realm. In closing the talk that began with the parable of himself as an untrusting patient, President Uchtdorf went on to remind his vast in-person and even larger virtual audience that

> As members of The Church of Jesus Christ of Latter-day Saints, we aspire to something unimaginably greater. It is exaltation in the celestial kingdom. It is life eternal in the presence of our Father in Heaven. It is the greatest gift of God. In the celestial kingdom, we receive "of his fullness, and of his glory." Indeed, all that the Father hath shall be given unto us. Exaltation is our goal; disciplineship is our journey. [42]

This statement may sound like the usual Christian speech promising an eschatological cashing out of humanity's status as being children of God. But it is something at once much more literal and much more expansive than that. This is because, for Mormons, Godhood is an achieved rather than ascribed status. As Joseph Smith preached in a funeral service given just two months before his death, "We suppose that God was God from Eternity, I will refute that Idea, or I will do away or take away the veil so you may see. It is the first principle to know that we may convers with him and that he once was <a> man like us, and the Father was once on an earth like us."[43]

More than the existence of the Book of Mormon, and perhaps even more than the Church's history of polygamy, this is the claim that does the most work to upset non-Mormon Christians. They view this statement as a heresy without historical or scriptural precedent. In saying that it is unprecedented, these Christian critics may be going too far: Mormon apologists have found plenty of examples of Christian claims of theosis (the becoming something closer to God) in the two-thousand-year history of the religion. But either way, within Mormonism, more is on the line than mere theosis. The promise of entering the third-highest heaven, the "Celestial Kingdom," and of being exalted to the highest degree of glory, comes with a promise of becoming a god and with the promise of dwelling forever with those with whom they are sealed. And in doing so, they are also doing that which was done before in other worlds: participating in an endless chain of creation where man is conceived, tested, exalted, and then deified to create himself anew on some other sphere. Mormonism offers salvation, but it offers something much more significant. It offers a chance to participate in the foundational chain of the universe, where gods create worlds after worlds, and where gods beget gods.

Second Series: Transhumanisms

§21

Zoltan Istvan has a problem. Like Elder Uchtdorf's problem, it involves health. Zoltan Istvan's health, specifically. The problem is that as he sees it, people are trying to kill him.

Naturally, he is battling this threat by driving across America in a giant coffin.

Unlike Elder Uchtdorf, whose position may be exacting but at least comes with a clear title and a recognized slot in a broader hierarchy, the nature of Zoltan Istvan's authority is somewhat hazy, and his self-assigned portfolio seems to be a bit wider. He's been a journalist with *National Geographic*, has written a self-published novel, and is also a self-styled philosopher. He claims to be the inventor of a new sport (volcano surfing). And he was also the 2016 Presidential candidate of the United States Transhumanist Party.

§22

As discussed earlier, transhumanism is not a political movement, or at least it is not usually a movement associated with the sort of electoral politics that the term *presidential candidate* conjures up. Transhumanism is something between a thought, a social movement, an aspiration, and an evolving fact. The shortest gloss for the term is that transhumanism is the positive anticipation of the possibility that increases in technology will allow *Homo sapien sapiens* to overcome their historic species' limits to such as degree that they become something else altogether. In one way, this is nothing new. As far back as Francis

Bacon, there was the hope of effecting all things possible so that increases in knowledge would better allow humans to control both their environment and their selves. And even scholars such as Benjamin Franklin anticipated a day where the limits of what it means to be human would be opened up to unprecedented degrees, a hope shown in his dreaming of being embalmed in a "cask of Madeira" so that in a future century he could be "recalled to life."[1] The speculative fiction of the nineteenth century, when not trading in cautionary tales or moral and political allegories, sketched out imaginary futures where human life was significantly different. Scenarios projected by various authors during that period for the distant year 2000 AD ranged from mastering weather by controlling the axis of the earth, to education by transmitting the content of books directly into student's heads through electrical wires, to an end of scarcity and the complete medicalization of criminality, to women (garbed in male drag) taking over all forms of productive employment while men wandered the world playing golf.[2]

After World War II, however, this sort of hazy futurism shifted gears. Technological research during the war gave birth to cybernetics, a meta-language where the human and the machine could be thought of as not separate domains but as similar and sometimes overlapping systems. This conceit served as an intellectual warrant for thinking of human limitations as a *technical* problem, and it gave a path for species improvement that wasn't the now-discredited eugenics of the early twentieth century. Government agencies such as the Defense Advanced Research Projects Agency funded programs that worked toward such ends as the removal of physiological limitations such as the human need for sleep, the mechanical enhancement of soldiers, and the creation of prosthetics so that human capital was not entirely lost by wartime injury. At a more prosaic engineering level, DARPA and its antecedent agency ARPA were responsible for ARPNET (the first instantiation of the internet), cloud computing, speech recognition technology, and self-driving cars, and it is currently working with NASA on the "100 Year Starship" project, an endeavor laying the groundwork for exactly what one might expect from the initiative's title.[3]

This efflorescence of experimental, defense-funded research came together in the early nineteen seventies with a counterculture-fueled Silicon Valley, and both were driven by some visions emanating from science fiction, the literary genre that inherited the legacy of nineteenth-century speculative writing. The internet was a technology that was a fruit of these synergies. But the internet also acted in its own right as an accelerant of other real and imagined breakthroughs. While the communicative advances that came with the internet allowed different groups to mobilize, the technological resonances that also came with the World Wide Web meant that it stood as both an example of a means to work toward rhizomic democracy and of future technological possibilities.

This sketch is a partial and overly Western genealogy of transhumanism's precursors; Russian speculation about technology, threaded at times through Eastern Orthodox mysticism, gave rise to a separate strand referred to as "Cosmism."[4] And this account also skips lightly over the means through which all these factors came together to cohere as a self-conscious community. Publishing, internet forums, organizations, speaking tours, and conferences were all means of bringing together people such that they could recognize themselves as transhumanists. But putting these caveats aside, the narrative that I've just recited is a standard origin story that most American Transhumanists would recognize.

Now, this is a story that steps lightly over *who* transhumanists are. Transhumanism is a hard community to canvas, as its boundaries are not clear, and its members do not necessarily make themselves manifest in any particular real or virtual locale. But drawing off of both surveys (albeit surveys with self-volunteered participants) and anecdotal evidence, it seems fair to say that we are dealing with a set of people who are mostly white, mostly male, and mostly in their twenties and thirties; they are also more likely to be single than the average American. They work in a variety of fields, but unsurprisingly, the technology sector is an outsized proportion of this population. And they are overwhelmingly agnostics or atheists (and often New Atheists).[5] This portrait is cast in terms of tendencies and likelihoods, and it misses the fine grain and the exceptions: gay, transgender, or polysexual transhumanists are not so rare, and there is a fair scattering of membership as far as race and class go as well. And this is only a picture of people who are active transhumanists. There are far more people that, rather than analyze or celebrate a transhuman moment, simply assume that it is our collective future.

§23

But what is striking about the transhumanists is not their demographics or their marked or unmarked nature, but their imaginings, even if there is no doubt a tight link between who they are in social and cultural life and what it is they dream. An example best illustrates this scope of transhumanist vision. Ralph Merkle is just one transhumanist, influential and perhaps exemplary, but not necessarily central. His *bona fides*, though, are unquestionably solid. He was a key innovator in the field of public-key computer cryptography, has worked on nanotechnology, and is a board member for one of the most important American cryonics foundations.

Merle is a heavyset but enthusiastic man, and a typical public talk on trans-humanism from him escalates quickly. He starts by laying out the current state of three-dimensional printing, drops down in scale to engineering at the atomic level, and then marries them, predicting a capacity in twenty-five or so years to our achieving the ability to freely arrange medical atoms. This development then becomes a promise for technologies such as nanocomputers so compact that one liter of nanocomputers would have the computational power of 10^{11} human beings. But nanotechnology is for much more than computing. We are promised innovations such as artificial red blood cells that would allow a person to hold their breath for an hour, robotic microphagocytes that effectively eradicate infectious disease, inexpensive and clean power, ultra-high-yield agriculture, the removal of CO_2 from the atmosphere, the undoing of anthro-pogenic warming by designing trees that grow diamonds ("How many people are concerned about global warming? Hands? I have good news for you: this technology will let us solve global warming"), zero-pollution manufacturing, and the cryonic suspension (for later resurrection) of the dead for millennia. He even foresees the availably of personal spacecraft that could tour the solar system in a few months. What is notable is not how wild these predictions are but how conservative this talk is in comparison to other transhumanist visions. Merkle's prognostications skip over additional common transhumanist, futur-istic touch points, such as a post-scarcity, post-capitalist future, uploading human minds into computers, the capacity to use advances in biology and bioengineering to redesign one's body in almost endless forms at will (some-times called *morphological freedom*), the capacity to make computer-virtual worlds that are indistinguishable from real ones and those that are immersed within them, and, finally, quantum archeology, a speculative technology that reads the disturbances in the universe itself as a means of retrieving, and some-times of salvaging and resurrecting, the otherwise irretrievable past.

§24

Expansive as it is, Merkle's presentation on the day I saw him speak did not touch on another vital lay line in transhumanist thought. He did not mention the singularity. This term is not originally of transhumanist coinage, and its use is not exclusive to transhumanism even today. *The singularity* was initially a mathematical term. As used in that discipline, one of the chief senses for that word is the point at which a function can no longer be graphed. It is where mathematical capacities for depiction collapse and where mathematical powers at imagining fail. The term in transhumanism means roughly the same thing. Though instead of being a generic term for a form of representational or logical breakdown, it refers to the hypothetical breakdown point of a single function. The singularity in the transhumanism sense is a term for the moment when, due to accelerating technology, human futures can't be foreseen. It is, in short, the point when computing becomes so powerful and self-aware that humanity would no longer be the dominant intellect on the earth. The engine of this moment that the singularity anticipates—or dreads, depending on what is imagined as happening during the unimaginable—is artificial intelligence. Specifically, what is dreaded or anticipated is the tipping point where artificially intelligent agents can design and execute subsequent artificial intelligence agents with greater analytic capacities than those possessed by the very intelligences that crafted them. In its negative form, this would be the point in the developmental curve when the AI parents, so to speak, will build children that surpass them, after which intelligence self-ratchets upward in scale and intensity.[6] To quote Verner Vinge, the mathematician, computer scientist, and science fiction writer who in 1993 first used *the singularity* in its current transhumanist sense, when the singularity occurs we will "enter a regime as radically different from our human past as we humans are from the lower animals." That would be the moment when these intelligences are no longer our tools "any more than humans are the tools of rabbits or robins or chimpanzees."[7]

As with all these other terms, the concept of the singularity has a prehistory. As used by Vinge, the singularity was an existential threat to the species, and this is a common understanding in subsequent elaborations of the concept.[8] Early technical discussions of creating "ultra-intelligent machines" considered the possibility with ambivalence.[9] Even after Vinge, this ambivalence continues. Some welcome the prospect of creating superintelligences that surpass humanity, even if it means the extinction of our species. The feeling here is that having a role in advancing toward a cosmic teleology heading toward intelligence is our purpose and perhaps our destiny.[10] But some anticipate it not with fear, or with a fatalist sense of tragic purpose, but with a bated optimism. A significant number of transhumanists, including Ray Kurzweil, inventor and Google's director of engineering, and one of the co-founders of Singularity University, see the singularity as a sort of Omega Point. For them, the singularity marks the moment where superintelligences will sometimes serve, and sometimes merge, with biological humanity in a way that effectively ends any limitation on the human.

§25

As presented by Kurzweil, the singularity is not a disruption but a sort of overcoming and completion. As was discussed earlier, for this reason, it's easy to see both the idea of the singularity, and of transhumanism more broadly, in essence as a *religious* thought. It certainly has its eschatological and soteriological overtones. Even the etiology of the term *transhumanism* has a religious genealogy. That is because the term *transhumanism* was not originally minted to discuss technological overcoming. There is a common misconception that the word *transhumanism* was originally coined by Julian Huxley, the evolutionary biologist (and also the brother to Aldous Huxley, author of *Brave New World*). The claim is that he coined the term when he wrote, "I believe in transhumanism: once there are enough people who can truly say that, the human species will be on the threshold of a new kind of existence, as different from ours as ours is from that of Peking man. It will at last be consciously fulfilling its real destiny."[11] His 1957 use of that word certainly was influential in the current understanding of the term. And his presentation of that word was rich with all the utopic resonances that the word *transhumanism* has since taken on. But whether or not Julian Huxley was aware of it, the prior art regarding the word *transhumanism* goes back much further. Before the technical term, *transhumanism* appeared in discussions of progressivist metaphysical philosophy. And before that, it was in what has been called the Standard Victorian Translation of Dante's *Divine Comedy*, where it was used in place of Dante's Italian neologism *tranumanar*. And the passage containing *tranumanar* itself has an etymological trail that runs back to Saint Paul's Second Epistle to the Corinthians, with Dante using the word to make "unambiguous references" to where Paul takes up the impossible task of describing the ineffable third heaven.[12] Following the full trail of breadcrumbs, we find Paul referring to himself, albeit in the third person: "And I know that this man—whether in the body or apart from the body I do not know, but God knows—was caught up to paradise and heard inexpressible things, things that no one is permitted to tell." Like the human contemplation of possible inhuman thinking machines two thousand years in the future, this was a moment when human reason and language broke down and became inadequate in grasping some other intelligence. It was a moment of transcendence.

§26

As was discussed previously in this book, this play with transcendence doesn't make transhumanism necessarily a religion. Etymology is not essence; after all, the word *secularism* also runs back to religious parlance and a theological division of the world if you follow the history far enough. But other filaments run from transhumanism back to the religious. It is common for both adherents and critics to refer to transhumanism as the "rapture of the nerds." (Although the reference is made with differing levels of sarcasm depending on the warmth that the speaker holds for the possibility of the singularity.) But even more to the point, theologians who address the matter of transhumanism sense some overlap between the two fields, or they at least see a contest between religion and transhumanism for some select spot in the contemporary imaginary. It is common for these theologians to also offer detailed Christian genealogies for transhumanism, invoking names that range from Pierre Teilhard de Chardin to Karl Rahner. Alternately, they may insist that the divination of humanity that can be read into some accounts of Christian intellectual history shows that Christianity was a transhumanism before transhumanism, as we know it today, was even a conceptual possibility. This does not mean that there is an across-the-board acceptance of transhumanism in academic theological circles. Some theologians critique transhumanism, claiming that it has a manqué, tacit understanding of Christian thought and does not make use of this supposedly rich aspect of its intellectual inheritance. Others see transhumanism not as derivative in its relation to Christianity but as deficient in its own right. These theologians argue that transhumanist visions of human thriving are narcissistic and shallow, or that transhumanism has no real grasp of, or remedy for, the problem of evil.[13]

Christian resistance to transhumanism can also be found in less academically rigorous varietals. American Evangelical and Fundamentalist apocalyptic thought has long been fascinated with malevolent, and sometimes outright demonic, instantiations of both technology itself and faith in technology. Supercomputers, bar codes, and computer chips embedded in the brain have long been apocalyptic evangelical leitmotifs.[14] Transhumanism's promise of freely reconfiguring the body, and of effectively unlimited transcendence of death and finitude, can easily be used to catalyze these already well-developed Protestant Luddite tendencies. This popular religious sentiment, in some ways, is a mirror of the theological debate about transhumanism, even though this apocalyptic discussion is carried out in a different register. Both these discourses, after all, see transhumanism as at once a twin and a rival to Christianity, or as Christianity's visage presented back to itself in a monstrous and inverted form.

That twinned sense of an identity and a rivalry between transhumanism and both high and low instantiations of Christianity does not mean that all critique of transhumanism is religious critique. When weighing whether or not transhumanism is religious, it is essential to note that religion is not alone in presenting critiques of transhumanism. As we have seen, even some of those within transhumanism who anticipate these transformations have had fears that the kind of technologies that are often grouped as being transhumanist might unlock existential risks to the species. Therefore, it should be no surprise that critics outside of transhumanism harbor the same worries. But there are also concerns that the damage done by these species-capacity-remolding technologies may be just as dangerous even if our species *isn't* extinguished, though in these scenarios the injury these technological capacities cause may be expressed in more subtle ways. Rather than coming in the form of a hard stop, the end of humanity could come as a strange transformation. For some non-transhumanist social critics, the threat here is that the contemporary social, cultural, and political balance is founded on an essentialist human nature. Therefore, injecting a new plasticity into that shared species' inheritance would throw off that balance and place into question long-running and inherently valuable forms of human thriving that are dependent on what we all supposedly share.[15]

We should note that these non-religious concerns are different in their logic from religious concerns about transhumanism. They are expressions of a kind of Burkean conservatism, an approach that sees the way to assuage their anxieties in the purposeful deceleration and deliberation of change rather than being a standpoint that sees transhuman life as inherently unassimilable or as a false telos. Even more importantly, these non-religious critiques do not coalesce into a community or an identity. Contemporary Luddism is a sensibility or an aesthetic that may be dependent upon current social and cultural formations for its existence. But this diffused Luddism doesn't install or expect the sense of a popular movement that we see in either religious belonging or secular transhumanism. Finally, in secular attacks on transhumanism, there doesn't seem to be that sense of a doubling that we see in some religious critiques of transhumanism, of a simultaneous identification and rejection with transhumanist thought. Secular critiques may be in opposition to transhumanism, they may even lay bare an enmity, but they do not express an anxiety about secretly being transhumanism's kin.

So, does this rivalry between Christianity and transhumanism mean that transhumanism is a religion, or at least is religious? Recall Abou Farman's argument that the appearance of a kinship between the two arises in part from some of the genealogical linkages between transhumanism and historic Christianity, and even more so from the fact that transhumanism has stepped into a void that was created by religion's effective withdrawal from the public discourse.[16] In this account, the advent of secularism meant that, on the one hand, physicalist accounts of memory, mind, and self have replaced those of the soul, and disciplines such as psychology and social sciences became the driving frames through which modern subjectivity was thought through. At the same time, due to the way that these academic arguments were structured, and the sorts of evidence and arguments that they used, these intellectual frames could not make determining public statements about the ultimate import of these issues and objects. Psychology might be able to explain paths toward happiness, or sociology might be able to describe what would make for a better functioning society. But neither could address why the Universe was here and what purpose it might have. It was not so much that secularism meant that religious understandings of ultimate issues could no longer be found: religion continued to exist, even if its 'good' form no longer had the right to set the terms of debate about these issues. In fact, religious articulations of these issues seem to have multiplied. That is because these ultimate issues became private concerns rather than communal understandings that gave a shared tenor to a collective life.

Furthermore, the imaginative possibilities opened up by transhumanist technologies have meant a return of these concerns to secular commentators who previously found them outside of their remit. The limits of these various secular social science explanatory frameworks remain. But the potentiality for new standards of time and ability means that these issues are now capable of being immanently worked through on a scale that overlaps with the lifespans (biological or otherwise) of present-day people. If, as many transhumanists claim, it is possible that there are now people living who will be alive forever due to transhumanist technology, or who could come to dominate the entire universe through the powers that this technology would bestow them, then we can speak about eternity and about the universe as practical concerns for the first time in our secular age.

Context is important in understanding the scope of Farman's point. Farman's narrative focuses mostly on the historical possibility of conditions for the appearance of transhumanism and the structures of subjectivity and desire that animate transhumanist thought at the moment of its advent. And we should remember that his project did not address the allergic reaction that transhumanism evoked in some Christianities that were already at odds with secular and modernist social arrangements. Even if Farman is correct about the genealogy of transhumanism, we should remember that kinship is relative, and one person's distant relation is another's close family. The other issue that is hanging in Farman's entirely plausible account is why this antipathy should often be mutual. Why should a transhumanism birthed from the open spaces in secularism have the same fraught relationship with religion that religion has with transhumanism? Given the contestation over the same domain of ultimate issues, and the way that many religious others react to transhumanism, it seems necessary to ask whether transhumanism exhibits any antagonistic tendencies when confronting religion. Is this a case of mutual schizogenesis, anthropologist Gregory Bateson's term where contestation between like groups ratchet up not despite their similarities, but because of them?

§27

The best vehicle (so to speak) for exploring the question of transhumanism's potential enmity and kinship with religion may very well be Zoltan Istvan's coffin-shaped Immortality Bus. Istvan's coffin-bus was not so much utilitarian transportation as it was publicity for his presidential campaign as the Transhumanist Party presidential candidate. Transhumanism as a political party isn't entirely novel or, for that matter, particular to America. There are transhumanist political parties in Australia, Austria, France, Germany, India, Korea, Canada, and the United Kingdom. There are even wholly de-territorialized virtual and global transhumanist parties. Altogether, there are reportedly transhumanist political movements in some nineteen nations; some of these groups are registered as political parties, some as political organizations, and some are still in the process of working toward some form of official status.[17] Despite these signs, transhumanism as an overtly political project is a relatively new turn. There are longstanding transhumanist advocacy groups, such as the Entropy Institute and Humanity+ (formerly known as the World Transhumanist Association). And as advocacy groups, they are working for social transformation and were thus engaged in a politics of a sort. But the scope of their activity did not encompass electoral politics.

In a way, Zoltan Istvan's political party did not really encompass electoral politics either. There were no primaries or caucuses through which Istvan was selected as the Transhumanist Party's candidate, he did not engage in fundraising, and when election day of 2016 rolled around, his name was not on the ballot of any state.[18] While on the campaign trail, he even said at times that he was likely to vote for Hillary Clinton, the Democratic Party's presidential candidate, though at other times, Istvan did insist he would cast his ballot for himself.[19] While Zoltan's coffin-bus was a way to build awareness for his candidacy, the candidacy itself was not an electoral project *per se* but an attempt to unify what he saw as a fractured social movement and build awareness for transhumanism as an idea. But this awareness was more than informing a broader world that transhumanism exists. The issue was not solely that transhumanism, and the technologies that transhumanism promotes or anticipates, was relatively unknown to a broader American population. Curing the public of this ignorance was an aspect of his project, as Istvan stated in his many interviews and self-authored essays. He labored mightily in laying out for a wider audience some of transhumanism's most breathtaking predictions of (positive) imminent change. He would quote a prediction by Aubrey de Grey, a gerontologist with a great deal of traction in the transhumanist world, that there was a "50/50 chance of bringing aging under what [he] would call a decisive level of medical control within the next 25 years." He would list technologies such as "gene editing, bionic organs, and stem cell therapy," and "[r]obotic hearts, stem cell technology, designer babies," and "3D printed organs" as being innovations likely to bring about that "decisive level of medical control." He would list established companies, such as Google's Calico, or startups such as Human Longevity, as evidence that these transhumanism projects were not just a theoretical possibility, or vapor-wear (to use programming argot), but were instead a tangible economic and social possibility.[20]

But Istvan saw bringing these developments to light as only the first moment in a much larger project. While this education had to be done, the real challenge was not merely building awareness but combating what Istvan saw as a deep-seated and already entrenched opposition to transhumanism. In one of his (many) *Huffington Post* essays, he laid out the threat that he saw facing transhumanism:

> Unfortunately, many people in America and around the world—
> especially those who believe in afterlives—are neutral or even oppose
> stopping biological death and aging with science. They feel it chal-
> lenges what is natural in the human species. Transhumanists call
> these people "deathists," those who believe and accept that death is
> a desirable fate.[21]

"Deathists" here is the name for those people who presumably long for death, and "deathism" is the mindset supposedly held by deathists, abstracted from any particular individual or group, and represented as an intellectual or moral position. As the essay suggests, these words are not terms limited to Istvan. These terms have been floating around transhumanist communities for quite a while and with a particular frequency among those segments of transhumanism that focus most of their energies in life expansion. In *Fight Aging*, a post in a long-running transhumanist life-extension blog, deathism has been glossed in the following way by the blog's author, a prominent anti-aging activist:

> When I say deathism, I mean a point of view or philosophy that pro-
> motes death . . . that almost always means death by aging: deathist
> views are those of technological relinquishment, apologism for degen-
> erative aging, and shying away from the medical progress that could
> eliminate the death and suffering caused by aging.[22]

As transhumanists imagine them, deathists do not fear the reaper. Instead, they supposedly see death as so organic to the fabric of human experience that they fear the reaper's absence. Death may not seem to be a phenomenon that would have a significant number of natural backers, or at least not under that name. It would seem that deathists (if they even exist) must understand themselves as being engaged in a different project.

Given this, who in the transhumanist imagination are the deathists? We are further told in one gloss on this topic that believers in "modern Malthusian environmentalism" and the "traditional religious cultures" are the largest deathist communities.[23] Of those two factions, it seems clear that, at least in Istvan's mind, *religion* is the chief adversary. Malthusian objections ultimately will be undone by technology not dissimilar to those which will bring about Methuselah lifespans: meatless meat, green technologies, environmental management, and an inevitable plummeting in the birth rate vitiates any overpopulation concerns. But unlike these Club of Rome style worries, which are presented as reasonable though mistaken objections to longer life, it seems that religious opposition is supersaturated by what is effectively madness. We are told by Istvan that

> [a]bout 85 percent of the world's population believes in life after death, and much of that population is perfectly okay with dying because it gives them an afterlife with their perceived deity or deities—something often referred to as "deathist" culture. In fact, four billion people on Earth—mostly Muslims and Christians—see the overcoming of death through science as potentially blasphemous, a sin involving humans striving to be godlike. Some holy texts say blasphemy is unforgivable and will end in eternal punishment.[24]

We should note that this is Istvan ventriloquizing the religiously observant, as can be seen by the presence of modifiers in what is presented as the discussion of an undifferentiated and somewhat anonymous block. These are not statistical presentations but sock puppets. But still, Istvan is fearful. And this fear is present because, for Istvan, what he is laying out is not just a cultural problem but a political one as well:

> Complicating matters for transhumanists is the fact that the 535-person U.S. Congress, the current U.S. President, and all members of the U.S. Supreme Court are 100 percent religious and believe in afterlives. Ultimately, this means the American government has little policy incentive to stop death or to put national resources behind life extension science to make citizen's lives far longer and healthier.[25]

In Zoltan Istvan's eyes, pie in the sky when you die is not just something that the master tells the servants to quell them, but it is also something that the master whispers to himself. And the master will govern himself accordingly.

There are possible objections. In Istvan's presentation, there is a quick stitching together of the potentiality of seeing life extension as blasphemous and of some texts that see such blasphemy as unforgivable. But it is not quite so clear that this is the same inferential chain that believers would make. It also assumes that these technologies would automatically be seen as both new and other, not as just one more instance of an already accepted open set of contemporary life-extending medical practices. There are a few, and often small, contemporary religious movements that have rejected either specific medical technologies or all of allopathic medicine writ large. But these religious groups have not been very successful in mobilizing the state's full monopoly on violence to further their prohibitions.[26] The jihad against prosthetic limbs and the crusade against angioplasty have yet to materialize. But at the same time, it is also clear that issues regarding where to situate the boundaries that cordon off birth, life, and death from one another have been flashpoints for social unrest and political conflict in the United States during the late twentieth and early twenty-first centuries. And it is not unheard-of for people to reject treatment for congenital conditions such as kyphosis, dwarfism, or deafness because the individuals bearing these conditions see them as being either essential to their makeup as a person or as traits given to them not by accident but by God.

Istvan's campaign may not seem worth taking seriously; viewed skeptically, it appears to be little more than a series of quixotic stunts. During the cross-country tour of the United States on his coffin-bus, he got chipped when he had an RFID implant surgically placed into his left hand at a biohacker event in Tehachapi, California. (He later makes a suggestion that Syrian refugees be chipped so that they should be monitored, opining that "[m]aybe Big Brother shouldn't be looked at really negatively . . . [m]aybe Big Brother isn't the bad guy—if he protects us from ISIS.")[27] The tour ended with him reading a so-called Transhumanist Bill of Rights at the steps of the Capitol Building, which occurred after his trying to post them to the building's doors as if they were a posthuman version of Luther's Ninety-five Theses. He ended up standing down when guards told him that the document would be torn down as soon as it was put up, and he ended up instead handing the manifesto to California Senator Barbara Boxer.[28]

But it is also arguable that Istvan should be seen as more exemplary of a position than as a mere ideological outlier. His position is not his alone. Remember, the term *deathist* is not particular to him, and it is possible to find statements like the following on major transhumanist forums:

> However, today the comfort provided by archaic religious superstitions impedes advancement and therefore should be set aside. We need to

grow beyond religion. But must we relinquish religious beliefs now, before science gives us everything we want? Yes. The most important reason to abandon religious belief is religion's opposition to most forms of progress. For the most part religion has opposed: the elimination of slavery, the use of birth control, women's and civil rights, stem cell research, genetic engineering, and science in general. Religion is from our past; it opposes the future.[29]

Like Istvan's, the essay that this quote is taken from goes on to argue for evolution as the warrant for progress as well, a claim that ultimately suggests contemporary religion is in its essence some kind of vestigial social organ. This discussion does not have the same air of militancy that can be found in Istvan's argument, but the logic of it is similar. Even if it is expressed at a slightly more temperate emotional gradient and still predicts a gradualist fading away rather than sudden extinction, an extinction of some sort is the imagined telos for the process.

It would be wrong to say that this opinion of religion is found throughout the entirety of transhumanism. According to a survey conducted by the Institute for Ethics and Emerging Technologies (the institution which was the venue for the quote immediately above), nearly eighty percent of transhumanists are not religious, and about thirty percent are libertarian. This large preponderance of avowed skeptics in the survey results matches the sense one gets regarding levels of unbelief in either religion or government when talking to secular transhumanists. But we should still be wary of assuming that all libertarians are also atheists or that all atheist transhumanists see religion as the enemy.[30] When thinking about transhuman intellectual diversity, it is essential to note that transhumanists can be mapped using distinctions far less crude than catch-all categories such as atheism or libertarianism. An example: one typology of transhumanism proposed by a transhumanist intellectual offered ten different "philosophical categories" of transhumanism, ranging from "democratic transhumanism" to "cosmopolitan transhumanism" to "hedonistic" transhumanism to "anarcho-transhumanism." To further disable any essentialisms, the author of this typology claimed that most individual transhumanists are best thought of as being informed by amalgams of these various schools rather than a fideistic total dedication to any one category in particular.[31]

But even this transhumanist diversity is reason enough to take Istvan seriously. This is because the contestations between various transhumanist sensibilities that Istvan sparks have had social effects. In what appears to be a classical instance of schizogenesis, during his campaign, the differences between Istvan's understanding of transhumanism and other transhumanist visions were exacerbated and, at times, weaponized. Mid-way through the campaign, the secretary of the Transhumanist Party publicly resigned, opting instead to align himself with the Open-Source Party, a different transhumanist-oriented political project. This move was caused in part because of the perceived wastefulness of Istvan's campaign. The resources squandered on the bus, as the Open-Source Party defector pointed out, could have been used to set up "16 clinics at $1,500 each in Sub-Saharan Africa, to provide about 2,400 children with free medical care, and prevent approximately 20 kids from dying of malaria." But this particular rejection was also motivated in part by Istvan's autocratic style. For this former supporter, Zoltan's out-of-the-blue suggestion that he would endorse Hillary Clinton was the final straw—a statement made without any consultation with the skeleton of officers who made up the non-profit organization that was the Transhumanist Party.

The issues put in play here were larger than individual allegations of malfeasance and mismanagement. The former Transhumanist Party secretary and now Open-Source Party member was also worried that Istvan was co-opting transhumanism and needlessly alienating religious believers;[32] he would also later allege that Istvan was in flagrant and open violation of FEC campaign laws.[33] Both critics and supporters held similar concerns about Istvan's campaign.[34] Speaking in his official capacity, the head of the United Kingdom's Transhumanist Party shared a comparable worry. He feared that Istvan had usurped a transhumanist voice and that Istvan's focus on immortalism was particular to him rather than something shared by all transhumanists. Finally, there was the claim that Istvan had "rejected all real party-building, due process, and even democracy itself."[35] As formulated by this UK party spokesman, transhumanism is multifarious, and progress demands cooperation and mutual understanding between different transhumanist collectives rather than something subservient to a single figure. Likewise, an online petition disavowing Zoltan Istvan's campaign received almost one hundred signatures. While this is not a large number, some of the people who appended their names were very prominent transhumanists, including Natasha Vita-More, a former president of the Extropy Institute and (as of the time of this writing) chairwomen for the board of directors of Humanity+, the latter organization being perhaps the foremost transhumanist organization in the world.[36] And more than one secular transhumanist has suggested in conversation that Zoltan Istvan was the reason that Nick Bostrom, Oxford University philosophy professor, best-selling author, and founder of the World Transhumanist Association (the precursor organization to Humanity+), had stopped referring to himself publicly as a transhumanist.

§28

Now, this moment of Zoltan-catalyzed political polarization is not the first time that a rent in the transhumanist social fabric has occurred. The history of transhumanist politics before the existence of transhumanist political parties is full of tensions between libertarian and leftist transhumanists, as well as clashes between and within different transhumanist organizations, list serves, and meetings. Zoltan Istvan, a mediagenic figure arriving at a time when transhumanists are taking the first hesitant steps into electoral politics, may be the latest and most public instance of this transhumanist propensity to schism. But these fractures have been present within transhumanism since it first started to cohere as a self-conscious movement in the closing decades of the twentieth century.

It may sound odd to have transhumanists battling over seemingly vaporware technology. True, surgical implants, prosthetic limbs, virtual reality, and self-driving cars are either now existing or at the cusp of being commonly accessible to select segments of Euro-American and Asian populations. But other technologies such as general artificial intelligence, the successful revival of cryonically preserved human beings, and the uploading of human consciousness onto computers all seem decades away at best. And to some, these technologies may appear perhaps forever beyond the reach of the possible. It is also not even certain that all transhumanists take these prognostications seriously. The self-presentation of some leading secular transhumanists often makes it seem as if they have a certain degree of comic/ironic distance regarding transhumanism. Leading figures like Max More (born Max. T. O'Conner—the husband of Natasha Vita-More), FM-2030 (the late and cryonically preserved Fereidoun Esfandiary), and R.U. Sirius (Ken Goffman) have all taken on science-fiction sounding names, in part as an object lesson to question ideas regarding the fixity of personal identity, but also as a playful provocation. Of course, there was also a pedagogical aspect to these new monikers: the former Max O'Conner chose the name Max More "in order to remove the cultural links to Ireland (which connotes backwardness rather than future-orientation) and to reflect the extropian desire for MORE LIFE, MORE INTELLIGENCE, MORE FREEDOM."[37] But still, playfulness remains. There is even transhumanist "fun theory" with careful discussions on how to create sufficient engaging novelty to ensure that effectively omnipotent, effectively immortal posthumans would not become bored as the millennia roll on.[38] If play is essential to the transhumanist endeavor, then there is always a suspicion that transhumanism itself may be a bit of a game.

Perhaps because some intuit this playfulness, or perhaps because of the wildness (from particular perspectives) of the ideas that they peddle, it is undoubtedly the case that transhumanists and transhumanism are often not taken seriously from the outside. There is a history of transhumanists being read as so strange that they are not capable of being either partners in or subjects of sober discussions. A *Slate* article documenting a 2007 transhumanist conference reads like a parade of broken toys: panels on "the self-demand amputation community," people advocating the use of "cyber- and biotechnology to elevate all animals to human status," and individuals who present themselves as "transsexual transbemanist" (because mere transhumanism is "too parochial").[39]

Alternately, a person could reasonably think that there is no need for rough-edged transhumanist politics because transhumanism as an idea has already won. The list of Hollywood movies built around transhumanist McGuffins is a long one. *The Matrix Trilogy, Ex Machina, Eternal Sunshine of the Spotless Mind, Ghost in the Shell, Brazil, Gattica,* and even *Avatar* and *Wall-E* all touch on transhumanist themes. And even if these productions are at times set in a dystopian key, they at least suggest transhumanist scenarios as apprehensible worlds, logically self-sufficient on their own terms. Commercials and other advertisements using transhuman imagery, such as white virtual spaces or eroticized female cyborgs, are even more ubiquitous in American culture. More to the point, it can be said that the presence and Bond-villain-like influence on American popular culture of figures such as Elon Musk and Peter Thiel shows that whether or not the futures that they are working for are likely, their transhumanist concerns have a level of considerable influence in public debate that was absent as recently as a decade earlier.[40]

However, the possibility that transhumanism is too playful to take seriously or too victorious to be vulnerable has not closed down transhumanist politics. Part of the reason these debates continue is the culture of Euro-American social movements, where contentiousness rather than consensus is the general direction of most social drift, absent a contravening force. But there may be a reason that is far more specific here: for transhumanists, too much is at stake. For some, the existential risks that come with the singularity are what feed this sense of transhumanism's importance, hence its political charge and value. For others, transhumanism is a chance to participate in a Silicon Valley–charged noosphere, where one can carve out a living (though not a fortune) through speaking events, workshops, and publishing. But for others, it is not what could be lost in the future, or what could currently be joined or captured, rather it is what will be gained that creates the real stakes of this socio-political tumult. It is a common view among transhumanists that the telos of the universe tends toward intelligence, complexity, and prowess (if we are to treat these terms as something other than synonyms of one another).[41] It is also common to see transhumanists chart the increase of intelligence as continual exponential growth. The upsweeping curve many transhumanists often draw to illustrate this growth starts with the appearance of life over four billion years ago. The line has a significant shift in the degree of rise-over-run as the curve reaches events such as the appearance of multicellular life and the Cambrian explosion, and it launches further upward with the arrival of humanity. But that is not the sharpest inflection point. There is a spot on the curve where intelligence switches substrates, leaving eukaryotes and animals. Starting there, intelligence is instead tracked by following Moore's law-driven exponential growth of computational mechanic capacity. There, the chart shoots almost straight up to nosebleed length, asymptotically approaching infinity. And for many transhumanists, this is only the beginning. Extrapolating this curve to cover future cosmic ages, they commonly claim that *all* matter will either be intelligent, or it will be a resource used by intelligence.[42] This is a heady future—and a weighty one. Because of this potential, even though this exponential growth seems to be hardwired into the warp and weave of the universe, it must still be guarded. There is a lot at stake when debating the possibility of overcoming the human and becoming something like Gods.

Third Series: Mormon Transhumanism

§29

Lincoln Cannon has a problem. He most likely does not frame it as a problem, and even if he did, it is not a problem that he would choose to escape. It may even not be too much to speculate that this problem created him, or at least created him as he is today. In some ways, his problem is the same problem that President Uchtdorf has. In other ways, his problem is the same as the threat that vexed Zoltan Istvan. And at the same time, though again he would not phrase it this way, it would not be entirely unfair to say that President Uchtdorf and Zoltan Istvan *are* his problems, if we think of problems in the generative and creative sense of the term as something that is productively engaged with.

We have met Lincoln Cannon before, speaking at one early-morning session of a conference where he outlined a future history in a lecture hall in the Salt Lake Public Library. Lincoln Cannon is a Mormon Transhumanist.

§30

He is not the only Mormon Transhumanist. He isn't even the only religious transhumanist. There are Buddhist Transhumanists and Christian Transhumanists. There are also Transhumanist Churches that are independent of ties to any larger religion or denomination, with their links to transhumanism telegraphed by names such as Terasem, The Church of Perpetual Life, or The Turing Church. (*Turing Church* being a pun on the Turing-Church thesis, a conjecture created by Alanzo Church and the well-known Allan Turing, that states that any real-world computation can be translated in an operation

conducted by a universal computer, or Turing machine). Among all these religious transhumanist movements, Mormon Transhumanists have a stature,
history, and size that makes them preeminent among religious transhumanists.
And most of all, Mormon Transhumanists benefit from the resonances, as well
as the gaps, between transhumanism and Mormonism as separate modes of
imagining the world.

Most of the original wave of Mormon Transhumanists did not know at first
that they were transhumanists, though almost every one of them always knew
that they were Mormons. These founders were in a place where being Mormon
was unexceptional: Provo, Utah. It is easy to imagine that all of Utah is Mormon, but that is an exaggeration. As of 2007, only about 60.7% of the state was
Mormon; Salt Lake City, which is often imagined by outsiders as being a re-
doubt of Mormonism, is only about half Mormon in population.[1] By way of
contrast, Provo, the point of origin for the MTA (Mormon Transhumanist
Association), is nearly 88% Mormon.[2] To an outsider visiting or staying in the
city, this high percentage of Mormons in Provo is not just a demographic fact
but also a palpable feeling. This is in part because of the monumental footprint
of Mormonism in that city. Provo is home to two Mormon temples. The more
traditional-appearing brick Provo City Center Temple, built out of the ashes
of the burned-down Provo Tabernacle, and the nineteen-seventies modernist
Provo Utah Temple, built with its back to the mountains that shoot up right
against the city's east side. Mormon temples are not usually subject to being
outshown, but both of these buildings are overshadowed by Brigham Young
University, the jewel of the larger set of Mormon higher-educational institutions. BYU was built on a bluff overlooking the city, and like other college
towns, both the university as an institution and the students and faculty associated with it seem to be inescapably woven into the life of the city.[3]

Mormon Transhumanism originally built itself out of a small circle of post-
college–aged men (and one woman), mostly recent BYU graduates still in the
Provo area after completing their degrees. Many first met through Mormon
boards from online sites such as Belief.net or other similar discussion forums.
(There was, for instance, a great deal of activity on a private internet forum
with the surreal title *Spock with a Beard*). Often working under pseudonyms
(various *nom de web* that ranged from an allusion to the Ancient Greek Language to Zen proverbs to Book of Mormon Saints), they built a small community centered on debates about philosophy, the relation between religion and
science, and the Mormon religion. Discussions around that last item were
varied and complex, with headings such as "Milk before Meat," "Warning
Voice From The BofM," "What is Atheism?" and "Brigham and Extraterrestrials." As these conversations evolved, they began examining Mormonism as

a set of doctrines and also as a ritual practice; the nature of revelation, visions, scripture, and prophecy were also frequent sparring points. There was even one running thread entitled "Adam is our God!!!," which was a long-standing meditation on Brigham Young's Adam-God doctrine, a nineteenth-century claim (which has long since fallen out of favor) that Adam, the Archangel Michael, and Elohim were the same figure operating under different names and acting in different modalities.[4] The Adam-God doctrine today has ties to Mormon Fundamentalism and to those interested in the speculative Mormon theology of the nineteenth century. As this group cohered, it certainly did not end up looking like any form of Fundamentalism. They were always in conversation with conservative sparring partners. But most of the participants in these early forums were suspicious of a literal Adam and Eve, wary of the Church's stance against homosexuality, interested in religious empiricism, viewed science positively without adopting scientism, and generally preferred intellectual abstraction over literalist readings. These shared conversations developed a particular grain. Eventually, this formed something like an informal salon. This larger group was not just likeminded people found online, but it was also cobbled together out of a network of friends, and friends of friends, who often found each other by being told about "some guy" who spoke in sacrament meetings in a nearby ward in the same odd way that they did.

While they had degrees in an eclectic set of disciplines such as linguistics, Portuguese, business administration, music, and philosophy, many of them found themselves working in Provo's rapidly expanding Silicon Slopes, the high-tech corridor that runs along the Wasatch Front. (The Wasatch Front being the name for the sprawling warren of cities, ranging from Nephi in the north to Brigham City in the south, that runs right up against the snow-capped Wasatch Mountain range.) The existence of the Silicon Slopes is no accident. The Slopes is an effect of both state and local government outreach as well as the result of long-standing local ties to the research wing of the American military. (For instance, the University of Utah was one of the four original nodes in ARPANET, the decentralized computer military and research network that mutated into the contemporary internet.) A history of computer-programming entrepreneurship is important as well: WordPerfect (the *de facto* standard for word processing software in the early days of personal computing) was designed by a graduate student and a professor from BYU. But the Slopes is not just the result of governing infrastructure, capitalist adventures, or a research legacy. It is also an effect of the relatively high level of education in the region. This educational level is proximately due in part to BYU's dominating presence, but more generally it's an effect of the higher educational achievements of Mormons demographically.[5]

But there is also a cosmological echo to be found both in the high levels of education and in the wider interest in intellectual labor that made this area of concentrated high-technology commerce possible. The Slopes, and more generally, the high levels of educated Mormons that populate the Wasatch Front, are reminiscent of the partially divinized place of intelligence and intelligences in Mormon cosmology. *Intelligence(s)* is a debated term sometimes used in the singular as one of the highest attributes of God, and sometimes it is used in the plural to refer to a host of pre-mortal beings, many of which would become incarnate on Earth as humans as part of their spiritual trials. Whatever the cause, technology has been thick on the ground in Provo and its environs for a while. Even today, Provo is the kind of place where you can see a roadside sign advertising robotics classes for Boy Scouts. So, it is not too surprising that this place is full of startups as well as established tech companies, many of which have set up large satellite offices in the region.

§31

It is also no surprise that many of the people who made up that first reading salon (that would become the nub of Mormon Transhumanism) shared something beyond a tie to the Church of Jesus Christ of Latter-day Saints. And it was also no accident that these same people should at times wonder if God's miracles and human science were akin: the distance between the technologically achieved and the miraculous is never that great in the Church of Jesus Christ of Latter-day Saints. This closeness between the two phenomena is one of the effects of an eschatology built around theosis. If God "once was [a] man like us, and the Father was once on an earth like us," then universal laws cannot be something created and sustained by God since that opens up the question of who or what was sustaining these laws before God was God.[6] While there are segments of the Church that would deny any account that questioned the ultimate supremacy of God, the solution to this potential aporia that has gained the greatest traction in Mormonism is that these laws *precede* God. This leaves God, even after his divination, as being subject to universal laws, including ethical laws (as it says in the Book of Mormon, "[n]ow the work of justice could not be destroyed; if so, God would cease to be God").[7]

This understanding of God as an exalted human who is locked into preexisting laws creates the problem of what then to make of miracles. *Miracle* is no simple term. There are a multitude of definitions for, and sensibilities concerning, miracles in Christianity. But the most common Anglo-American Protestant understanding, running from David Hume's rejection of the miraculous to C.S. Lewis's endorsement, is that miracles are "an interference with

Nature by supernatural power."[8] In this reading, miracles are, in essence, an interruption of nature's laws by a divine being. According to a strict account of Mormon metaphysics, however, this understanding of a miracle is not a cogent proposition. If the law comes before God, then God is boxed in a way that forecloses any interference with nature that would suspend the natural order.

This is not to say that miracles are absent. Depending on the strain of believer, there is enough providence and divine intervention to satisfy the most supernaturally intoxicated forms of American Evangelicalism. It is just that Mormons take a different approach, not erasing the miraculous with the known laws of the universe but reconfiguring the metaphysical mechanics through which miracles are effectuated. Thus, despite the ontological limitations, miracles are still a part of the Mormon imagination; one could even argue that a place for the miracle is mandated in Mormon scripture, not just by example, but by command. As presented in the Book of Mormon, on his deathbed, the colonist, king, and prophet, Nephi, warns his future descendants that to deny the miraculous power of God is to deny God himself. Regardless of the scriptural warrant, miracles are thick enough on the ground in Mormonism. Some of the more florid miracles associated with folk Catholicism or Pentecostalisms are not a feature of the religion. But Mormonism is full of the sort of unexplained healing and providential coincidences found in much of Protestantism (and healing, as one of the priesthood blessings, is given a heightened degree of attention, though it rarely takes the form of an instantaneous absence of a recently quite palpable condition). But more to the point, whether or not the wonder is understood as a sudden change or a slow recalibration, these miracles are not seen as God suspending natural laws. Instead, the miracle is seen as God *leveraging* those laws to his own ends. The acts of God appear miraculous not because they are beyond any possible understanding but because of a lack of understanding by the merely mortal recipients as to how they were accomplished. As put by Brigham Young, "The providences of God are all a miracle to the human family until they understand them. There are no miracles, only to those who are ignorant. A miracle is supposed to be a result without a cause, but there is no such thing. There is a cause for every result we see; and if we see a result without understanding the cause we call it a miracle."[9] Not all of Brigham Young's theological speculation has made it into the twentieth century (recall the Brighamite speculation that conflated the identity of God and Adam), but this aspect of his thoughts has been given a more solid reception by the contemporary Church.

This makes miracles, in effect, just another example of technology and an almost perfect anticipation of the last of the science-fiction maxims called Clarke's three laws: "Any sufficiently advanced technology is indistinguishable from magic." This understanding of the miraculous is common (I've never heard a Church member dispute it when I've put it forward), but that does not mean that it is always salient. For most Mormons, this understanding of the miracle is just a bit of ontological scaffolding, a sort of cosmological spandrel that has no effect on how they think of day-to-day miracles or on the kind of emotional charge or spiritual import that miracles bring when they occur. For many Saints, it is the concept of the miracle as beyond any human understanding rather than of the miracle as being a part of a chain of cause and effect that does most of the heavy lifting here, ensuring that the miracle retains the same patina of wonder that also marks the Protestant, Catholic, and Orthodox miracle. But for other Mormon believers, the metaphysics of the miracles-as-techne says something important about the nature of God—and also about the possible future capacities of humans.

It was this idea of miracles as technology that informed the discussion group. Already saturated in internet startup culture, again and again, they hit upon the idea that anticipated new technologies could make Mormon eschatological longings something that could fall within human reach in a not-too-distant future. The millennium's end to sickness and poverty, the renewal of the Earth, the construction of a new Jerusalem, the inauguration of Methuselah-like life spans, and, most of all, the resurrection of the dead "in the twinkling of an eye" just might be accomplished through human technology and not divine fiat.[10] This is not to remove any divine responsibility: this world was purposefully created such that these technological wonders were achievable by advanced human beings who were sufficiently intellectual and spiritual. Nor was this imagining unique to this moment or this place. I've had many current Mormon Transhumanists talk about what one member has called the "proto-transhumanist moment," an instant when they imagine a scientific accomplishment of religious marvels while reading texts like a science fiction story, comic book, children's Bible, or while daydreaming in seminary, the Mormon name for the often-boring religion courses given to youth during secondary education. I've also heard of members jokingly refer to themselves as having been "dry" transhumanists before finding the MTA—this phrase being an allusion to "dry" Mormons, a term for people who have the ethical values and lifestyles of Mormons, but who are not members of the Church. (The *dry* here indicating that they are unbaptized.) Other Mormon doctrines also played a role in the founding of the group, making these proto-transhumanist moments of thought overdetermined. Mormonism teaches that spirit is simply a more refined state of matter. And that in their way, all entities, including God, have a material body. While this is not the same concept as the sort of materialist assumptions that are given methodological (and often ontological) weight in discourse about science in Euro-American circles, these two ideas do rhyme with one another, and this allows for a sort of quick shuttling between frames without too much cognitive turbulence.

§32

The logic of transhumanism does not necessitate a belief in or concern with extraterrestrials. But the fact that Mormonism assumes an open cosmos has importance as well when considering Mormonism's resonance with an intellectual and social movement that is open to the possibility of non-human, extra-solar intelligences. Founded on the American frontier in the nineteenth century, Mormonism is a genuinely post-Copernican religion. The ancient Israelites wrote the Bible with the understanding that Earth was at the center of creation and that the sun, moon, and stars were attached to the firmament, a great vault over Earth that separated the waters below from the waters above. Smith worked at a time when the idea of each star in the heavens was thought to be a sun like ours, surrounded by possibly inhabited planets. This idea was common enough in Smith's America to be referenced by Thomas Paine in his attack on Christianity and to be used by the Protestant minister Thomas Chalmers in Christianity's defense.[11] In fact, the idea of a plurality of worlds, with earth as one world among many, is again inscribed into the very sacred texts of the religion. The Book of Abraham, a scripture supposedly translated by Joseph Smith from some scraps of papyrus purchased by the Church in 1835, acknowledges a multiplicity of worlds and suns, including Kolob, the "governing" sun that is "nearest unto" the Lord.[12] Similar claims can be found in the Book of Moses, a part of Joseph Smith's translation of the Bible that included sections that "had been taken from the Bible, or lost before it was compiled." In this text, there are passages where God grants Moses a vision of the endless swath of worlds, all of which had "inhabitants on the face thereof."[13] Apart from these explicit claims made in Mormon scripture, Joseph Smith also is reported to have taught that "the moon was inhabited by men and women the same as this earth, and that they lived to a greater age than we do—that they live generally to near the age of a 1,000 years." These lunar men were "averaging nearly six feet in height, and dressing quite uniformly in something near the Quaker style."[14] Similarly, Brigham Young had hypothesized that the sun was inhabited because he believed that such a celestial body "was not made in vain."[15] (These kinds of lay astronomical claims were common in the nineteenth century, even among orthodox Protestants who saw the existence of these aliens as evidence of God's power and beneficence.)[16] This Mormon doctrine of the plurality of worlds even appears today in strange moments, such as in a folk theory that utilizes the multiplicity of earths in defense against creationism. I have been told more than once about some eccentric seminary instructor or odd uncle that argued that dinosaur fossils were not traces left over from Earth's ancient past. They were merely remnants of some other world that God reused when he fashioned this world from whatever spare matter was lying around at the time of creation.[17]

While none of these folk-astronomical claims are central postulates to contemporary Mormonism, the idea of a plurality of worlds is central to Mormon ritual. During the temple endowment ceremony, which to a large part is a reenactment of a particularly Mormon understanding of Genesis, there is a reference to both worlds that preexisted this one as well as to references to other (seemingly contemporaneous) worlds. A plurality of worlds is even presumed by what is arguably the most defining instance of Mormon iconography. Mormon art often refers to or incorporates the *Christus* statue, a Danish nineteenth-century marble statue of a robed Christ gazing downward with his hands spread in what looks to be a gesture of benediction or welcome. The Church has used reproductions of this statue several times, including in several temples as well as in the 1964 New York World's Fair (where its presence was "intended to help visitors understand that the Latter-day Saints (or Mormons) are Christians").[18] It is common to see smaller reproductions of this statute in Mormon homes. (The Church-owned religious bookstore, Dessert Books, sells nine-, twelve-, and nineteen-inch reproductions.)[19] But the first *Christus* statue acquired by the Church is the one that is most commonly referenced in the visual culture of the Church. These photographs of the sculpture, which are located in the north visitor center in Temple Square, Salt Lake City, focus as much on the expansive mural that surrounds the statue as on the statue itself. Located on top of a spiral walkway, the statue is set inside a cavernous dome that is painted to look like planetary space. Amid a field of stars and large, suspiciously cloud-like nebula, there is a multitude of planets (including a canal-streaked Mars). One of the most defining images of the Mormon Christ is not just a Cosmic Christ but an Interplanetary Christ as well.[20]

§33

Mormonism also has a utopian edge that plays a role in laying the groundwork for these proto-transhumanist moments. Early Mormonism is most often understood as a religion complete with scriptures and rites. But it is also possible to frame Joseph Smith's initial attempts to found a Holy City as another instance of the utopian communities that were so common to the American nineteenth century. When the Saints first ingathered to Zion in Kirkland, Ohio, Joseph

Smith had a revelation that, under what was termed *the law of consecration*, Saints should consecrate their property to their bishop so that a United Order could redistribute it to end poverty. Even after this proto-socialist endeavor failed, Brigham Young would continue to set up various other collectivist United Orders in the inter-Mountain West. Joseph Smith also drew up utopian plats for the various planned instantiations of the city of Zion that came together as discrimination and violence slowly pushed the Saints to the edge of the continent. This totalizing civic plan, which was posthumously realized in Salt Lake City, is built around a single central temple, with outward shells of concentric orthogonal plots. (The plan looks something like the map of a contemporary microchip.)[21] Even the early form of "celestial marriage" carried out by Joseph Smith can be thought of as being akin to the sorts of sexual experimentation seen in "complex marriages" in places like the Oneida colony.[22] Utopianism is broader than transhumanist imaginings of societies perfected by technology, and it cannot be reduced to science-fiction portrayals of societies without flaws. But there are deep overlaps between these categories, allowing for synergistic moments where Star Trek–like post-capitalist techno-societies are shot through with religious visions of the lion lying down with the lamb.

But there is one element of Mormonism that makes the religion potentially simpatico with both proto-transhumanisms and with a fully realized transhumanism as well. It's fair to say that this element is central to the Mormon Transhumanist Association's understanding of itself. And this same element is almost certainly the root of post-Copernican, cosmological, and ontological Mormon religious visions as well. This element is not a specific piece of doctrine or scripture, but it is instead a tradition, an aesthetic, a mode of thought. It is the late-nineteenth- and early-twentieth-century Mormon practice of religious speculation.

Early Mormonism's break with established Protestant doctrine, as well as its adoption of an evolutionary and perhaps infinite cosmos, not only opened up horizons but raised questions. If God was once a man, was there another God before him? Are there multiple gods? Is there a community of gods, and if so, then what does a society of gods look like? If God has a physical body, what is the nature of that body? Where is it housed? If science and miracles are not separate categories, then what is the proper attitude toward science for sincere Saints, and what sort of truth does it hold for them? Are there hierarchies of scriptural and scientific truth, or is truth all of one substance or order? Speculation could be seen as a response to the sorts of theological statements about materiality, divine origins, and the engine of the miraculous. But at the same time, it could also be seen as the conditions of possibility for these statements to be articulated and have any sense in the first place. Therefore, speculation was not just a proposition about God or the nature of man, nor was it merely a topic for discourse or a mere exhaustion of listing the various supernatural forms that might be imagined in a post-Copernican, unbound Mormon universe. It was instead an ethics of thought and a tolerance for diminishing degrees of certitude. Small decisions about opening up for iteration previously foreclosed questions (Why not another scripture? Why not earlier Gods?) created virtual vistas that could be endlessly explored if one is open to what these fields of untrodden potentiality might bring.

While this tradition of thought does not enjoy the pride of place it once did, now that the Church is arguably more concerned with presenting an outward veneer of Protestant respectability, it was a facet that was there from the faith's beginning. This line arguably started with Joseph Smith, both through his use of revelation and through his founding organs such as the School of the Prophets, which worked to synthesize religious experience and secular knowledge. It can be seen in moments such as Brigham Young's promulgation of the Adam-God doctrine. And the same ethics of speculative thought was certainly found in early Mormon thinkers and Apostles such as Parley Pratt, author of *Keys to the Science of Theology*, as well as Parley's brother Orson Pratt, a self-trained astronomer and avid pamphleteer who frequently resorted to philosophy and science in his defense of the faith. This speculative tradition continued into the early twentieth century with B.H. Roberts, who attempted to fuse then-cutting-edge astronomy, scientific cosmology, and, perhaps most importantly, evolutionary theory to create a comprehensive account of Mormon doctrine.[23] This speculative tradition should not be taken to be of one mind or even to have been particularly irenic. Brigham Young and Orson Pratt fought bitterly with one another, and Roberts' magnum opus *The Truth, The Way, The Life* had its publication blocked by Roberts' fellow apostles, and it only came out posthumously over a half-century later.[24] Much of this thought was concerned with working out the metaphysical conundrums that Joseph Smith's vision raised. But because of the post-Copernican nature that was shared by both Smith and his later Mormon interpreters, this work also necessitated a deep engagement with the physical and humanistic sciences.

It was this speculative line of thought that was remembered by those Mormons with a taste for the forgotten corners of their religious inheritance. And this included the founders of the MTA. The centrality of this speculative strain to Mormon Transhumanism has been acknowledged both by members of the group as well as by outside Mormon scholars who have commented on the Mormon Transhumanist thought, such as Richard Bushman. (Bushman, being an individual who is in many ways responsible for a new wave of Mormon history, is also a key figure in a more open and less pugilistic style of Mormon apologetics.)[25] It was this nineteenth-century intellectual tradition that was informing the circle of Silicon-Slope employed and adjacent recent college graduates. They turned to this intellectual tradition as they read and debated with one another; they used it to play with the sober metaphysical minimalism of American pragmatism and with an inebriatedly open potential Mormon technological miraculous. As this intellectual tumult went on, one of the members of this group, Lincoln Cannon, stumbled across a paper by the transhumanist philosopher Nick Bostrom. Bostrom's paper asked, what was the likelihood that this universe was not foundational, but merely one of a great many computer simulations being carried out by some more original, non-simulated Universe? Almost simultaneously, a friend of Cannon's brought to his attention Ray Kurzweil's life-extension-oriented website. As Cannon explained it, "As I studied the two, I started to connect the dots and realized that both of these men were part of a community that had a lot in common."[26] He had discovered transhumanism.

For Cannon, that a secular community such as transhumanism even existed was a surprise. Paradoxically, he was also surprised by how unsurprising these thoughts were. "These are the ideas that I've had myself, for many years, and I didn't know that there were secular people who had similar ideas, and that they were called transhumanists. I'm a transhumanist!" This transhumanism was not a replacement for his religious beliefs but rather something logically entailed in the idea of religion and particularly in the idea of his religion. In his words,

> Humanity is not static. Never has been, never will be. And so, as we are changing, as we are evolving toward these superhuman, post-human projections that we've always had, and sometimes we've called them Thor and sometimes called the Zeus and sometimes whatever, but as we evolve toward them, and even if it's merely passively or if we actively chose to try to become like that, and if choose wisely and well to try to emulate the more compassionate ones and less the more lightning-bolt throwing ones, then we change, and we evolve and we become more like them, and we build tools that make us more like them, and I anticipate that as we continue to try to do this, and as we continue to choose the better Gods and not the worse ones, which if we emulate we will just destroy ourselves, that we'll become increasingly like them, and that's—in my mind, that's what Mormonism and Transhumanism are both about. And honestly, that's what the Gospel of Christ is, at its very heart, about, for me.[27]

Soon many of the other people in Cannon's circle would make the same realization. In rather quick order, an initial group of fourteen wrote a constitution and set up the Mormon Transhumanist Association as a 501(c)(3) non-profit corporation in 2006. That same year they were accepted as an affiliate in H+, the world's largest transhumanist umbrella association (though that membership would later be briefly interrupted for a few months in 2010 when an H+ board member objected to the very idea of Mormon Transhumanism). The next year the MTA collectively authored a document that was half article, half declaration; they quickly had it published in *Sunstone*, a major magazine in both progressive and academic Mormon circles.

By 2020, the Mormon Transhumanist Association had grown to almost one thousand members and was the largest religious transhumanist organization in the world.

§34

What is the day-to-day life of a Mormon Transhumanist? This is a natural question and one that is also flawed from the start. As will be shown, there is no paradigmatic or representative Mormon Transhumanist that can stand for all others. Issues involving a tightly mixed series of independent variables–involving gender, sexuality, and their relation to and history with the Church–means that a Mormon Transhumanist's day-to-day life can vary widely. And besides, not all members see the MTA as affecting their day-to-day lives; I've heard a member say that their lives aren't different at all because of their membership.

But still, generalizations can be made about the way that Mormon Transhumanism is actualized in their lives. There is just the bare fact of a thriving on-line community of nearly 1,000 people, which is animated by friendships that have gone on in some cases for over fifteen years, though this is not a phenomenon unique to Mormon Transhumanism. Effects more specific to Mormon Transhumanism can be sorted on a continuum running from concrete forms of self-care to abstract cosmic orientations. For some, Mormon Transhumanism is a catalyst for an enchained regimen of habits and dietary practices. One member credits Mormon Transhumanism for his intermittent fasting, a technique that has root in Mormon religious practice, and in some life-extension-oriented techniques drawing from laboratory research on the effect of caloric restriction in mice; the use of nootropics (smart drugs and cognitive enhancers) and geroprotectors are not unknown. They are sometimes taken in the form of prescription drugs but most commonly as commercially available dietary supplements.

For others, their transhumanism is expressed in ambitious, large-scale personal projects. One member, a computer programmer for a large entertainment corporation, has, as an individual, recreational project, used mathematical graph theory to analyze astronomical data sets. This analysis was used to map possible pathways between stars that could potentially be used as routes for human (or extraterrestrial) travel through the local region of space—with *local* meaning the 2,000 stars closest to the earth's sun. This project ended up being presented as both a publication and as a session at GraphConnect, the most prominent annual conference dedicated to the use of graph databases. This work, in turn, led to a collaboration with a postdoc SETI researcher. Later work expanded on the prior project and used a much more recent astronomical database to map potential networks between stars with known exoplanets; this project was intended to assist scientists looking for technosignatures that could index the presence of alien life.

For others, Mormon Transhumanism brings a different relationship to technology, sometimes with an expansive understanding of technology. This often takes the form of an appreciation of ready-to-hand technologies that are so ubiquitous that their nature as technologies are forgotten: clothing as artificial skin, cars as mobility-expanding mechanisms that (temporarily) merge with the body, food as not natural products but the result of sometimes centuries of conscious human manipulation and processing. When it comes to much of the unthought communication technology that is now common throughout large swaths of the developed and semi-developed world, there is a day-to-day realization of the sometimes world-spanning infrastructure necessary for it to operate.

Compared to body practices or large-scale projects, this continuing reassessment of already achieved human imbrication with technology is more of a change in consciousness than conduct. But it is still worthy of note. This is because such attention to technology is also an expansion of religion. Work *with* technology—including not just the usual work in science and computer programing but work conceived of expansively to include fields as diverse as medicine, education, and art—also becomes what is sometimes described as spiritual work. It allows for an expansion of religion from the sites where it is traditionally located in Mormonism (the meetinghouse, the temple, scripture study, and domestic rituals such as the weekly family home evening) to a broader horizon. As one MTA member put it, this dilation of religion by associating it with technology "means I see the LDS church as simply one environment to practice my Mormonism and Christianity rather than it being what I reduce my religious identity to." I've heard members report feeling a "burning in the bosom" at times during Mormon Transhumanist meetings, using a phrase usually indicating the affective experience of being moved by the Holy Ghost during sacrament meetings and temple ordinances.

For some Mormon Transhumanists who already have problematic relations with the institutional church, this expansion allows for a break with Mormonism and a shift to a sort of vague technospiritualism not unlike what is seen when secular transhumanists think at the level of cosmological futures rather than near-contemporary technologies.[28] But for many others, it allows for a *greater* engagement with the institutional Church, or at least the avoidance of any diminishment. The Church, no longer forced to bear the entire weight of religiosity, instead becomes a plank in a greater program. Relativized and contextualized as simply a moment in the long arc of a greater techno-religious quickening, these transhumanists find ways to hold on to their Mormonism, even as they realize institutional Mormonism while making Mormonism anew.

§35

The challenge facing Mormon transhumanism is familiar. On the one hand, it is the challenge that tests Zoltan Istvan: how to midwife the birth of technologies that promise not just to prolong or nurture human life but to transform it. On the other hand, Zoltan Istvan is the problem. The New-Atheist edge of much of transhumanism cuts against the grain of Mormon Transhumanism. The organization has to continually suffer not just transhumanists who view the MTA as just another implausible sample of weirdness to be shelved away as another curiosity in the futurist *Wünderkrammer*, but it also suffers those who think of the Association as a deathist fifth column within the transhumanist body-politic. And while many transhumanists are not so extreme, we can see in Istvan's language and work that, at least at the level of the imagination, the sanctions visited on religion and the religious could be quite extreme.

Mormon Transhumanists also share President Uchtdorf's concerns, even as they, in a way, are an example of Uchtdorf's concerns. Uchtdorf and the rest of the leadership of the Church are a (generative) problem for the MTA as well, not in the sense of the official organs of the Church serving as an opponent for Mormon Transhumanism but rather in that Church anxieties complicate the MTA's project. The MTA is not comprised only of faithful adherents who are in good standing with the church, but it is also comprised of those who are either a product of, or who have thoroughly internalized, the inter-Mountain West Mormon culture. While believers in good standing are an important part of the MTA, MTA membership is not limited to the orthodox, compliant, male, white Mormon. The association also has members who have left (or been expelled from) the Church. It is also a home for Christian Transhumanists who, until recently, had no organization of their own. The MTA even houses some atheists who are drawn in by the way that, unlike other transhumanist groups, issues of transhumanism and religion are often foregrounded in Mormon Transhumanist discussions, and it also houses those secular transhumanists who are interested in the MTA's openness to charitable endeavors. This diversity is positively seen by the leadership of the MTA, and it is something that they work to maintain and, at times, to expand. (While not mapping onto the same axis of religious diversity, this primarily male organization has been concerned for several years about working toward something closer to gender parity in membership.) But still, the MTA is an organization that is primarily comprised of, intended for, and informed by believing Saints. And this is important because it allows Mormon Transhumanism to do a significant part of its work: that of producing alternative, yet still faithful, Mormonism. While the MTA often presents itself as being merely an advocacy group for Mormons interested in technology, the MTA also works to make Mormonism seem more plausible to some doubters by recasting the religion as something not only in harmony with natural science but as something *implicit in the promise of* natural science, and perhaps as an inevitable expression of the cosmic and teleological forces as they are understood by natural science. Like Uchtdorf, Mormon Transhumanists wish to build a Mormonism that is immune to the skepticisms and schisms that are being catalyzed by the internet. And they wish to do so not by redirecting the attention of doubters to literalism and orthodox understandings but by laminating that orthodoxy to the rationality of science and to the hope of transhumanism.

Because of the MTA's very nature as a geographically distributed community that is interested in the cutting edge of information technology and carries out a large part of its social activity through internet-mediated means, the association can be seen as yet another heterodox faction within the Bloggernacle. The intimate ties between the MTA and the Bloggernacle can't be denied. In fact, the very person who coined the term *Bloggernacle* has also served as one of the past presidents of the MTA. MTA members are often guests on various Mormon podcasts. Sometimes they are speaking in their capacity as members or officers of the Association. But at other times, they are talking about issues that do not immediately seem particular to religious transhumanism, such as child-raising, biblical hermeneutics, or gender and sexuality as it relates to the Church.[29] Likewise, many Mormon Transhumanists are regular presenters, or even organizers, at *Sunstone* symposia, a series of conferences put on by a progressive-leaning Mormon studies periodical. And finally, while the organization takes no official position on issues like women's ordination, gay marriage, or the status of LGBTQ Saints, the sentiments of many MTA members–and also of most of the members with the greatest internet presence–lean more toward the position of the progressive internal critics of the Church, a common view in the Bloggernacle.

The MTA is an outsized and organizing presence among religious transhumanists and, increasingly, in the Church itself. The question is whether this is despite or because of its positioning at the intersections of so many fault lines. While some Christian transhumanists accept them, many Protestants dismiss Mormon theology, and therefore Mormon Transhumanists, as unbiblical. Even though Mormon Transhumanists sincerely anticipate, and often work for, a post-human future, many secular transhumanists dismiss them, stating that their religiosity makes them, at best, self-deluded or cognitively impaired. And finally, while there has been no disciplinary action taken against anyone for their affiliation with the MTA, or because of their transhumanist beliefs, they are non-conformists and free thinkers in a time of Mormon religious retrenchment and general institutional panic over the internet. Many Mormon Transhumanists feel that disciplinary action is a constant threat or perhaps even an inevitability. Starting in 2018, some hardline Mormon apologists have begun to notice the organization as well, and their attention is not always kind.[30] Yet despite all this, the MTA not only endures, but it continues to grow, and its members turn their imagination and efforts toward what they feel their highest calling is: realizing the essential function of the universe, which is a machine for the making of gods.

PART II
Mormon/Transhuman

Fourth Series: Kolob runs on Domo

§36

Interstate 15 is a ribbon of American Cold War infrastructure, a freeway running from the United States' southwestern border with Mexico, to its northern border with Canada. Part of this larger arc formed by this thoroughfare is threaded through the state of Utah, bisecting Salt Lake City, and its penumbra of suburbs and commuter cities, all while shadowing the Western edges of the Wasatch Mountains, the peaks that are so much a part of the Salt Lake vicinity's skyline. The part of Interstate 15 that goes through Salt Lake City does double duty, serving as a conduit for interstate travel and also as an important route for the daily in-city commute. Despite the stark, desert beauty of the region, and the sometimes diaphanous, sometimes imposing sky above, the view is not that different from what might be seen in other major American commuting freeways. There are the usual suburban homes (though it should be noted that in Utah, due to lower land values, suburban homes tend to be much larger than those in other parts of the United States), and also the occasional landmark buildings. For instance, the software company Adobe has a large glass structure in Lehi, a commuter town south of Salt Lake City proper; the building looks not unlike a giant crystal Lego lit up from within. (Less exaltedly, there is also a nearby miniature golf complex built to look like Mount Rushmore.) And on the interstate, there are billboards. The billboards here are no different architecturally from advertisements elsewhere in the United States. Their content, however, is not quite identical to billboards elsewhere; there are innumerable ads for plastic surgery, for instance. (Utah is notorious for having a much higher per capita count of plastic surgeons compared to most other regions in the United States, which led one major American periodical to refer to Salt Lake City as "the Vainest City in America.")[1]

Sometimes, though, it is even harder to know precisely what a billboard is promoting. Such was the case with one that appeared on the side of Interstate 15 in the spring of 2017. The far left-hand side of the billboard had a minimalist white-and-blue square logo, and the rest of the sign was taken up by the gnomic declarative statement—set in a giant but standard Helvetica-like San-serif font—that "Kolob runs on Domo."

On its face, the statement is puzzling. It is even more puzzling than it might appear. That is because this message is pulling its language not from one cant, code, or technical vocabulary, but rather from two. The second term, "Domo," is the name of one of the larger startups in the Silicon Slopes. "Kolob," the first term, is from a different argot. Immortalized in both scripture and hymn, Kolob is the star or planet where God lives.

While it did not rise to the level of being a *cause célèbre*, something about this sign hit a nerve in the greater Salt Lake metropolitan region. One Mormon public intellectual was so annoyed by the sign that he joked on Facebook about setting up a GoFundMe so he and his family could escape from the Utah Valley. A Twitter user had a much more succinct reaction, appending a picture of the sign with the exclamation "Good hell." Others thought that it must be some snide attack on the Church, as suggested by a discussion chain on the ex-Mormon subreddit entitled "'Kolob runs on Domo.' Which one of us filthy apostates did this?" On another ex-Mormon subreddit discussion chain (the billboard birthed more than one thread on that website) some people were so put off by it that they simply asked: "Is that real?" This query triggered a challenge of the whole thing in the form of someone else exclaiming, "Holy fuck. No WAY that's a real billboard." Another commentator at once tried to jokingly summarize, categorize, and dismiss by calling it "[j]ust a little blasphemy in a high density mormon [sic] area." Eventually, the owner of Domo himself had to chime in. Replying to a tagged tweet regarding the billboard, the founder and CEO of the company replied that "I thought it was good-natured fun but we've had a few complaints. Thinking we are making fun of Mormons. . . . I'm one so why do that?"

§37

Whatever the original intent of Domo's owner, it is evident from the intensity and confused negativity the billboard generated online that the pairing of these two languages was disjunctive, and perhaps even a bit wrong. Something about pairing a tech company and a scriptural proper noun triggered either laughter or derision, and sometimes both. One way to consider this reaction is to say that the offense lay in a juxtaposition of the sacred and the worldly. The problem with this analysis is that it ends too soon. We are told one thing is sacred, that something else is worldly, but we stop before thinking about what traits give each discourse its character and what effect those traits have. To put it another way, we do not ask what the features of sacred and worldly languages are. A subtler way of reading the gap and hence the tension between the two languages is to observe that these languages operate in different ways. One is arguably about concealment and revelation, while the other is a technical vocabulary that, while not secret, is still effectively obscure to those outside its discourse community. Consider this contrast: Mormonism is a religion rife with secrets, secrets present at the very foundational moments of the Church. The most central scripture, the Book of Mormon, was purportedly taken from gold plates engraved with characters in "Reformed Egyptian" that had been purposely secreted under the ground. And one hundred and sixteen of the pages from these engravings, along with some apocalyptic visions of the entire history of mankind, have been sealed from the world by divine command.[2] Even after they were uncovered, the plates that constitute "Joe Smith's Golden Bible" were kept continually obscured by thick cloth while in the custody of the American prophet, who is described as using seer stones to decipher them.[3] As a project, Mormon secrecy goes far beyond even this, inserting itself deep into daily life, especially in the Wasatch Valley.

By way of contrast, the technical language of the internet, alongside other dialects that could be considered virtual regional variants—such as those of computer engineering, cryptocurrency, and technically oriented futurism—may have a high buy-in cost and may occasionally have a few words or phrases that can appear annoying or even euphemistic. (Consider, as an example, all the work done by the term *disruptor*.) However, these are not languages of disguise per se—even the language of cryptocurrency, which has cryptic right in the name, is a language not of secrecy, but instead of an open and distributed processes. Technical languages allow an actor to speak *technically*, that is to say, to speak strictly, accurately, and factually (even if the import and nature of these facts are ultimately socially shaped and determined). Technical languages glory in their referential, not their poetic, power. And as we will see, this capacity of technical languages is even greater in transhumanism. Transhumanism can first serve as a switching point and gathering house for all the different technical registers used in the various technologies in which transhumanism is interested. But just as important is the second change transhumanism works on these languages: that in looking forward to a complete transformation of human life through technical means, and in envisioning many of the determining existential facts about our humanity as technical problems, the imaginative scope of these discourses becomes expansive.

The differences in these modes of speech, when paired, have novel effects. The pairing of a language of secrecy and a technical language can at once work toward an erasure of secrets, but also to the proliferation of secrets. The technical language allows a response to a series of enchained crises involving history, race, and gender, all of which are, in large part, caused or catalyzed by secrecy, which is encouraged by or attributed to the Church. This allows for those believers who make use of technical language to, in effect, obscure serious differences between what *they* believe and what is believed by more conventional Mormons. But the pairing of these two languages creates a paranoid proliferation of accusations of secrecy from some non-Mormons as well.

The kind of secrets that two such languages can create is not, though, of the same order as that created by a religious language of secrecy alone. The language of religious secrets, at least as far as its Mormon instantiations go, are languages of known unknowns, where the obscured target domain is in ways pretty open many of those who truck in the language.[4] And the use of technical language in a religious domain, which is in large part a response to the Mormon economy of religious secrets, is either a tacit or accidental secret, fed by the receiver's assumptions about orthodoxy and her inability to decode this language. The discernment of secrets can itself take the form of secrets, and those secrets are even harder to discern because they are at once overcoded and unmarked. At the same time, these other secrets open up the operation of secrecy by encouraging others to fabulate all sorts of possibilities regarding what *might* be concealed. And this can even open up the doors to secrecy's hermeneutic shadow: paranoia.

While this is not the only lesson we can take from "Kolob runs on Domo," it can also serve as a meditation on the ways that the internet can catalyze doubt in relatively insular religious communities.[5] But at least here, this other facet of the discussion only reflects the paradoxical light that open secrets shine.

§38

As noted earlier, a great many MTA members have worked or do work in the technology sector. Which is to say, they work at places like DOMO.

This DOMO-adjacent interest in the technology sector shapes many online MTA discussions. But this interest in technology affects more than what is discussed; for quite a while it also shaped the venues in which these discussions took place. In its early days, the MTA would realize itself through then-novel forms of internet mediation. Members would take advantage of email listservs, a minimalist technology whose MTA expression hobbles on to the present day. They would struggle to connect through awkward and then-new video-chat technologies. Even in early days, membership scattered across various American and European time zones would meet in virtual rooms with strange buzzing sounds and wildly fluctuating volume settings. These video-chat feeds would contain nothing more than live-cam talking heads wearing bulky earphones with built-in mics, projected on various virtual screens in the video-chat clients.

For a period, the Mormon Transhumanist Association had a presence in Second Life, a then-thriving virtual world full of wandering, talking avatars tethered via the internet to the home computers through which these avatars were controlled. Some early MTA members would go to Second Life's transhumanist spaces to talk to the other transhumanist avatars about the MTA, sometimes even giving planned talks on the subject, complete with PowerPoint slideshows that would play in the background as a lecture ran its course. These MTA members in Second Life were, in the words one MTA member, virtual missionaries for a virtual world.[6]

MTA heralds were not the only such missionaries, though. At the time when Second Life was at its peak in both reputation and active users, multiple other unofficial Mormon presences wanted to use this space to share their own Mormon testimonies. These Second Life users built spaces with names like Deseret (the title of a new-world land occupied by ancient middle-eastern immigrants fleeing the aftermath of the Tower of Babel), Zarahemla (the name of a Nephite metropolis which, due to its "iniquity and abominations," God burnt to the ground by fire and lightning), and Adam-ondi-Ahman. (The latter space was a virtual island named "after the site where, according to Mormon beliefs, Adam and Eve resided after their expulsion from the Garden of Eden and where, before Jesus Christ's second coming, Adam will judge his posterity, receiving all of the keys bestowed in each successive dispensation, preparatory to turning them back to Christ.")[7] In these virtual spaces, virtual Mormons would build virtual recreations of both the Salt Lake City and the Washington DC Temples. There were virtual Sunday meetinghouses constructed out of virtual versions of the red bricks commonly used for these structures in the physical world. These virtual meeting houses would contain virtual pews, virtual hymnals, virtual organs, virtual Sunday-School rooms, and even virtual basketball courts for virtual after-school activities. There were virtual billboards that took advantage of their virtual nature by containing whole chunks of text (including the "Proclamation on the Family," a Church document advocating for a recognizably heterosexual-leaning take on marriage and gender that would never fit on the billboards in non-virtual 'meat-space').[8] There were virtual reconstructions of the Christus statue, of the Family History Center, and even of the Brigham Young University–Hawaii's central administrative building.[9] There were virtual Church of Jesus Christ of the Latter-day Saints kiosks that, when activated, would stream videos of talks by Thomas S. Munson, who was Church President at the time. There were avatars who wore BYU t-shirts, and there were avatars who shaped their faces to resemble Joseph Smith. There was even a virtual dance club named Brother Brigham.[10] The Mormon cyber-space volunteers who built these enclaves were not the technology-oriented Silicon Slope sort of Saints; they were more representative of the Church of Jesus Christ of Latter-day Saints: disabled retirees, or librarians, or housewives. They certainly were not transhumanists.

For two months, the Mormon Transhumanist Association rented the top floor of a modernist office high-rise located on one of the Mormon-oriented virtual properties. But the arrangement fell apart, and the MTA was (virtually) evicted. The reason given for the eviction was that the MTA's virtual landlords felt that by having the word *Mormon* in the group's title, the MTA was insin-uating a nonexistent official status or link with the institutional Church. (The irony of being told this by the owners of a virtual real estate property called Mormon Island was not lost on MTA members.) But what was even worse in the eye of the virtual landlord was that, in their capacity as MTA members, some of the individuals in the Association had presented at Sunstone—a mod-erately progressive Mormon conference associated with an equally progressive Mormon journal—and attending a Sunstone symposium was (supposedly) incompatible with a temple recommend. To this cyber-rentier, Sunstone, and hence (thanks to the transitive property of heresy) the MTA as well, smelled of apostasy.

Luckily, virtual worlds like Second Life were not the MTA's only venue.

§39

While it did not always carry the importance it does now, Facebook is, without doubt, the most trafficked bit of MTA's current internet real estate and one of its oldest venues as well. The MTA's Facebook presence dates back to 2007, the same year that MTA members first started making presentations to secular transhumanists on Second Life. The page, now reformatted as a discussion group, is still active in 2021 (though this may not be for long—there is a great deal of interest in moving to, or even setting up, a decentralized form of social media, outside of corporate control). Not everyone who is a member of the Facebook discussion group is also a formal member of the MTA. There are several reasons for this. A small number of the people conflate the MTA's

Facebook presence with the MTA itself, believing that activity on the former automatically enters them in the latter's rolls. Others are religious transhumanists, including Mormon religious transhumanists, who feel for idiosyncratic reasons that they cannot sign off on the Mormon Transhumanist Association's six-point "Affirmation," but who also want to engage with others on the overlap between Mormonism and transhumanism. There is also a smattering of very conservative, orthodox members of the Church of Jesus Christ of Latter-day Saints who are suspicious of what they see to be a progressive lean in the group's politics. Some non-members active on the discussion group are Mormon intellectuals who do not endorse transhumanism but travel in the same Mormon circles as those of MTA members. And some group members, usually influential figures in the world of secular transhumanism, have no ties with the group beyond their status as past MTA-conference keynote speakers.

The Mormon Transhumanist Association's Facebook group has multiple uses. Until an obligatory switch to Zoom, the Facebook group was often used to livestream the monthly in-person meetups from the Utah Valley local MTA chapter, so that those in other locales could participate. (And while this speaks more to the level of stability of the various platforms and devices used to broadcast these meetings than anyone's futurist bona fides or their programming acumen, MTA members are amused that these transmissions are often cut short, or sometimes wholly unsuccessful, due to technical difficulties.) Neither transmitting nor publicizing these meetings, though, were the original purpose of the Facebook page; they were merely a subsequent use. These in-person group meetings were instituted in 2008, after the original Facebook page had been set up. In fact, according to one founding member, the impetus for starting these meetups was that while Facebook conversations were going well, they could stand to be supplemented by more traditional interaction: "transhumanism was cool and all, and maybe we'll get mind-uploaded eventually, but in the meanwhile, it's nice to meet real people in the flesh from time to time." The Facebook page is also used to publicize other upcoming events. The annual conference, for instance, is given a lot of attention, but so are the annual family get-togethers, in-person meetups, and even an occasional theater expedition to see a futurist-leaning science fiction movie.

The primary use of the MTA's Facebook presence, though, is the same as the primary purpose of most of the other group- or topic-centered pages on Facebook: posting hyperlinks to various websites, and then discussing the hyperlinked material in the comments. The topics of these posts, usually centered on some news story or blogpost, vary. There is as likely to be a discussion of developments in the tech sector as there is to be a discussion of transhumanist futures, or of Mormon doctrine or Mormon politics. It is fair to say, though, that discussions of current and near-future technology constitute a preponderance of posts. And these technological discussions are often *quite* technical. To give one example of how into-the-weeds these conversations can get, there was a long-running debate on the MTA's page about whether or not SpaceX, the private space program run by Elon Musk, was placing too much responsibility for its software on the software's developers, and not enough on internal quality assurance engineers who were employed to test the software. As the trailing chain of comments grew, this discussion turned into a debate about the general importance of quality assurance engineering; about whether quality control engineers should be engaged in manual testing or should write code to test the software (that is, be SDETs, or Software Development Engineers in Test); about the use of continuous integration and continuous delivery (CI/CD) pipelines to automatically test each iteration of code during the development cycle; and about the value (or lack of value) of agile programming (a style of software writing that focuses on software development being done by physically co-present and egalitarian teams who work in short spurts to encourage the software to be more responsive to changing user needs).

The SpaceX discussion is inside baseball, to use a particularly American expression; that is, this one thread presumed an already existing expert knowledge for those following the details of the conversation. An example of this is a post that started as a commentary on whether the term *Artificial Intelligence* should be disambiguated. The discussion ended up descending into a somewhat prickly back-and-forth about whether it was even theoretically possible to simulate human beings on a Turing Complete computer—*Turing Complete* meaning a general-purpose computer that, in theory, has the computing flexibility to simulate any other Turing machine by running that different machine's data manipulation algorithms on its own hardware. While these two threads were unusually thick discussions, oftentimes, both the back-and-forth and the topic addressed on the page are less opaque. Still, even the more comprehensible discussions can be quite technical. An instance of this is the repeated arguments concerning Moore's Law, the observation that the computing power of chips tends to double every two years. A common point of contention in the group is whether or not Moore's Law is still good, or if it is running up against limits mandated by the laws of physics, such as issues of waste-heat, and the limitations resulting from having to design and manufacture chips at an effectively atomic scale. Other times, issues such as the ethics of decision-making in self-driving car programming are batted about. While the topic of self-driving cars is one that is easy to grasp in concept, discussions of such ethical quandaries drift into questions of how ethical decisions could be technically implemented.

To misuse Hegel's *bon mot* that the mysteries of the ancient Egyptians were mysteries to the ancient Egyptians, this tendency to turn to technical aspects, as well as the diversity of different potentially relevant technical domains, means that many Mormon Transhumanists have trouble understanding what they are discussing as Mormon Transhumanists. Not all members, of course. The closer one is to the information-technology-sector-employed prototypical member, the more likely one is to approach a complete command of the various technical vocabularies and competencies discussed. However, not everyone will have competence in every topic of discussion, and so even the prototypical member will not have the knowledge to follow every conversation in the forum. It is hard to control a string of technical languages for topics as diverse as neurotropics, genetics, computer programing, space technology, cryonics, and so on.

Further, many members are not prototypical, do not engage in these discussions, and at times have suggested that these interactions often escape their understanding completely. There are after all Mormon Transhumanist artists, Mormon Transhumanist educators, Mormon Transhumanist charitable NGO professionals, and Mormon Transhumanist housewives, drawn to the group either by the breadth of the transhumanist vision or by the degree of autonomy from some of the more orthodox aspects of the Church that the MTA affords. These individuals bring their own insight, and they are not in any way shunned by the prototypical group members. Some have even been on, or at the time of this writing, currently are on, the MTA's board, or have otherwise taken on essential responsibilities for the running of the association. But that does not mean that the prototypical group members police their interest in these technical domains for the benefit of non-prototypical members, either.

Again, this lack of competence in all the possible technical argots that can come up on the MTA Facebook page is a point worthy of qualification. First, members at times try to leverage this disparity in comprehension by posting a link and stating the equivalent of "I don't understand this." Sometimes this sort of post requesting clarification about a new scientific claim, or engineering or programming issue, floats uncommented on. But at other times, people will try to either explain the phenomenon under discussion or at least try to give some sense of its possible broader importance. At least on technical issues, this is a rather peaceable community. (Discussions of Mormon politics and culture, as we will see, can be a bit more sharp.) This generosity brings us to a second point: the category of technical issues is sometimes so approachable that despite the technical challenges, anyone can enter. Often discussions of topics such as the social impact of a technology are cast in such a way that there is no technical-language-competence barrier to joining (even if it is always subject to falling down into some technical rabbit hole). And third, this category of technical issues does not exhaust MTA online discussions. In addition to the technical discussions and the announcements about events and meetups, there are usually links to topical humor (often to internet cartoons such as XKDC or Saturday Morning Breakfast Cereal, regular series that touch on fields such as science or religion). But still, even at their most approachable, the MTA discussion threads are always just a conversational turn away from deep waters.

§40

In addition to technology, there are two other common topics of discussion. One item is transhumanism—and particularly religious transhumanism—as a programmatic philosophy (as opposed to the other ways that transhumanism can be addressed: an aesthetic, an ethic, a community of interest, a research agenda, and so on). The other topic is the institutional crisis in Mormonism. These topics are more closely associated with one another than they may appear, but it will still take some heavy lifting to flesh out the intricacies of the relation between transhumanist thought and contestation about the Church.

Let's take the institutional crisis first. As they lean progressive, MTA members tend to be more sympathetic to dissidents on these issues, but they are not of one voice. And their debates—or just as often, metadebates about the frequency and progressive lean of these debates in MTA-affiliated social media— are heated and common on Facebook. The heat is because there are both real stakes and a perceived trajectory to progressive Mormonism. The path that many orthodox Mormons imagine is that entertaining progressive Mormon concepts often leads to unbelief. The logic goes as follows: if you question the Church on the immorality of homosexuality, or on the veridical nature of the *Book of Mormon*, or on the Church's history of polygamy, other aspects of the Church start to look suspicious as well, and skepticism can drift toward apostasy. This is a perceived trajectory, not necessarily an actual trajectory. Some people who fall into the progressive Mormon world of blogs, podcasts, print media, and online social networks do end up rejecting the Church of Jesus Christ of Latter-day Saints. But other progressive Saints will try to recode religious belief in different ways, such as seeing the truth of the Book of Mormon not necessarily as *revealed* but instead *inspiring*, or they see value of Mormonism's history and cosmology more as mythical and edifying than as informative and veridical.

Some more progressively minded believers work toward a liberal-but-still-committed position because they find it more intellectually satisfying, or because they find such a stance parsimonious. But another reason to stop this side of apostasy is that in Mormonism, unbelief comes with a cost. Given this, many Mormons who no longer find their faith either compelling or convincing often work to hide this fact from others, and even, in a way, from themselves. This is so much the case that there are many church-attending Mormons who are to all effect atheists, believing little or none of it, but not willing to place their position or their family relations in question.

§41

This meditation on trajectories toward unbelief brings us to the other discussion topic: transhumanist apologetics. But to understand the role that transhumanist apologetics plays in both the perceived or actual risk of nonbelief in progressive Mormons, and in work done by the sort of resistant technical vocabularies discussed earlier, something else must be attended to first: the greater Mormon economy of secrets. Religion is capable of numerous definitions, a fact that has led some to question whether any one definition could be sufficient, or whether we should even trouble ourselves with definitions in the first place.[11] This has led some to suggest that the focus should rather be on what religion *does*, or to put it slightly differently, on the various ways that religion can express itself.[12] And one common phenomenon in religion is a play of markedness and absence, a play of secrecy and revelation.[13] Whatever the value of this theory is in the abstract, both external circumstances and their religious imaginary have made Mormons double down on secrecy and revelation as a form of religiosity. These are not absolute secrets, to be sure, in the sense that they are unknown; they are secrets that most people know but are not allowed to articulate openly. These secrets, which range from a history of polygamy to the series of ritual temple initiations undergone by most adult Church members in good standing, are things of which effectively all Mormons are aware (albeit with varying degrees of clarity). But because of various taboos and prohibitions, these open secrets still work to particular effects.

Part of this is a legacy that results from having to keep secrets not from other Mormons but from outsiders. As has been shown by linguistic anthropologist Daymon Smith, the necessity of concealing polygamy shaped much of the nineteenth- and early twentieth-century Mormon speech norms.[14] This was at first concealment of the existence of the practice of plural or *celestial* marriages. But after the lynching of Joseph Smith and the exodus of the main body of Mormons to the intermountain west, the focus of concealment shifted. Speech norms became about keeping secret from outside law enforcement officials not merely that polygamy was occurring, but also who specifically was engaging in it, so as to protect these parties from the legal repercussions of their status.

Crystalized in a pithy saying sometimes referred to as the Mormon Creed, the phrase "mind your own business" became a watchword for Mormonism as early as 1844. As a slogan, it has a life in Mormon song, poetry, and homiletics. It was even memorialized "for future generations" in "an ornate back-painted glass fixture" in a major Utah LDS temple.[15] As a moral paradigm, the creed was seen as encouraging an individualistic reliance while also rejecting the questioning of authority.[16]

As an ethics of speech, "mind your own business" was chiefly expressed as a proscription on asking too many questions about a fellow Mormon's personal affairs. But it was also a refusal to speak either of one's affairs or the affairs of others, regardless of the inquiry's subject. The logic was that one never knew whether one was either talking to an agent of the United States or to someone who was an informant for a US agent. (As waves of prosecutions for polygamy during this period attest, this was not mere paranoia.) Adherence to this imperative was often taken to an extreme. Married couples were encouraged by the Church to purposefully obscure the dates of their weddings. While giving testimony during polygamy or cohabitation trials, wives and mothers would express confusion as to when they married, who the father of their child was, what the child's family name was, whether or not the child was married, and to whom (if anyone) their child may or may not have been married. Conversations with others on these subjects were inevitably forgotten as well. Sometimes people expressed confusion or ignorance as to whether or not they even *knew of* the accused.[17]

Tactics like these allowed polygamous marriages to go on long after the Church announced a ban on the performance of such marriages. Many believers, including some in the Church's hierarchy, assumed that the original institutional disavowal of plural marriages by the Church in 1890 was merely a statement given out to satisfy the pressure of gentiles, and not an actual change of positions. For almost a generation afterward, many in Church leadership continued to officiate at polygamous marriages, insulating other Church members from any complicity through "minding their own business" and not informing them.[18]

As a maxim, "mind your own business" is, for the most part, a historical artifact, and not referenced during contemporary church gatherings. (I was once told by a Mormon friend about an occasion when it was brought up during a priesthood meeting—a meeting of adult and older juvenile Mormon males in good standing with the Church—and someone commented skeptically that "this doesn't sound like us; no one minds other people's business more than we do.") This older maxim, which worked to separate belief from action, and to encourage a consequentialist—as opposed to representational—language

ideology, eventually faded away during the early twentieth century. This transformation coincided with polygamy being stamped out as a Mormon practice
(except for a few fundamentalists), and with Mormons adopting an ethics of
speech that, in large part, prized sincerity.[19] The qualification *in large part* is
an important one, though. Unlike the prevailing Protestant language ideology
that advocated a direct, expressive relation between a logically prior mental
state and a subsequent linguistic expression, there is even today a vital set of
linguistic prohibitions that deeply affects Mormon religious speech.[20]

§42

Unlike previous limitations on language intended to disguise domestic practices, the speech prohibitions of the current moment are rooted in the Mormon
Temple. Three Mormon particularities, specifically an identification with Old
Testament Judaism, the early influence of various forms of folk occultism, and
the early introjection of Freemasonry, have in the temple given Mormonism
a set of hermetic ritual practices that are without any clear Protestant precursor
or analog.[21] The inheritance from the temple includes many essential familial
ritual practices, such as celestial marriages and the sealing of children (both
rights in effect permanently incorporating the participants into the postmortal
family). The temple also houses certain initiatory rites (referred to as *endow-
ments*) and the various rituals for the dead (including, but not limited to, the
practice of baptizing the dead). These can only be performed inside of a consecrated LDS temple. It is also notable that many of the rituals there must be
kept, in the language of one prominent late-nineteenth/early-twentieth-century
Mormon leader, "secret from the world."[22] This is to a large degree an ethical
imperative and not an absolute block to sharing knowledge; these rituals have
been described, and even surreptitiously filmed, numerous times. Further, not
all elements of the temple are under a ban, and certain parts of these rituals
can be freely discussed. But still there is an understanding that disclosure of
some of the core elements of these rites is reprehensible, and it is shocking
when they are made publicly.

As observed by the anthropologist Bradley Kramer, this secrecy does not
lead to a practice of invisibility.[23] Instead, it gives rise to a complex play of oc-
culting and disclosing information about temple rites and one's experience of
them, and also of one's general status within the Church. Through the unmarked
incorporation into extratemple speech of portions of entextualized temple
material, one can simultaneously allude to temple rites, attempt to channel
some of the sacrality of the temple, and also index status as a temple-going
member to those who are also so-experienced and so-situated. Being known

to be a temple-going person does further work, as not everyone has temple access; in addition to meeting specific age requirements, recall that one must also have a temple recommend—a certificate that states one is complying with various dietary and behavior prohibitions, has endorsed the validity of the upper echelon of the Mormon hierarchy, and has a testimony regarding religious truths such as the truth of the Book of Mormon and Joseph Smith's status as a prophet.[24]

Even the formulaic questions used as part of the temple recommend can be inserted in language as entextualized snippets, such as in the case where some Mormons criticize public figures using elements of temple-recommend questions:[25]

> For example, Mormons who do not like Mitt Romney (a small but vocal minority among American Latter-day Saints) regularly criticize his reputation for truth-telling not merely by calling him dishonest but by accusing him of not being honest "in his dealings with his fellow men." Even progressive Mormons who would otherwise eschew archaic gendered language retain the "fellow men" here, because that is the language of the Temple recommend question: "Are you honest in your dealings with your fellow men?" Thus, the criticism is meant to implicate not just Romney's honesty but his temple-worthiness.[26]

A few words are slipped into conversation, and the scene doubles; in what is an ordinary moment of watercooler-conversation-cum-political-critique, the speaker is now also a phantasmic bishop, probing to see if a virtual Romney is worthy of the temple, and the object of critique also becomes a virtual ward member, potentially too unclean for entrance to the ritual heart of the Church.

§43

Despite their utility, these series of entwined indexicals have some limits. For example, not all temple communication is fair game. As mentioned, there is an unwritten rule that places some of the communication at the center of the rituals off-limits (Kramer lists "the imparting and reception of holy knowledge, signs, keywords" as being particularly taboo).[27] But putting aside these particular elements, this still leaves the temple-experienced with a code, allowing surreptitious communication.[28]

Then there are secrets of other sorts, such as the ritual temple garments, first issued during temple endowment ceremonies, that are worn such that they are hidden by conventional clothing (the *underclothing* that is occasionally derided by outside critics of Mormonism as *magic underwear*). It is apparent from the literature on them that these garments tend to act not so much as cryptic statements, but more like implements in regimes of self-subjectification, a constant reminder of one's status, one's community of belonging, and one's vows.[29] And it is true that especially of late, the Church has been open about these ritual garments, going so far as to post pages on an LDS-operated website (and even a video on YouTube) to explain the garments' relevance to non-Mormons.[30] However, because of their hidden nature, temple garments open up a possibility for hermeneutic play that is not unlike the surreptitious incorporation of temple speech into everyday language. A line of clothing showing through a shirt or blouse can communicate righteousness. A Mormon skeptic or former Mormons can signal apostasy by wearing clothes that, due to the depth of the neckline or the height of the hem, indicate that they have either not been issued temple garments or have chosen to not wear them.

§44

There is a different category of occulted information: secrets which are not open in nature but are hidden from the mainstream of Mormon believers. These secrets come in three categories: (1) information that escapes access due to the social site at which it is located, (2) information that the Church purposefully withholds from general access, and (3) virtual secrets—information that is accessible, but which appears to be secret as an accidental effect of these two other modes of secrecy.

The highest echelons of governance in the LDS Church are the organs that must be affirmed during temple recommend questions: the "President of the Church of Jesus Christ of Latter-day Saints," the "members of the First Presidency and the Quorum of the Twelve Apostles." These are individuals who have both shown loyalty to the Church by their work in lay-positions such as stake president, mission president, or the like, and who have shown some success in their chosen careers (usually business, law, or occasionally medicine). The deliberations of these men, who were until quite recently entirely white and almost entirely Utah-valley-originated, are secret;[31] but despite this secrecy, because these men often visit various Mormon wards, and (most importantly) regularly give public speeches (especially at general conferences), there is a sense that people *know* them. This knowledge often extends to creating phantasmic opinions on these apostles' imagined attitudes toward current religious difficulties. Individual Mormons will often claim to have a sense of the positions individual apostles hold on some of the various challenges that the Church faces. It is not uncommon for a church member to be able to quote from memory particular passages from apostles that they admire (and frequently to quote lines from apostles for whom they have much colder feelings).

Despite this imaginary individualization of the apostles, when the Church announces a new policy, that policy is presented as the unanimous, divinely inspired position of Church leadership. This possibly fictive unanimity leads to what might be called a Mormon Kremlinology, a fabulation of the role that each apostle played in the discussions that formulated that policy, and particularly imaginings as to what each apostle has militated for the various real or possible outcomes. This speculation reaches a fever pitch when a particularly divisive policy touching on some social issue comes to the surface. One example is a 2015 position classifying adults in same-sex marriages as apostates, that is, having them share the same status as individuals who have committed "murder, rape, sexual abuse, spousal abuse or intentionally cause[d] serious physical harm to others, as well as those who engage in adultery, fornication or abandon their family responsibilities."[32] This policy also prevented children of parents in gay marriages from receiving Church-sanctioned baby blessings, being baptized, becoming priests, or volunteering for the traditional young-adult mission work until after they are legal adults. People speculated on whether this was something that Elder Uchtdorf or one of the other high-ranking members stood for, or if some member was backed into this position once it became clear that he would not be able to sway a sufficient number of other apostles to his side.

Many of these controversial positions are formally announced, often through an organ such as the Church newsroom, but not all of these positions arrive with a great deal of formal institutional heralding. The policy on baptism and priesthood ordination for the children of same-sex marriages, for instance, became public knowledge only because it was leaked (as pages from the *Handbook of Instruction*) on Facebook by a prominent podcaster who had himself been excommunicated for apostasy, though many believe that the real reason was his championing of the rights of LGBTQ Mormons.

The *Handbook of Instruction* is a guide that is supposed to inform Mormon bishops, stake presidents, and other Church officials with more than notional oversight duties how to carry out their responsibilities.[33] Not only does it contain a script for carrying out rites, it contains capsulized recitations of some of the core of Mormon doctrine. It also includes instructions on how to shepherd, and at times discipline, the members of the ward. The *Handbook* comes in two volumes—a blue "volume 1" and a red "volume 2"—which differ not only in color and their respective ordinal positions in a sequence but in that only the latter has historically been available to the public. Copies of the former were previously only available to a circumscribed set of Church officeholders.

(In 2020, the Church consolidated the two handbooks into a single text, available online to anyone.)

This uneven availability had effects. It has been claimed that since the second volume was made available in 2010, the sacredness of the volumes has, to a substantial degree, been vitiated.[34] From talking to my informants, my sense is that even a diminished sacredness is not the only affective charge associated with these books. Though most regular Church members know the rough outline of the regulations contained within the first volume, the fact that information can be leaked from the text suggests that there are still unknown contours of the book. Especially because bishops often exercise considerable discretion in disciplining Church members—with some bishops being notably more lenient, and others often far stricter (a phenomenon that is referred to as *bishop roulette*)—understandings of the exact rules contained within the first volume are often hazy. And while the cognitive stress this occasions can be overstated, this haziness, in tandem with the Mormon Kremlinology discussed before and the Church's historical tendency to be stingy in giving outside access to potentially embarrassing historical documents, can sometimes lead to a suspicion that the Church is *hiding* things from the main body of its believers.

§45

This brings us to our last category of secrets: *virtual secrets*—things that appear secret when they are not, either because they are not actually occulted, or because they are secrets that do not exist. Rather than being actual esoteric knowledge, they are the potential to imagine secrets, created by gaps in the record and a learned sense of Church patterns of occlusion. Virtual secrets are a logically subsequent formation (though not necessarily a temporally subsequent one)—an effect of the suspicion that the Church is hiding something. These secrets are virtual in two different ways. One is in the most contemporary and base sense of the word: these are secrets that are generally unveiled through the set of mediate distance technologies we have come to call the internet (podcasts, blogs, social media, and email-circulated PDFs). The advent of these technologies has allowed individuals to take elements of Church history the Church neglects and to bring them to the fore in ways that, say, Mormon continuing education or Church periodicals (such as *Ensign*, the official periodical of the LDS) would not.

To be frank, Mormonism as an institution both cannot and will not cover everything about the history of the religion that would raise eyebrows: Joseph Smith's polygamy, apparent contradictions in different accounts Smith gave of the first vision that started his prophetic career, Smith's earlier career as a

treasure hunter, Smith's use of folk magic like seer stones to translate the Book
of Mormon, problems with the historicity of Mormon scripture such as the
Book of Mormon (which in the eyes of critics is full of anachronisms and
arguably has not had any archeological or population-genetics confirmation)
and the Book of Abraham (which, recall, was translated by Joseph Smith from
some Church-purchased papyri, a document which is currently understood
both by secular and many Mormon Egyptologists as *not* containing a hidden
testament from the patriarch Abraham). Nor can the Church address all of the
now-embarrassing aspects of Church history that occurred after Joseph Smith's
leadership: withholding priesthood from male believers of African descent, the
violence that was unofficially sanctioned by the early Church in both the Nauvoo, Illinois era and in the later Western intermountain years. Not only would
a full discussion interfere with the pedagogical narrative history that the Church
likes to present, but to be honest, this is material that a great many Mormons
do not care about. Day-to-day problems of keeping a ward operating, fulfilling
callings (volunteer positions that one is asked to take up by a bishop or stake
president), taking care of extended family and profession, handling marriage and
parenting, and attending to all the other quotidian problems that come up through
the course of a day, exhausts most Mormons' attention.

Hence the relative silence on this front; assuaging doubts that do not exist
can only feed those doubts. Rather than assuage anxiety by addressing these
issues ahead of time, speaking about questionable events that most members
were not asking about generates anxiety by bringing it to the attention of Saints
who previously were not concerned. Furthermore, arguably none of this information is *hidden* in the first place. All this material is out there in the public
record, and many Mormons with some interest in these matters already have
a familiarity with many of these facts. But the advent of a critical mass of
Mormon-related material on the internet, which is to say the advent of the
Bloggernacle, has substantially dropped the opportunity costs of attaining this
knowledge, and more importantly, has given space to a virtual community that
can disseminate this information. Discovered online after one is often deep
into one's Mormon career, this information often feels like a revelation, specifically a revelation of material that the Church does not want one to know.

There is some very typified language in this broader community to describe
the arc traveled by many of these newly minted Mormon skeptics. These *faith
crises* (the usual moniker used for this transformation) builds on some initial
injury in the phenomenology of faith, a sense of the hollowness of faith, of
God not being there for someone, or the sense of emptiness that follows be-
reavement. Or it takes the form of political alienation, such as an uncertainess
with the Church's role in politics, and particularly in the Church's recent overt

participation in campaigns against same-sex marriage. Or it can be an empathic injury resulting from their perception that someone is being hurt by the Church, usually by being treated shoddily or coldly by a bishop or by fellow believers. In the wake of the injury, people will turn to the internet to look to answers and will end up following a rabbit trail of information that they say they have been unaware of; the result is that rather than receiving solace from the Church, their doubts are often magnified exponentially. Individuals in a faith crisis will still prize some forms of doubt over others, focusing, for instance, on feminist critiques of the Church, or highlighting problems with historicity of scripture or with the behavior of early Church leaders. But whatever the initial occasion of the crisis is, or form the crisis takes, once the crisis takes root their specific concerns can now be set into relation with other domains of religious skepticism and critique and present something that looks like a global indictment of the Church. And the general language used is of a prior innocent faith in the Church that falls away as secrets that the Church hid are unveiled, and they "open their eyes."

§46

There are some interesting resonances between Mormon Transhumanists and individuals who go through faith crises. While it is by no means a prerequisite for being a member of the MTA, many of these Mormon transhumanists also report undergoing the equivalent of a faith crisis, often articulated in almost the same language. For a few, this crisis in faith ends up making them unbelievers, and the few remaining ties to Mormonism that make being a member of the MTA are primarily social and cultural. Others, though, present this crisis as a temporary stage—more than one has described it as a phase of being *effectively* an atheist—and present their becoming transhumanists as *returning* them to Mormonism. As it is said in Mormonism of someone who can speak with religious conviction about the truth of the faith, being a transhumanist allows them to *have a testimony*.

This postcrisis position, though, is not a return to the same form of Mormonism that they left. It is one where the task of becoming God is a *practical* and indeed *technical* project, and one where the weight of the labor devolves onto the believer. This is also a faith in a much more *naturalist* God. Mormonism has always leaned into a certain kind of naturalism, giving God a physical body, and speculating that natural law preceded God, rather than having been instituted by him. But in most Mormonisms, this naturalist aspect was a minor theme, and the aesthetics and cosmology of the divine-human relationship did not differ significantly from that of most other Western Christianities, with

God being in effect a wonderworker who wrought recognizably Biblical miracles. The God of the Mormon transhumanists, though, is something else. (S)he[35] is perhaps a space alien, designing a world or overseeing a pocket-cosmos, or (S)he is perhaps a programmer running a simulation. In short, God is the kind of being or beings that could be situated in some specific, socially recognizable space, albeit a space of technology far in advance of ours. God could even be located (though this is not something frequently discussed by the MTA) on or near Kolob.

There is the possibility that God is simply the name for a certain kind of emergent capacity inherent in us, a future entity that we will craft out of ourselves, rather than a presiding entity that already stands as extant and actualized. The possibility that there already is a God, though, is assessed as rather high by most Mormon transhumanists, despite their deep naturalist inclinations and faith in human-self-actualization. Many Mormon Transhumanists hold to what they call the *New God Argument*, a post-Copernican syllogism which reasons that if we are capable of achieving something like Godhood, and if there is no particular reason to believe that we are special, then it is likely that some other entity has preceded us in making themselves into a divinity.[36] And if these entities were able to overcome the lure of violence and self-destruction (something evident through their capacity to achieve Godhood in the first place), then they are compassionate. And if they are capable of creating worlds, and desire to do such in a compassionate manner, then they most likely forged this earth as well as an exercise in something like love.

Those Mormon Transhumanists who are convinced by this argument have a real faith in an actual creative, benevolent God. They can affirm their religious identity as Mormons and classify themselves as believers. But this shifting to divinization as something achieved through technological labor has had another effect. The Mormon Transhumanist vision is a teleological vision of improvement over time, rather than one of adherence to an always-available ethical and spiritual standard. This processual understanding allows for ethical error in the past to be simply that—error—and does not create issues of legitimacy of the Church. This is because human striving faced toward the future, rather than divine foundations, forms the backbone of this rearticulated religious logic. If humanity is slowly and collectively approaching theosis, it would make sense that the errors of prior generations of believers would be cured over time by those generations' inheritors. This frame enables these often culturally progressive-leaning believers to come to terms with the problematic inheritance of the Church while also presenting themselves as fully believing Mormons without experiencing cognitive dissonance.

But the specificities of the language of transhumanism are obscured by the language of Mormonism when it comes to barebones declarations of faith. And when more complex conversations ensue, the dip into the technical language of near futurism leaves non-transhumanist listeners behind. Many Mormon Transhumanists report feeling pushback from fellow members of their ward because of progressive political views, but I have never heard anyone complain about animosity because of their reimagining of Godhood as a technical achievement. In short, *this capacity to use the technical language of humanism disguises the fact that they are no longer conventional Mormons, and that they view God as a position to work through via engineering and not an entity dispensing salvation.* A crisis of faith, triggered by virtual secrets, is averted and disguised by the secreting work achieved through the impermeability of a near-future technical argot. This is to say, the disruptive frisson of these two languages does not just strip away; in the right circumstances, it builds.

And what of "Domo runs on Kolob"? The problem with the billboard was that it put together two languages that (outside the MTA) are not usually conjoined. But this juxtaposition also caused puzzlement: about the meaning, about the motives, and about the author. Sacred language problematizes its origins and appropriateness; technical language occults its meaning. At once a shibboleth and an enigma, it can only be read, as that one Reddit commentator said, as "[j]ust a little blasphemy in a high density mormon area [sic]."

§47

This is not an operation without some sort of remainder. These accidental or formal secrets of the MTA may not be known, but they are felt. Gilles Deleuze and Felix Guattari once observed that the contents of secrets are in a sense never actually secreted away from all knowledge; that observers may not identify the secret per se, but that they will inevitably notice the semiotic container that holds the secret within it, and the knowledge that there is a secret within that space, even in the absence of knowing the actual import or substance of that secret, will have social effects.[37] Fellow ward members, along with bishops, stake presidents, and other Mormon institutional figures capable of either discipline or opprobrium, may not respond to or even notice the way that technical language can work to obscure changes in meaning of the occasionally secretive language of Mormonism, at least when spoken by these transhumanists. Others, though, do rise to the challenge. And like skeptical Saints whose momentum in their faith transition is accelerated by imagining possible Church secrets, internet critics—not being privy to the actual nature of the secrets of

Mormon Transhumanism—confabulate their own paranoid truths regarding the association.

Speculation about, and criticism of, Mormon Transhumanism can be found all over the internet in various subreddits, blogposts, and the like. Given the brevity, velocity, and antagonism that shoot through Twitter, it is not surprising that site gives the purest distillation of Mormon-Transhumanist-oriented suspicion. In the wilds of Twitter, discussions of Mormon Transhumanism are discussions of a group that is in league with the Antichrist, aligned with secret cabals of Mormon political figures such as Mitt Romney, and wishing to use technology to bring back half-human, half-divine monsters such as the Nephilim of the book of Genesis. All this is supposedly being done on the downlow, but somehow these Twitter critics have managed to see through the lies and uncover the conspiracy.

To a degree, this always-bubbling paranoid speculation amuses most members of the Mormon Transhumanist Association. To their eyes, there is a yawning gap between their average annual conference full of deli-sandwiches and PowerPoint presentations and the imagined global conspiracies in which they are supposedly deeply implicated. That gap is so vast that they reasonably find these charges comedic. But in a sense, the paranoia of their critics is understandable, even if it is not excusable. Part of it is simply a conspiratorial desire to link secrets together to make them all legible. This explains the tendency to link together Masonry, with its secret initiation rites, Mormonism, with arcane temple practices, and transhumanism, with its apparent confidence and clear-sightedness about a future that others cannot imagine. The shuttling between a language of faith, redolent with secrets, and a language of technical processes highlights the degree to which meaning always remains slightly unmoored—something that, when mixed with preexisting animus toward Mormons and overly primed eschatological imaginations, must act as a red flag to those who, influenced by a hyper-literalist referential Protestant language ideology, attack them online.

It may be that in time, the working language of the MTA, which now exists as a shared sense of communicating through code-shifting between two languages, will, in turn, evolve into a sort of pidgin, where both coding systems could be merged, immunizing them from these sorts of attacks and insinuations. However, such change would undo what is achieved now by the combination of a religious language of intentional concealment, and a technical language of accidental impermeability. Or rather, it would undermine what is done by the ways these languages approach one another and yet fail to combine. This is because this combination of different registers of speech, when taken collectively, is not a language used to conceal one's state of belief or hide any breach with orthodoxy, though it may accidentally work to that effect. Rather, this form of speech is ultimately used to expand what Mormonism means while granting Mormon Transhumanist the freedom and power of both religious and speculative-technological domains that otherwise would not be mutually comprehensible at all.

Fifth Series: Discipline, Belief, and Speculative Religion

§48

In a world where religious skepticism is networked and distributed, individual doubt does not remain individual for long. Circulating both information of alleged misdeeds and accounts of personal frustration and embarrassment, online consociates can quickly become comrades. As noted earlier, the advent of the internet created an explosion of relatively easily accessed moments of problematic church history. And with the heightened accessibility also came criticism of the Church leadership on stances such as the status of gay and lesbian Mormons and on the ordination of women (ordination into the priesthood being a status enjoyed by all adult male members in good standing). This openness also brought renewed attention to long-running debates regarding the veracity of texts that are given scriptural status in the Church.

Most importantly, it allowed for the organization of protest movements on a national and international scale. Examples of collective dissent include the Wear Pants to Church day protests, which challenged LDS doctrine on gender through a collective, purposeful, Church-wide breach of unstated Mormon sartorial norms including the demand that women only wear dresses or skirts to religious ceremonies. Originally held in 2012, and repeated in 2013, it urged Mormon women to wear pants to church services held on a particular December weekend.[1] Protest has gone beyond clothing choices. Recall those Mormons associated with Ordain Women, a web-centered Mormon feminist movement, who attempted to attend the historically male-only General Conference meetings in both 2013 and 2014; both years they were denied tickets to the event. In

2014, they publicly protested their exclusion, though they were shunted to a free speech zone in Temple Square.

Shortly after these protests, the Church responded. In June of 2014, the Church excommunicated *in absentia* Kate Kelly, the lawyer and Mormon Saint who had founded Ordain Women. This was not the Church's only rejoinder to increasingly mobilized dissent. In January 2015, John Dehlin, a prominent Mormon podcaster who had advocated for LGBT Mormons, was also excommunicated. In each case, apostasy was the stated grounds for excommunication. (In the case of John Dehlin, the apostasy charges also claimed that in addition to giving teachings that questioned church authority, he had also stated that The Book of Mormon and The Book of Abraham were fraudulent.)[2]

This is not the first time that the Church has gone through a Thermidorian Reaction after a period of apparent openness: similar shifts in media have caused similar thaws and then similar countermeasures. In the 1970s and 1980s, the establishment of two independent print magazines dedicated to Mormon thought (*Sunstone* and *Dialogue*), and a brief period of greater access to the Church archives for trained historians, lead to an intensification and greater visibility of academic-informed debates about Mormon history and Mormon conceptions of gender. In 1993, Boyd K. Packer, the President of the Quorum of the Twelve, warned during one of the semiannual general conference meetings of "the dangers . . . from the gay-lesbian movement, the feminist movement (both of which are relatively new), and the ever-present challenge from the so-called scholars or intellectuals." Shortly afterward, six Mormon academics (who came to be known as the September Six) were excommunicated; many of them at the time were working at the church-affiliated Brigham Young University. Working in diverse areas such as Mormon Church history, Hebrew Bible interpretation, and feminism-oriented scholarship, the sole thread that seemed to run through all of them was publicly circulated academic work that ran against the grain of Mormon orthodoxy.[3]

The 1993 excommunications created an atmosphere where similarly situated Saints feared that they would be next.[4] It was much the same after the excommunications of Kelly and Dehlin. Rumors of other immanent excommunicators started to circulate, both in person and online; and over a dozen individuals are known to have been excommunicated for online statements in the wake of Kelly's ouster. On Mormon-themed online forums, church members claimed that they had been threatened with excommunication by their bishop for saying anti-Mormon things online. Many of these reports suggest that the bishop lacked any knowledge of how the internet functions; one Reddit post on the ex-Mormon subthread was titled "My bishop threatened to excommunicate me because of my blog. I don't have a blog."[5]

There are no reports of any member of the MTA having been excommunicated because of activities in the MTA. Nor have any MTA affiliations led to lesser disciplinary actions, such as being disfellowshipped or losing temple recommends.[6] But anxieties about excommunication still have effects. The first is the danger of individual expulsions and sanctions that still may be coming. This danger is particularly an issue because on issues of gender many members are progressives, at least as understood on the American political spectrum; many also have an outsized presence on the internet, which is the current locus of church disciplinary activity. Again and again, in interviews members expressed anxiety that their web presence could lead to attention from their bishop.

This proliferation of excommunications also raises a second problem: how the MTA should collectively respond to the challenge of excommunications of Mormon dissidents. Senior MTA members sometimes talk of changing the governance structure of the organization by reorganizing it along the line of blockchain (a form of distributed democratic governance molded after bitcoin). Meanwhile, the MTA has a rather traditional form of governance that comes with its nonprofit tax status, which means that its decisions must be made centralized. Furthermore, there is a tendency for leadership to unintentionally give the appearance that they are speaking on behalf of the entire MTA, even when they are merely expressing a personal opinion. Given the high internet presence of current and previous heads of the MTA, and the variety of views of MTA members, this means that the justice of each of these high-visibility excommunications end up being vigorously debated after they are announced in the wider Mormon public. Complaints usually consist of demands that the MTA take one posture over another (often a stance that would align with the excommunicated Mormon dissidents). Making statements that do not openly contradict current church doctrine, while still aligning with the political desires of members, often results in quite careful parsing of words as well as strange combinations of actions and statements. An example of the latter came when some MTA members marched in a Gay Pride parade as allies. At the same time, the MTA leadership tried to square the circle by simultaneously expressing support for LGBTQ individuals but also deferring to church teachings on LGBTQ issues.

§49

This mass of disparate views suggests that while the MTA does not discipline its members, as an organization and a collection of individuals, it is both indirectly and directly shaped by the potential disciplinary care of the Church. Mormonism matters, particularly Mormon oversight and character formation. It is easy to imagine the Church as a religion of discipline. For instance, local bishops keep track of which members of the ward are tithing the obligatory 10 percent, and have a keen eye as to who is attending meetings and for how long. There is also a tradition of monthly home visits by elders to Mormon houses within the ward. While these visits primarily are an opportunity for teaching, whatever the intents of the home teachers are, its function as a form of surveillance is undeniable.[7] There is also an obligation for members, and particularly for youths and young adults, to confess immorality, particularly sexual immorality, to their bishop during one-on-one interviews. And while it is by no means a formal practice, there is also a tradition of what can be called ward informants. These are fellow members of the local ward who volunteer reports about questionable activity to their shared bishop. The impetus for reports is usually the moral judgments of fellow believers—perhaps even personal animus in rare cases. However sketchy its origins, this information still circulates through the ward's capillary system, and this information as well as the knowledge about this kind of information being circulated, has its effects. And even something as small as an immodest Facebook profile picture can be the object of ward-level disciplinary attention from a bishop.

These capillary structures of control are not all that different from many other hierarchical, bureaucratic, and ethically oriented American institutions such as the Boy Scouts or the PTA. However, this strain of Mormon discipline also consists of active work on the body. Dress, particularly *female* dress, is noted, though there is an implicit pressure for all members to live up to Mormon sanctuary meeting dress codes. But discipline goes more in-depth than religious sumptuary laws. Also on the disciplinary side of the ledger are the various dietary prohibitions arising from the Word of Wisdom, a revelation received by Joseph Smith in 1833 (*Doctrines and Covenants* 89:1–21). The specific meaning of the original prohibition is not entirely clear, and some aspects of it, such as a mandate for an effectively vegetarian diet, are ignored. However, as developed since the twentieth century, it is understood and commonly practiced as abstinence from tobacco, alcohol, black and green tea, and coffee. Again, these dietary restrictions are in flux: in 2012, the Church released an official statement to the effect that caffeinated sodas are allowed—though many Mormons still avoid them.

There are also what might be called various "tests of endurance." Ward sacrament meetings can be a physical trial unto themselves. Hours of repetitive hymn singing and amateur homiletics—all while wearing traditional Sunday best and struggling to contain fidgeting children—are far more taxing than might be imagined or perhaps even conveyed. This is not just the opinion of an outside ethnographer. Mormons themselves regularly comment on how tiring these meetings are, and I have heard them described as spiritual exercises not despite this pressing dullness, but because of it. The idea behind this understanding is that these meetings function as virtue-building ordeals when properly approached. (I have even had Mormon friends apologize to me for the dullness when I attend sacrament meetings—and an apology for the boredom of a core ritual is something I have never before experienced as an anthropologist.)

§50

This would make the Latter-day Saints seem to be a contemporary example *par excellence* of what might be called, for lack of a better name, Asadian religion.[8] As part of a larger project attacking foundationalist accounts of religiosity, the anthropologist and critic Talal Asad famously has argued that anthropological accounts of religion were too centered on the concept of belief. He argued that belief as a category was not a religious universal but merely one religious mode that happened to be favored by contemporary, secular-ensnarled Protestantism. The costs of this anthropological privileging of belief were that forms of religion that instead stressed some other aspect, such as practice or submission to religious authorities, were often depicted as "bad" religions. As a counterexample, Asad documented how a different religious western formation, the Medieval Catholic Church, saw belief not as primary to the faith but as a secondary product. This framing of belief followed from the Medieval Church's capacity to make determinative statements about what is true, as well as the Church's ability to discipline those who would break with such statements. Punishment is not the only aspect of Asadian religion; the training of believers is a part of the disciplinary mode of religion as Asad sees it. In a move that would be particularly influential to later scholars, Asad argued that Medieval religious subjectivities were also the result of bodily practices that honed the sensorium and trained affective responses. The idea that there were modes of religion that could operate through both the foregrounding of authority and the voluntary submission to regimes of discipline that act on the body has gone on to become a mainstay of the anthropology of religion, particularly in conversations regarding reformist branches of Islam as well as in certain strains of Pentecostalism.[9]

The analogy to Mormonism should be clear. What is and what is not proper Mormonism is the result of institutions with a capacity to punish and of physical prohibitions or trials that Saints are compelled to undergo. The contemporary LDS also seems to be a perfect exemplar. Teaching both in wards and at home is informed by *Ensign*, the official LDS magazine and a host of Church-produced videos; there is also the training that occurs when adult males go through the Missionary Training Center complex in Provo, Utah. There are also regular reenactments of events in Mormon history, such as the handcart-trek to Utah, that trains the senses while concretizing Mormon history.[10] Continuing education is hard-built into the church ritual calendar, with one out of three hours at each weekly chapel sacrament meetings being dedicated to these pedagogical efforts.[11]

There are even moments when surveillance, ordeal, and pedagogy converge; Mormon adherents in good standing with the Church are asked on a semiregular basis to give talks to the ward; the topics and times are chosen by the bishop and are particular to the Saint asked to speak. The talks themselves are often a hybrid of biographic narrative as well as research and reflection on the requested topic. Here, adherents are at once forced to both display their knowledge and their capacity to self-present as a particular performative style informs ward talks. (Suddenly breaking down into tears at the appropriate time is not a strict requirement, but it is an indication that one has internalized the genre.) They are designed as pedagogical exercises for both audience and performer. There are limits to this. If either a planned talk or a short, improvised testimony shared in front of the ward starts to go south, there will usually be no intervention. One of the few exceptions to this rule is when the talk or testimony questions or accuses Church hierarchy. Here, other Saints (and often other Saints close to local leadership) will go up and spontaneously feel moved to give testimony about the faithfulness and service of local leaders. But generally, when a talk goes off into some seeming tangent and starts to introduce material about UFOs or Native American new-age spirituality, things are left to play out. But these things are also remembered.

§51

An aside: What else happens during a sacrament meeting? This is what it was like during fieldwork: We walk through the parking lot; inside the cars people are assembling themselves, fiddling with their ties, checking their makeup, looking their children over to make sure that they are presentable. If this is a fast and testimonial type of sacrament meeting (usually a once-a-month event) there may even be a little grouchy sniping in the car before the whole family goes into the chapel. With adults skipping meals for most of the day, and often the older children abstaining from food as well, the traditional prechurch testiness of many religious families can sometimes feel more elevated during this type of meeting. This parking lot flurry of activity may be occurring early in the morning, but it isn't necessarily always the case; careful economizers, the different wards in the stake often share the same chapel building for their separate sacrament meetings. Each ward is given the building for an allotted slice of the day: sleepy early mornings, hungry late mornings, languid early afternoons.

The clothes worn and fussed over are distinctive, though not bizarre. Dress is white shirts, sober jackets, khakis, dark patternless wrinkle-resistant suits, tightly knotted ties, and dress shoes for men. For women: solid blouses and sweaters, skirts, long dresses, flats, close-toed shoes, and understated jewelry. Boys may have clip-on bow ties; girls may be wearing party dresses. No one will stop you at the door if you aren't dressed appropriately at the chapel. (When I first started attending sacrament meetings, tweed blazers, intricate, florid ties, and blue button-down shirts were the uniform I wore when going there before someone did me a favor and took me aside. I was oblivious to the unspoken norms I was stumbling past, but others obviously did notice.) While there will always be a few members who dress idiosyncratically, and the exceptions are noted and excused, the shade of formality in the clothing, and the degree of sloppiness with which it is being worn, are rich ways to signal (either accidentally or on purpose) that a Saint is tending more towards righteousness or apostasy, someone who can be trusted with responsibilities (a calling) or someone who should not be relied upon.

We leave the car, making our way to the building. People gather at the chapel entrance, meeting each other, shaking hands, giving hugs. The chapel behind them usually looks very church-like: red brick or southwestern faux-adobe with a church tower punctuating the profile of what is often a low, one-story building. Some older chapels were designed and funded by the stakes members on their own and have the architectural wildness and variability that one might expect from full local control. But these expressive chapels are increasingly rare, and many of them have slowly drifted towards the default ascetic found in the institutionally funded buildings of the contemporary Church. The chapel may read as a church from the outside, but the inside of the average contemporary chapel is not laid out like a church, or rather, it is not *entirely* laid out like a church. Yes, there will be a room that is set up in a familiar manner, with pews, hymnals, a podium, and a church organ (or a convincing electronic church organ). This is the room that most people will drift towards after they make their way through the gauntlet of greetings that occur among the cluster forming in front of the main entrance. But there are also hallways leading to a warren of meeting rooms, classrooms, offices, and sometimes even a gym or a half basketball court. When the first hour of sacrament meeting is over, most of the people will drift to one or another of these rooms for classes on gospel doctrine, for meetings of the all-female Relief Society, or for the priesthood meetings that are limited to males who have been ordained to at least the Aaronic priesthood (that is to say, almost all men who are roughly teenaged or older). And some Saints will inevitably go instead to the exit, to either make their escape by driving off or to at least take shelter in the car while other family members attend these later sessions.

But all this movement and room-hopping occur *after* the first hour. That first hour formally commences with a greeting, usually by the bishop (again, a volunteer position of someone also shouldering the obligations of some other profession as well). Then the first hour starts in earnest as a hymn is sung. To ears accustomed to the breathiness of contemporary Christian music, there is something antiquated about that moment when the organ starts to swell and the members flip open their hymnals; it is like the return of some lost Protestant low-church age, some piece of the nineteenth century anachronistically erupting in the present moment. Many of the hymns sung today are classical pieces of the nineteenth century's low Protestant hymnody. But many are not, even if these more distinctly Mormon usually sound like those other familiar Protestant hymns to untrained ears. They are hymns of the Church's own invention, most of them penned during the Saint's tumultuous first century. Whatever the source, the hymns are quite challenging, though many of the adults are surprisingly up to these sacred songs, thanks to performance skills they have habituated over a lifetime (a Mormon enthusiasm for choirs does a lot of work). That said, even the most inspired children will squirm while the hymn plays and the sacrament meeting runs on, playing with toys discreetly handed to them by a parent when the child becomes fidgety or perhaps is just left gazing into the modern-day seer-stone that is a smartphone.

There are (inevitably) announcements, usually the sort of dull organizational information that constitutes church or fellowship announcements all over the United States. Sometimes something more happens during this announcement period. Every so often there will be someone taking on a new supporting role in the ward, something like the ward clerk, the executive secretary, the temple preparation seminar teacher, the librarian, the temple and family history consultant, or the ward music chairman. In these cases, the nominated person will come to the fore of the church, and then there will be a vote to see if the Saint given this calling will be sustained. These are merely *pro forma* democratic moments, a mere ratification for someone already preselected by the bishop or stake president; anything other than a unanimous show of hands here is a rare and offputting turn. Immediately after the vote, a quick blessing by someone or someone will occur, the person giving the blessing also stating that in this act of "setting apart" this new officeholder, he is "acting under authority of the Melchizedek Priesthood," which is another piece of the liturgical boilerplate. At other times, someone will be released from a calling. These releases are usually a subdued moment of spiritual bureaucracy, inevitably involving the bishop or some other ward or stake authority expressing gratitude for all the work the released person has done. While thanking someone for their service may occasionally be a tender moment, for the most part, the sustaining, setting apart, and releasing of these men and women are just movements in the operation of the worn-smooth social mechanics of an organization acting entirely through the volunteer labor of a priesthood of all male believers and their spouses.

Another hymn commences, and as it plays out the sacrament is blessed by two (usually youngish) male volunteers. Other young men distribute the blessed bread on polished metal trays with handles (though older men sometimes have to supplement the ranks of these deacons depending on the demographics of the ward).[12] Every time that I've been to a sacrament meeting, the sacrament had both the taste and the feel in the mouth of sandwich bread, but I have heard of wards resorting to other kinds of bread. Similarly, polished metal hand trays are used to hold the "blood" as it is distributed. The blood is water in small disposable paper or plastic cups that look perfectly sized for measuring a dose of cough medicine.

After the sacrament comes the speaking. During most meetings, the remaining time will feature a few talks by individuals who have been preselected by the bishop or some assistant and who almost always have been given a stated topic of focus. These talks are a little like religious Toastmaster presentations, brief homiletics on faith or forgiveness or on some plank in the "plan of salvation," the Mormon term for the eschatological narrative scheme of their cosmology. Whatever the topic, there is almost always an autobiographical beat, a moment when the speaker makes things approachable by recounting the struggle with this problem or seeing some familial figure, fellow ward member, or close friend exhibit that virtue. These talks are only rarely the spoken equivalent of an academic treatise. (In fact, the rare moments of such arid and purposefully intellectual talks were, back in the foundational days of the MTA, how a few Mormon Transhumanists found each other; after sacrament meeting, someone would come up to one of the presenters for that day and compare a talk just given to another talk they heard when they were part of a different local ward or stake). But the talk is a feature of only one type of sacrament meeting.

If this is a fast and testimony meeting, no one has been asked to speak in advance. Instead, people volunteer themselves, coming up to the podium to give short accounts of whatever it is that they feel the spirit compels them to share. The feeling of a driving urge or a pressing obligation, perhaps even the classically Mormon "burning in the bosom," are the usual phenomenological cues that the spirit wishes you to speak. But for some, *any* desire to speak is enough to suggest that they are being prompted. A person might tell of God's presence in their lives, or recount a time that a prayer was answered, or give a spontaneous elegy for the temple and covenants in marriage, or talk about the importance of reaching out to their gentile friends to show them the character that comes with being a part of the Church. For that matter, people can bring up any number of other topics or events. While sometimes themes emerge on their own during the hour of testimony, or are caused by the shadow of an impending National Holiday or recent news event, the uneven levels of confidence and skill in public speaking, and the sometimes disjunctive jumps in topics, give the whole thing an air of a call-in radio show. Often these testimonials will be shot through with clichés, passages from various general conference talks, or other essays penned by influential figures in the Church (these quotes are also common to the prepared genre of talks as well). These strung-together quotes are the Mormon equivalent of Homeric epithets such as "wine-dark sea" or "swift-footed Achilles," turns of phrase that serve to give the speakers time to have their words catch up with their thoughts and to help index the genre of speech in which the speakers are participating. But even at their most clunky, these testimonials are not *all* clichés, and when present, these clichés are summoned up in the service of something like a spontaneous production of an affective prose-poem, particular to only one person, at only one place, at only one time. And as such, they are laden with emotion, joy, tears, gratitude, or sorrow—sometimes in seriatim, and sometimes all at once. Like all other forms of sacrament meetings in the Church of Jesus Christ of Latter-day Saints, these fast and testimonial sacrament meetings are often tedious. But at moments when someone shares something that comes across to Mormon ears as a little raw or emotional, these meetings can become something else, something uncomfortable, celebratory, or uplifting. And quite often they can become very personal.

§52

Without taking anything away from the disciplinary work being done by the Church, it seems that the sort of speech that makes up most of a sacrament meeting cannot be seen as a simple porting of Asad's account of the Medieval Catholic Church onto twenty-first-century American religion. The kind of religious belief that Asad was trying to parochialize through juxtaposing it with the Medieval Church is undoubtedly present in the Church of Jesus Christ of Latter-day Saints as well. Mormonism is a religion of belief. Church members who are asked to give talks during sacrament meetings are often advised to speak from the heart, and as we have seen, the implicit demand for an autobiographical aspect suggests more than mere replication.

In fact, these talks bear all the hallmarks of the modern, sincere forms of speech associated with anthropologist Webb Keane's idea of an agency-centered and sincerity-prizing Protestant Semiotic Ideology.[13] Mormon Studies scholar and trained anthropologist Daymon Smith has argued that the mainstream LDS is now a thoroughly modern institution, with contemporary ethics of speech prizing both transparency in speech and agentive responsibility.[14] Fundamentalist Mormons, who still practice polygamy and have rejected the contemporary LDS's bureaucratization of the Church, focus on disciplining the body, and they reject any sort of cartesian dualism. The LDS, however, sees its mission as instilling faith in eternal truths in the minds of Saints; this suggests a gap between the body and the intellect. The Church's commitment to an expansive concept of freedom and agency reflects a focus on the mind. Mainline LDS is without question a religion built around concepts of free autonomous will; the purpose of life on earth is for individuals to test their will through repeated exercises of agency. This sincere, modern Mormon neoorthodoxy, in the words of sociologist Kendall White, Jr., is notable in that it affirms "the fundamental doctrines of the sovereignty of God, the depravity of human nature, and salvation by grace . . . [and] may be closer to Protestant fundamentalism and neoorthodoxy than to what I and others esteem to be traditional Mormon thought."[15]

This combination has seeming discontinuities built into it; for instance, the focus on a cartesian dualism runs against the touchstone of Mormon theology, which is, in large part, centered on the claim that everything, even God, is a material body of some sort.[16] But there are specifically anthropological conundrums raised by this split between body and mind. This marriage of belief and power as modes of religiosity seems to go against Asad's formulation. This is to be expected, as belief is the religious index of choice in the age of Euro-American secularisms. That is because belief is the religious dispensation that comes with modernity and secularism. In the era of belief, the Church is limited in its power to compel submission by the presence of institutions such as the state, and it is likewise hemmed in its capacities to make truth claims by competition from institutions such as the sciences. Unable to touch the body, only mind and belief are left. This is because original Mormonism was never secular in the first place.[17] One could also suggest that this contradictory copresence of sincere belief and religious discipline in a secular age is, in part, explained by the concentration of Mormons in the western reaches of the United States. This demographic fact creates areas which may be formally secular, inasmuch as concepts of freedom of religion are acknowledged, but where the thick presence of the Church gives it an ability to compel—similar to that which Asad associates with presecular Christianities. But there are problems with this thesis. It treats secularism merely as a political arrangement and not a social vision that creates divisions when it believes it is merely cutting up various conjoined institutions at the joints. But a more telling flaw is that assuming a critical mass of believers in the intermountain West hollows out secularism fails to explain the grain of Mormonism outside of these redoubts. Only 35 percent of American Mormons live in Utah. And the hypothesis that concentration allows for coercion is also putting aside the observation that nearly 57 percent of Mormons worldwide live outside the United States. This is important because, at least as far as the formal aspects of the religion go, these extra-American Mormons exist under a system that is effectively identical in theology and internal organization to that which is found in the United States. I have heard accounts of American Mormons who, attending sacrament meetings in places as far away as Papua New Guinea, were shocked by how similar these services were to the sacrament meetings in Provo or Salt Lake City. (There are, of course, minor differences, the inevitable result of localizations.)[18]

This seeming contradiction could possibly be cured by the claim that belief is what follows discipline, an argument that again would be in harmony with the Asadian model of repeated training and submission to inculcate a particular subjectivity. The only difference is that the disciplinary mechanism is used to install a different Protestantized belief rather than a medieval one. But this again doesn't quite hold. Affective upwellings, in the absence of training or discipline, as a response to encounters with what an earlier generation of anthropologists would call symbols, seem baked into mainline Mormon religiosity.[19] The idea of spontaneous affective states, resulting without any causal priming, is important in the LDS. It is a touchstone of Mormon missionary and apologetic rhetoric, regardless of the degree or length of experience with the Church, that if one prays on the veracity of the Book of Mormon, one will receive an answer. And this answer is an event that would occur well before the subject has become enmeshed in the larger Mormon disciplinary apparatus. There is also a tradition of personal revelation in the Church; Saints are told that petitioning for such revelation is a central part of the religion. Church literature does suggest a series of practices that are supposedly preparatory, such as fasting and bible study, and it lists a host of forms that response can come in, including voices and even visitations from God. But these preparations seem threadbare when compared to the discipline honed Pentecostal and charismatic modes of subjectification as depicted by anthropologists, and these Mormon exercises don't seem to be systematically undertaken.[20] Instead, these revelations and other related phenomena, like the presence of the Holy Spirit, seem to take the form of Geertzian moods and motivations rather than of bodies or minds trained into an automaticity in behavior and sensory apprehension. This should alarm us because Asad used Geertz as his example of an anthropological theory of religion that was supersaturated by late-Euro-America belief privileging sensibilities. We are left with something that, while not rising to the level of a contradiction or even a paradox, is at least a puzzle: we have a religion which seems to fit both a theoretical model intended as an alternative to belief, and which also fits what the very model of religion-as-belief that discipline was intended to stand in contrast with.

§53

The Church of Jesus Christ of Latter-day Saints, then, is a combination of belief and discipline. And this admixture runs orthogonal to contemporary anthropological typologies and causal narratives of religion. What then is the MTA? Are discipline and belief major modes of religiosity in the association? The MTA's own sensibilities regarding religion appear to suggest an answer. Members of the MTA have a vision of religion as primarily about practice rather than belief: it is common for founding members of the association to refer to religion as a "social technology" for installing "strenuous moods," a term they borrow from American pragmatist philosopher William James (pragmatism being an outsized influence on many of the original members of the group).[21] This is a picture of religion that they often invoke when tussling online against New-Atheist narratives of religion as a series of irrational beliefs. Focusing more on social effects and the purposeful inculcation of shared sensibilities, the MTA's account of how religiosity is operationalized is consonant with an Asadian narrative. It is not belief, nor adherence to an intellectual system, nor an internal fidelity. It is an introjection of an aesthetic and an ethic by way of the myriad practices and socialities that constitute quotidian religion.

But at the same time, when disaffected ex-Mormons (including some who retain membership in the MTA) or agitated outsiders come to criticize Mormonism, discussions on the MTA boards tend to drift into a kind of apologetics. There are subtle apologetics to be sure, focused not so much on the veridicality of scriptures but on the character of the religion. When anti-Mormon critics crash into the servers to accuse Joseph Smith of immorality due to polygamy (which is often presented as adultery and borderline pedophilia), the picture that many MTA members offer up as a defense of Joseph is not like the usual plaster-saint portraits that mainline LDS members offer up of their prophet. For much of the MTA, rather than denying these accusations, Joseph is instead presented at times as a deeply flawed man who intended well, as a religious genius who was reaching for the truth. Similarly, when attacks on The Book of Mormon or The Book of Abraham are hurled at the group, rather than a literalist version of circling the wagons, instead the defense is usually that the sorts of flaws being alleged (including the inclusion of supposedly plagiarized material from other nineteenth century texts, or from the King James Bible) are to be expected considering the context of the writing. It was a different era, with different senses of intellectual property and of the importance of originality. Similar statements are made when anachronisms in the Book of Mormon are pointed out (such as the oft-commented presence of horses and elephants in the pre-Columbian New World). The argument is that Joseph Smith could only interpret from language and terms familiar to him. To expect otherwise is to have a vision of the truth that is anachronistic if not ahistorical. But most importantly, such criticisms miss the point. It is not whether there is a literal truth contained in these books, but rather that they are true in the sense of inculcating proper moral character and indexing a sense of the divinity that can be worked with in both day-to-day life as well as in the opening up of future ages.

While it would be wrong to say that these defenses give the group a fideistic edge, these apologetics are repeated expressions of a qualified, hesitant, but still quite real faith in Mormonism. And we could say that this faith in some ways carries more weight than the endorsement of religion as a praxis, specifically because these defenses offered by some MTA members are endorsements of religion as a practice. It is not a belief in the tenets of a religion, but it is a belief in the efficacy of religion. This means that their views about religion not being concerned with belief *are* an expression of belief. This is not the only mode of religiosity for many MTA members. Many are indifferent to this question, and there are those who are angry with the institutional church after having a negative personal experience, or after having to break with some political or social position of the church. And it must be remembered that for most of the MTA, what is done online does not exhaust their full engagement with religion. But to a large part, in the online spaces that constitute the social center of gravity for the MTA, these MTA members don't seem to be engaging in praxis but instead declaring their belief in praxis.

Still, it is important not to overvalue the claim that the Mormon Transhumanist Association is about belief, because this only touches on a small part of MTA activity. Defenses of Joseph Smith, of the folk-magic elements of early Mormonism, of polygamy, or of the veracity of the Book of Abraham are not the purpose of the group. They speak of other things. Some of the discussions are about the group's charitable endeavors; working in conjunction with secular transhumanist groups, for years the MTA had supported an orphanage in Uganda. And the association has also seriously discussed opening up a combined medical clinic and research center in the United States (though concrete steps have yet to be taken). As shown earlier, discussions of technology take up most of the group's energies. Some of this is speculation about near-term changes, technological innovations, or incipient cultural shifts. Here we have somewhat banal discussions of grab-bag technological novelties, of new silicon chip architecture, self-driving cars, new genomic medicine, in-process and planning-stage space missions—as discussed previously in this work. Then there are questions about the social echoes of this. For instance, if we manage to add twenty or thirty years to the current life span, what would the resulting gerontocracy look like? (This is a very real question for progressively inclined members of a religion that is governed by a set of quite literal 'elders').

Extend the temporal horizons of these hypotheticals just a bit, and a tipping point is reached as cosmological issues are touched on. Examples abound. A new video game is announced in which an entire galaxy is simulated, and someone inevitably raises the question: is this how God made the world? Is this how we will make worlds ourselves as we become Gods through transhumanist technologies? Can wormholes be used to create entire universes? Will direct neurological communication between individuals allow them to slowly converge into one essentially godlike superentity? As these broader questions have a temporal stutter built into them, this is also an inquiry into whether and how God accomplished these feats in the past—when he instituted the current order of things. As anthropologist Abou Farman has observed, transhumanism in its secular form is about the gap between what *is* and what *might be* possible, and that the arc of this investigation is left running in the direction of the cosmic.[22] The MTA also inhabits the gap where desire meets anticipation; they just resort to a different vocabulary than that used by secular transhumanists when their thoughts close in on these ultimate concerns, and they see their discussion of transhuman eschatologies as possible discussions of the genesis of the universe.

And it is in these discussions—rather than in their defenses of Joseph Smith, The Book of Mormon, or their subjunctive-mode refusal to entirely reject celestial marriage as a historical practice—that their oft-repeated kinship with the more speculative tradition of nineteenth-century Mormonism becomes clear. Starting with Joseph Smith and running into as late as the 1930s, there was continual experimentation with speculative accounts of how the cosmos was structured. Relying not on revelation directly but on analogical reasoning from propositions that were rooted in revelation, high-level Mormon leaders such as Brigham Young, Orson Pratt, and Parley Pratt would speculate how far back goes the lineage of Gods ancestral to our God, whether and what gods ate, if they socialized with other gods, whether they were omnipresent or if their materiality necessitated their being situated in specific locales (among other such issues). These were post-Copernican imaginings, a plurality of gods for a plurality of planets in the universe, and they also referenced contemporary astronomical and mathematic discoveries. The similarity to the MTA's practices should be obvious. This similarity is not lost on the MTA nor on outside observers, such as Richard Bushman, a noted Mormon historian who has collaborated with MTA members and spoken at an MTA-organized conference.[23] Though now introjected with a different vocabulary due to both changes in science and engineering and the development of Silicon-Valley entrepreneur capitalism, this is a continuation of a mode of Mormon thought that goes back to the first instances of the faith.

§54

This kinship is also important to the anthropological question of religious discipline and belief. Consider the discussion of belief and discipline in the mainstream Church of Jesus Christ of Latter-day Saints. Recall that they seemed to be unsteady; both modes of religion seemed inescapably present, and it was uncertain what connection or ranking between the two might be. But that points to the nature of the relationship. In the Church, the link between belief and faith is not causal (even though it is easily thought of that way) but structural. The reason it is thought causal is that belief is seen as the product of discipline, at least in classical Asadian reading of the Medieval subject. There are certainly numerous ethnographies by anthropologists—such as Rebecca Lester's study of Catholic nuns, Sabah Mahmood's study of the women's Da'wa movement in Cairo, and Joel Robbins on Papuan confessions to Ayala Fader on Hasidic education—that show some kind of reliance on faith. These works document the sort of faith that anthropologist Joel Robbins would call a belief *in* rather than a belief *that*, and they testify to this belief becoming much more intense as a result of prolonged submission to disciplinary exercise.[24] This leads to the illusion that, in this arrangement, belief follows discipline.

But belief cannot be seen as entirely a product of discipline. French philosopher Gilles Deleuze, writing on modernist cinema, has stated that belief in this world is necessary, and it can be supplied by cinema but also through

Catholicism (and speaking in a French context, it seems clear that Catholicism is supposed to stand for the broader category of religion).[25] He specifically states that he is discussing "belief in this world," but as both his reference to Catholicism and his references to Kierkegaard suggest, this belief is not so much an explicit rejection of any transcendence, but rather it is some faith that allows the grasping and navigation of the immanent plane of existence that (at least in Deleuze's metaphysics) exhausts reality. This somewhat abstract philosophical statement has been given ethnographic rigor in its use by Richard Baxstrom's depiction of the rapidly churning city of Kuala Lumpur.[26] His discussion of both Islamic and Hindu adherents' attempts to foster this belief in solidarity with a condition of radical uncertainty shows us that this belief can come in religious forms. The degree to which they struggle also suggests that this faith can come in different intensities, and that its absence, while not impossible, is pathological in the sense of being an unhealthy and ultimately unsustainable state of being. Belief does not necessarily precede action. And it does not always retain the same form throughout the life of the individual. But belief of some sort is always already there, and if events cause it not to be, then the subject's coherence and capacity for action are at risk.

This makes sense. Except in cases of physical coercion, submission to religious discipline has a voluntaristic edge; in most of the ethnographic cases with disciplinary modes of subjectivity, inasmuch as they are comprised of both techniques of the self and some form of group accountability, there is a sense that choosing not to opt in to the regime is always a possibility.[27] These disciplined adherents belong to clerical orders, particularly demanding denominations, or specific study groups. For most disciplinary forms of religion, people are not born into its mechanisms but have to volunteer to be a part of the disciplined community. And how can we understand that initial choice except as an expression of something like belief? But then at the same time, it is also true that we can say that discipline is also always already present. These individuals are not acting in a void; they are making the choices in situations prestructured by religiously inflected institutional power relations and bodily habituation.

It is not that we have discipline extruding belief—or belief birthing discipline—but belief and discipline as mutually implicating one another. With this understanding, instead of belief summoning up discipline or discipline summoning up belief, both the ratio and quality of belief and discipline are affected by one another, though in different ways and in different circumstances. It is this transformation, the continually moving tangents of the intensity of belief and discipline, that makes it look like one process mechanically produces the other. This is not to undo Asad's dispensationalist account of belief and discipline as having both transformed radically at the onset of secularism, rather it's to view it from a different perspective: a qualitative shift in ratios rather than an utter transformation into an untranslatable structure. Nor is it to make belief and discipline fundamental elements; under other sociocultural constellations, it is easy to image constructions of the person where the divide is not between belief and discipline but between other aspects, activities, or relations. But this does raise a question. If we do not see belief and discipline as causing each other in seriatim, but as different sides of the same coin, as some kind of mutually dependent forces cohering into some amalgam or composite, does this mean that we can talk about the work that belief and discipline do, in conjunction, on other modes of life?

§55

To understand the work of the belief-discipline composite, we have to look in two different directions. The first direction is outward. We should ask what the belief-discipline composite does to forces that are formally exterior to religion as a sort of system.[28] We can see an example in discussions of nineteenth-century Mormon history. Much of the current Mormon belief-discipline process is a way of combatting or preempting extra-Mormon threats. Twentieth-century Mormonism tries to produce a picture of itself that will not trigger the genocidal, violent response from Protestants that shaped much of the first century of Mormon history, when there were both direct (through vigilante) and indirect (through state and federal government persecutions) attempts to eradicate both the religion and the community. There are other extrareligious forces that the belief-discipline composite must act on, such as the biology and appetites of human beings; these are forces that can exist without religion but must often be shaped in particular ways if religion—particularly a religion that hypertrophies kinship and marriage—is to operate.

We ask this question about the outside of religion, even as we conceded that another mutual implication, that of the secular/religious divide, means that it will be difficult to assess what is inside and outside of religion. The barriers are permeable, shifting, and unmarked; in some moments, there will be no barriers at all. In cases where religion is part of a total system, such as Mormonism, this challenge becomes more significant. Kinship and religion are conjoined in this case—along with a great deal of economic activity, ranging from business ties to multilevel-marketing essential oil sales. To the degree that the belief-discipline composite affects them, is it acting on material *outside* of religion or *within* it?

But there is also the possibility that the belief-discipline composite works to shape the outward face of the faith as well as the borderlands where religion bleeds into something else. It might be doing self-regulatory work as well. This work is why we have to ask what it is inside of religion that is being operated upon by belief-discipline. It is evident that the composite is working to fight interior centrifugal forces. But it is also apparent that a centrifugal force be extant, or at least continue to be extant for as long as the particular belief-discipline composite that acts upon it exists. This observation is just Foucauldian metaphysics 101, the axiom that power and resistance presuppose each other. One does not produce mechanisms of control unless there is a population or a force whose actions or mere existence threatens to resist control. But that presupposition does not mean that the relationship between power and resistance is always the same. Sometimes—rarely—power so occludes resistance that resistance appears to be absent; in other times, power allows for heterotopic spaces where some form of resistance can flourish. And there are times when power can try to make resistance its ally, even as it works to delimit where this alliance can occur.

But what interior force can the belief-discipline composite be working on? Speculation is a good place to start in order to give some value to the empty space of interiorized resistance. In fact, speculation may be the preeminent internal mode of religiosity that the belief-discipline composite has to govern. Belief-discipline always works towards a telos, be that a determined set of capacities, the installation of a particular image of the thought, or both. But speculation is about a virtual production of multiple teloi, arcs that lead out in new directions that would demand new capacities and different images of thought. It is an internal multiplicity, all the more unsettling of religious fixed points due to speculation's phantasmic nature since, unlike fully concretized religious apparatuses, it needs no infrastructure before it is realized as a concerted project. Worse yet, speculation, even as it decenters teloi, cannot be eradicated. Speculation is a mode of religiosity that is as necessary as it is dangerous. It is necessary as institutions can only survive change to the extent that they are able to reimagine themselves under new conditions. But at the same time, speculation can also produce arcs that would either mutate the institution beyond recognition or cause its dissolution. But it is never absolutely clear in advance what form of speculation we are dealing with, whether we are weighing something constructive, corrosive, or (perhaps most dangerously) something that is both. Therefore, the institution must, in its awkward and distributed manner, make a retrospective judgment as to whether speculations are transformative or poisonous. Only once speculation is actualized as a concept can it be weighed.

Because of the *après-coup* nature of the decision, leniency is one of the names that we can give the flexible membrane that at times dilates to allow discipline and belief to constrict speculation, and at other times it contracts to let speculation in. Leniency, conceived of as a purposeful recalibration of thresholds, is ultimately the deciding element in shaping the ways all the constitutive elements of this religiosity express themselves. If leniency is framed as the degrees of variance from the telos accepted by the system, then it is the general predilection for leniency that controls the degree to which the system is opened up for entropy—but leniency also controls the chance of creating a "hopeful monster," a term that a previous generation of biologists used to explain a mutation of the system which is more suited for the current climate and which can be laterally propagated throughout.[29] Leniency eats at institutionalized religion's internal cohesion, but it also allows for new modes of that religiosity to incubate. The development of new forms has happened before with Mormon's slide from nineteenth-century experimentation to quasiprotestant neoorthodoxy. Perhaps it will happen again.

§56

The Mormon Transhumanist Association appears to have engaged in potentially positive, nontoxic speculation. At the border of the Church of Jesus Christ of Latter-day Saints, MTA members took the wealth of accessible information within the Bloggernacle about nineteenth-century Mormonism, and they synthesized it with current discussions regarding advances in technology, creating a new mode of religious thought that is neither chiefly belief nor discipline, but invention. By way of contrast, those who used other forms of reason to imagine different regimes of gender—or a different relation with what are now scriptural texts—have been judged as engaging with poison.

This is not an automatic judgment. Speculation that plays with dangerous forces, even forces that result in censure in other instances, might not be challenged by the belief-discipline composite if they stay on the virtual side as much as possible, consisting of pure problematics without any real actualizations. An example is Taylor Petrey's essay, "Towards a Post-Heterosexual Mormon Theology."[30] Petrey is an associate professor in the department of religion at Kalamazoo College. His academic work focuses on Christianity and Judaism in antiquity, but he also writes on issues of Mormon theology. Petrey's essay was published in *Dialogue*, a popular progressive Mormon print venue, and it received a great deal of attention in the Bloggernacle. Furthermore, it touched thematically on many of the third-rail issues that Kate Kelly and John Dehlin played with. But Petrey's essay merely asked what problems a postheterosexual

Mormon theology would have to deal with if such a theology were to exist. While these problems touched on some of Mormonism's theological distinctiveness—such as divine sexual procreation, what relations can be sealed in Temple rituals, and the eternal nature of gender—Petrey did not present any reformulations. He merely identified them as issues that could be addressed through the use of existing theological and speculative material. In short, unlike the speculative work of Dehlin and Kelly, Petrey kept on the side of generative problems rather than that of concrete actualizations or unalloyed truth claims. In other words, he dwelled in speculation. And as such, the Church as an institution felt that judgment could be withheld—at least for now.

Much like the violence of the nineteenth century may have forced discipline to be more elaborated than in other forms of American postsecular religion, the relative recent history of Mormonism may make speculation more visible. Born in the Burned-over District of New York, the frontier territory serviced mostly by itinerant and self-authorized pastors, the initial disciplinary apparatus was too attenuated to prevent a speculative mutation of Christian primitivism born from the crossfertilization of folk magic, historical speculation, and new scientific pictures of the world. This speculation was originally granted a *de facto* internal leniency, but in the wake of a shift in the Utah basin from charismatic to institutional authority, the first generations of Mormons made it *de jure* as they tried to imagine what Mormonism would be. The intense bubble of nineteenth-century speculation that followed Joseph Smith's death can be seen as necessary, as there was an intense desire to reject the conceptual and ritual aspects of a Protestantism that had itself violently rejected Mormonism. Rather than see itself as a religion of institutional control, it was a religion of freedom, the highest degree of leniency. Leniency marked Mormon practice as well. Even the bodily disciplines seen as essential to contemporary Mormonism were not obligatory in those early days. For example: after receiving the "principle" of the "word of wisdom" in 1833, the prophetic mandate inveighing wine and "strong drink," Joseph Smith sold liquor at his tavern and asked for wine hours before his Nauvoo martyrdom. Mormons built distilleries in the Utah Valley, and Brigham Young discouraged smoking during sacrament meetings but did not forbid it. Mormonism had to invent itself, and constrictions on experimentations with practice and thought would not facilitate that work. The speculative capacities utilized through these acts of invention remain a potential aspect of Mormonism today. All this is to say that historical forces have made speculation an outsized component of Mormonism and perhaps with other fast-growing nineteenth and twentieth-century cosmological religions like Cao Đài,[31] one of the speculative religions *par excellence.* This suggests that if we wish to better understand not just discipline or belief but the wider religious ecology that these modes are a part of, we may wish to also keep an eye on religions like Mormonism and groups like the Mormon Transhumanist Association to understand the future of speculation as a religious modality, and perhaps to understand the future of religion itself.

§57

And what is that future? Moments of speculation, without a doubt, but also of contestation, as vying speculative spaces opening up from various different positions and cosmologies start to buffet one another as they expand. In the case at hand, Mormon imaginings crashing into transhumanist fantasies, and religious transhumanism pushing against them both. Myths of the future were created. When these expanding horizons of speculation and mythic thought reenter into the world of *doxa* and *praxis*, of belief and technique, their future will be shot through with moments of hardening and regimentation as this thought is translated back into the less malleable realm of group consensus and material practice. Dreaming of living forever, either after the singularity or during the millennium, will demand that people train bodies and thoughts, master rites, gather proficiencies, produce technologies. This will be done all in the here and now, all in very specific and controlled ways. And these actualizations can never have the pliability as the inspiration that first animated them. There will be moments of oscillation, too, as speculation's technical and cultural realizations, in turn, open up opportunities that were unseen before as possible intermediary concrete act or habit. This is when speculation's inauguration as something done or attested to in the here and now rather than as something dreamed opens up the speculative inflationary landscape yet again. If we freeze the dead for resurrection, does this technical fix become another mode of commemoration? If we bury the deceased faithfully, and in harmony with the guidelines of our religion in hope of their resurrection, what does that act open up in the imagination of life, death, and the body? And there will be moments of collapse, when speculative horizons are shut down either because they lose their symbiotic (and sometimes parasitic) relation with the belief-practice composite . . . or when speculative horizons are crushed by outside forces or internal enemies. While not always set in a paranoid key, both Mormonism and Transhumanism can be nervous universalisms, opening their hand to everyone in theory but in practice apprehensive about what they imagine to be their competitors and their enemies.

And finally, every so often, very rarely, we will see something like speculation rush ahead of its borders, in a way that might be likened to the inflationary bubble that some astronomers speculate grew the early universe. Those will be the moments when the speculative horizon becomes so vast that the belief-practice composite cracks open, and, for a few seconds, through these future myths the light of a virtual eternity shines directly into thoughts, bodies, practices, language, institutions, and technologies . . . doubling back into concrete forms, and thereby changing lives.

PART III
Science Fictions

Sixth Series: Freezing, Burying, Burning

§58

We have introduced the three movements, forms, or modes in Mormonism, transhumanism, and Mormon Transhumanism. Two of them—Mormonism and transhumanism—rhyme with each other, for lack of a better word, in their insistent materialism, their naturalistic understanding of the fantastic, and in their expansive vision of what humanity could ultimately become. They also rhyme in that they have universalistic aspirations, but at the same time feel threatened by forces outside of their particularistic communities. While neither of the two movements are creatures of the internet, the internet has left a mark on each of them. In transhumanism it has accelerated a certain quarrelsome tendency, directed out toward religion, but also inward, as various transhumanist instantiations debate one another. For Mormonism, the advent of the internet gave birth to the Bloggernacle—like transhumanism, a virtual community of discussion and debate, but also an accelerant of the circulation of information that for some deeply problematized their relationship with the Church, sometimes to the breaking point.

Mormon Transhumanism at one level is merely a response to the resonances and homologies between Mormonism and transhumanism, a simple positing of what it would mean if the eschatological promises of Joseph Smith's restored gospel were intended to be carried out through technical means and human agency rather than divine fiat. But at other levels, it is something else. Mormon Transhumanism rearticulates what Mormonism means in a way that enables a different relationship with the institutional Church, both historical and contemporary, such that Mormon Transhumanists can escape literalism and a parochial conservatism while still seeing themselves as genuine, faithful Mormons. Dancing between and combining together different codes and languages, they find something like a freedom instead of what would—to them at least—feel like a compulsion or a submission. And the freedom they find is a mode of religiosity that is distinguishable from both elements of the belief/discipline composite and, thus, in some ways, its other: pure speculation without structure at all, religion in the subjunctive mood, a wild proliferation and exploration of potentialities.

Speculation is not unique to either Mormon Transhumanism, Mormonism, or religion; Mormon Transhumanists build on a nineteenth-century Mormon speculative tradition, just as they borrow wildly from transhumanist imaginings. And due to genealogical linkages and shared existential conditions, speculation among these three objects sketches out the contours of similar problems: how will the dead be preserved, how will they be raised again, how might new worlds and universes come about, and on what sort of intimacies will we rely in bringing all this into being? To the degree that these are narratives that go beyond the empirical and engage in producing concrete expressions of "transcendental deductions" regarding the shape of things to come, they are myths, in the sense of the term used by Levi-Strauss.[1] These are cosmological concerns that thus draw on mythical modes of thought, whether those myths are religious or scientific-technical. And due to how these groups are at once so close and (at times) so opposed, these mythic exercises in speculative thought read as strange variants of one another: at times as almost identical, at others as twists and inversions of one another, where agency is foregrounded or erased, where fabric sheets are juxtaposed against metal sleeves, where gender is a chasm or a continuum, where God is a singular or a composite being.

The next set of series explores these spaces. Now that the actors have all been sketched and the hard sociological work done, the variations, structural involutions, and interference patterns for these three imagined solutions can be charted. And what is more, we can attend not just to the proliferations and differences amid this virtual thought, but also to the way that these solutions at times double back and reshape concrete human arrangements, reconfiguring situations, transforming problems, and at times opening up vistas for further speculative leaps.

We begin with the dead.

§59

They feel cold to the touch, but not as cold as one might think, considering the low temperatures they keep, and what's inside them. They are metallic silver in hue, stainless steel in composition, so polished that they mirror everything around them, including all the other near-identical containers (*near-identical* because they come in two sizes, one set slightly taller than the other). They are clean. In fact, this entire area is exactingly clean, though it does not read like the cleanliness of a thoroughly run doctor's office or that of a laboratory, even though both of these descriptions could be adequate analogs for what is done here. Rather, it seems instead to be the cleanliness of an industrial site or a tech business, which could be adequate descriptions, too.

The generic name for the half-device, half-container I'm touching is *dewar*; this particular model I now rest my hand against is the *bigger dewar*, called this "because of the large masters at the bottom" of the container (since my visit, they have also added the *SuperD* or *superdewar* to their inventory).[2] All dewars, by definition, contain cold liquids, in this case, the sort of liquid that boils off well below the freezing point of water: liquid nitrogen.

They also contain what most people would call *corpses* (up to four of them) and *severed heads* (up to forty).[3] Here, though, they use different terms. The intact corpses are called *whole body patients*, and the heads *neuropatients*. Thought of this way, they are not dead, regardless of what might have been said by an attending physician, or for that matter by the State. To the minds of the people who work here, and often (but not always) to the family of the deceased as well, they are instead in what might be considered the most perilous and extended instance of intensive care.

The plan is that these *patients* will all live again.

§60

ALCOR is located just outside of Phoenix, Arizona. Even more precisely, it is in an office park in a sparse and somewhat desiccated part of Scottsdale, a part of town at the edge of things, full of dun-colored faux adobe office buildings and storage areas.[4] It is literally within sight of the Scottsdale airport, a facility ALCOR uses to fly patients in, usually by the sort of small, chartered flights in which airports of this size ofttimes specialize.

When discussing the airport, ALCOR and its spokespeople usually go to some pains to mention that while they are extraordinarily close to the travel-hub, it is not in the flight path. In a similar vein, ALCOR representatives will acknowledge that while there is some irony in keeping bodies frozen in liquid nitrogen in the middle of the Sonoran Desert, Phoenix has an extraordinarily stable and consistent climate. There are no blizzards in Scottsdale, Arizona, there is little rain, and (despite what struck me as the oppressive heat there) the Scottsdale area has what is described by some ALCOR literature as "generally pleasant weather."[5] Likewise, ALCOR stresses that in comparison to Riverside, California—the operation's original home before a 1994 move—Scottsdale has very little seismic activity. Together, these descriptions paint a picture of Scottsdale as not just extraordinarily cozy (pictures of golden sunsets and the Scottsdale downtown Riverwalk appear on one ALCOR webpage, along with the claim that "Scottsdale is also a nice place to live, retire to, or visit") but as extraordinarily safe.

This is not the logic of an individual, family, or business thinking of where they should live for the next few years; making life decisions based on these desiderata such as the lack of seismic activity and the absence of system-disrupting extreme weather events would come across as neurotic. But this environmental and seismic stability is necessary to ensure the continuity of two essential goods: electricity and liquid nitrogen.[6] There are other concerns, most of which have a similar cast as focused on long-term security. If you were to go into the ALCOR building, you would walk past a plastic chemical hazard notice sign on a glass door (the door also has "Thank you for not smoking" stenciled in polite white lettering) and enter a lobby that looks much like what you would find in many small businesses: faux-marble tile floors, soothing blue walls with some modest brown wood trim, a matching blue couch next to some modernist chairs and a glass table, covered with promotional material. And behind that couch, a shoulder privacy wall made of shiny metal that reflects light in a rippled grid pattern; that wall also includes the word "ALCOR" in bold relief, spelled out with overly large blue letters. But then you would be ushered into an antechamber on the side. One wall of the antechamber is a bulletproof window that opens to a room reinforced with metal plating and Kevlar. That is the room that contains the dewars. Safety, all this seems to suggest, is not just *a* concern, it's *the* driving concern.

If you ask the people there who they expect to repel with metal plating and Kevlar reinforcement, they will say that they do not know. Perhaps it will be religious fanatics, terrorists, looters, or something similar, but the sense is that this metal bay is set up to thwart a danger which has not developed yet, and which they cannot imagine. But just like the need to plan for geological and climatic safety, just like the need to ensure access to electricity and liquid nitrogen, ALCOR has to prepare for a future that may stretch well behind the (first) lives of anyone who works for the nonprofit. ALCOR is not thinking in human scales of time, or of passing human wants, needs, and fears—or at least is not thinking of the sort of wants and fears that are measured in the life of just one generation. It is hoping that it will make those under its care, both whole body patients and neuropatients, into immortals. But such hopes may not be realized for decades. They may not be possible for centuries.

§61

There are tours at ALCOR. Interested people can sign up in advance for these tours, to have cryonics explained to them, to be shown videos of things such as heat maps of brains as they are cryonically preserved, to have the steel- and Kevlar-reinforced doors opened for them so they can walk through the dewar floor. The tour is usually given by the same man. He is an amazingly, almost unbelievably, muscular Englishman, with a goatee and red hair cut short on the sides and slicked back on the top of his head. It is hard to gauge his age, but when looking at him, you have the sense that whatever that age is, he looks very young for it. When I saw him in person during my tour, a jangling at his wrist repeatedly caught my attention as I saw flashes from a metal bracelet next to a more sober-looking biomonitor. I later learned that the bracelet he was wearing was an ALCOR emergency ID tag, a way of alerting aid workers who come across an ALCOR member in an emergency about what to do with the wearer's remains. Not in person, photograph, or video, have I ever seen him wear anything other than all black. He is a confident speaker, and his language does nothing to hide his Oxford education or his subsequent American PhD, but it is not purposefully showy or academic either. It is not the breezy language of someone wanting to close a deal or a car salesman's patois. Rather, it is the language of someone trying to speak as clearly as possible so that his points are understood.

Despite the guide's measured demeanor, the tour is unabashedly evangelical; the hope is that people who have been shown the premises will sign up to become regular ALCOR members, filling the form out and paying the three-hundred-dollar application fee for a first-time member. If one joins, there will be other fees in the future, of course; there is an annual fee of just over five hundred dollars (though the annual fee for additional family members and minor children will be considerably less). There is also an expectation that members will have a certain reserve of funds on hand for when cryopreservation inevitably becomes necessary. ALCOR suggests that new members have ready two hundred thousand dollars for whole-body cryopreservation, or eighty thousand for neurocryopreservation. There are additional possible expenses if one requires a standby team; that is, if one needs ALCOR specialists to show up when one is "hospitalized or in a terminal condition," flying out "to [one's] bedside to begin its procedure upon legal death."[7] An ALCOR representative will also find a way to mention that they can put new members in touch with financial planners or life insurance salespeople that can help meet the expense. There is no push for the high-ticket option; during the tour, the guide will inevitably mention that he plans on preserving only his brain. He will explain that he expects to be old and wizened when he dies, and would, therefore, prefer a new body at the time he is resurrected.

Given that the tour occurs more or less twice a week, despite the ease of the presentation, there is a settled, rote aspect to it. The guide will stop to answer questions, but unless he is thrown off the beat in some way, the expositions flow into set conversational channels as he walks through the facility, commenting on the contents of each room. In the patient bay—that is, the large warehouse full of the mirror-like dewars all reflecting one another—he will describe the dewars' design, the way nitrogen is added, the way ALCOR personnel check to see that a certain temperature is maintained. In what looks like an operating room, he will stand next to a mockup of a hospital bed (albeit one with a hammock instead of a mattress) and, using a resuscitation-dummy-like plastic body surrounded by plastic ice cubes—he will illustrate the way a patient's body is cooled during preparation, even as circulation is artificially continued in a sort of post-declared-death version of automated CPR. He will talk about how they drain the body, how they replace blood with vitrifying chemicals injected through the veins, how they sometimes have to crack open the chest-plate bones to make sure the chemicals are coursing through the body. He will present the device they use to sever neuropatients' heads and inject the coolants into the carotid arteries.

Throughout the facility, the hallways are all decorated in the same way. On the otherwise blank white interior walls, there are rows of framed photographs set at about shoulder height, their frames identical in size, though a few of the photographs are set out in landscape mode, so that the image runs wider than it is tall. The pictures vary: some are in color, and some are in black-and-white. Some are candid photographs, action shots of the person running, or scaling a mounting, or boating. A great many are professional studio portraits. Many of the photographs are of individuals, though there are a few photographs of couples, and the very occasional photograph of a dog or a cat. Each framed photo bears a small plaque, and while the specific information varies, the form is always the same. Each plaque has a name, the month, day, and year that the person (or persons, or animal, or animals) in the photograph was born, and underneath that the month, the day and year they were cryopreserved. ALCOR only had about one hundred and fifty patients at the time of this writing, and many patients are preserved *anonymously*, that is, without ALCOR acknowledging their presence to the outside world, in accordance with that patient's posthumous request. So there can only be so many of these photographs. But at some points in the tour, after seeing photo after photo after photo, one can get the feeling that the number of these photographs is endless.

§62

As stated, there is a pattern to the speech given in the tour, not a script, but more along the lines of grooves made through habitual wear, answers and expositions for the predictable questions, performances polished by repetition. The same material heard on the tour can be heard popping up again and again in videos people make of their visits to ALCOR—there are many such videos, as there is something about cryonics that fascinates both technophiles and technophobes. This is to say that something about cryonics demands documenting, a distributed informational archive to complement the archive of bodies.[8] But every so often, a question from a tour member will shake the monologue, throwing it into some unexpected area.

Such is the case with a question in one particularly striking video. Our redheaded, black-robed guide is standing among the dewars, saying something that seems like another piece of boilerplate, until the thought slinks off track. "You do occasionally have visits. We have, actually, every year, an annual meeting of directors," he says, and then (apparently answering an unheard question, given the shift in topic), "his first wife is cryopreserved, so he'll come in here and say hello to her." As he does this, our Virgil puts one hand on a nearby cylinder, the way I stretched out my hand to touch a dewar during my tour, as he reenacts a dewar-side greeting. "So yes, occasionally he visits, and says 'hi.'" He stutters slightly with that last sentence, rushing his response. "It's a tricky emotion," he adds (after an editing jump-cut, so there is, again, no real knowing what this is a response to), "because, if you—well, let's say you don't believe in an afterlife, you're saying goodbye forever, and maybe in time you can get over it. But in this case, you know," he turns to regard a nearby dewar for just a second, before turning back to the interviewer, "we don't know for certain we're bringing people back—we hope they come back—so you could still be saying sayonara, but it will be a lot later than you see them. So, it will be difficult."

Another cut, and the conversation has moved a bit further along, though everyone is still standing among the dewars: "Uh, sometimes what people do—I've suggested this to a number of people—they might write a letter, once a year, to say, your wife or your husband or your father." There is another quick editing jump, returning us to the starting camera position, showing rows of dewars at his back as he continues, "Write them a letter once a year, it keeps them alive in your mind emotionally, and of course, they actually get to see what happened when they were cryopreserved and catch up when they come back." During the conversation, his face, while not grave, is matter-of-fact. There are only two times when he breaks into a smile: an incongruous small

smile when he says "you could still be saying sayonara to them" and a full
smile, bordering (despite his measured tones) on the enthusiastic when he
speaks about the time when a cryopreserved person would hypothetically "catch
up when they come back."[9]

§63

In the first decade of the first century of this second millennium, the anthropologist Tiffany Romain did fieldwork with ALCOR members and other cryonicists, who were located both in the Bay area and also in Scottsdale. At one
point she described the stark functionalism of the typical cryonicist's home:

> Many of the cryonicists' homes I visited appeared as if the inhabitants
> had just moved in, often open spaces with sparse, functional furniture,
> one or more computers, a small display of photographs, a television,
> and, often, a large collection of books. These spaces hint at a life lived
> in the virtual space of the computer and in the imagined space of
> literature.[10]

This minimalism is not just a decorate aesthetic. It is mode of human connection, or rather, of deconnection. One former president of ALCOR, in conversation with Romain, describes the "typical cryonicist" as "not very social," as
"often single."[11] True, there were some cryonicists who "described themselves
as grabbing hold of life by going on adventures, having parties, or developing
deep emotional relationships."[12] But these deep emotional relations are not
with their direct descendent. Romain recounts:

> Most members of the cryonics community are childless, and while in
> many cases this may be coincidental, many choose not to have children
> as a matter of principle based on the importance of tending to the self
> through ongoing education, financial achievement, and other forms of
> self-improvement . . . cryonicists tend to adopt particular values, ethics,
> and goals. For many cryonicists, having children is considered an un-
> necessary diversion of resources that can and should be devoted to the
> self, especially if one is to achieve immortality. Phil, one of the few cry-
> onicists I know with children, once said to me, "They're good kids. But
> if their moms hadn't wanted them, they wouldn't exist." He did not see
> much value in passing on genes or creating new generations and pre-
> ferred to work toward a world in which people no longer need to pro-
> create since the extension of human lifespans would maintain the
> human species. Indeed, I have heard some in the community theorize
> that having children is an evolutionary byproduct that could very well

become vestigial as humans come closer and closer to becoming immortal. I have also heard several lay theories within the cryonics community about genetic or brain structure differences between men and
women that cause men to favor life-extension philosophies and women
to favor procreation and the conservative maintenance of cultural
traditions.[13]

One need not live on through one's children if one has the possibility of living
on oneself. Perhaps this is why our guide only describes the letters to the cryopreserved as being written to wives, husbands, and parents—not children.

That is not to say that there are no children. Member A-2789 was only two
years old when she became ALCOR's one hundred and thirty-fourth patient.
After her death from pediatric brain cancer, she was *field preserved* in her home
in Thailand and shipped in dry ice to the United States, where the neurosep-
aration was performed. ALCOR ends the official announcement of this preservation by saying that the patient's "family, extending well beyond her mother
and father, were supportive and have said they plan to also make cryopreservation arrangements with Alcor. No doubt being surrounded by familiar faces
of loving relatives will make the resumption of her life—as we hope and expect
to be happen [sic]—easier and more joyful."[14] The statement also takes time to
mention that this was ALCOR's first cryopreservation in Asia.

While it is not stated in the announcement, elsewhere, our spokesman
mentions something else distinctive about this case: the whole family watched
the cryopreservation procedure. "That's pretty unusual," our ALCOR spokesman says in the same video where he discusses writing letters to the patients,
"most people don't want to observe the procedures." He looks down for a second, and adds, "who wants to see surgeries in a hospital, either?"[15]

§64

Recall ALCOR's concern with long-term preservation, with continuing access
to electricity and chemical coolants. This anxiety is an exercise in foresight,
but there is a bit of hindsight in it as well. The hindsight here is not because
natural disasters have broken the continued free access to these goods in the
past, nor because of prior incidents of anti-immortalist vigilante violence.
Rather, this is a response to institutional cryonic memory, an anxiety left over
from a particularly deep-cutting trauma the cryonics community experienced
decades ago.

This trauma is sometimes referred to as the "Chatsworth disaster"; at other
times, its magnitude is downgraded and made more social by referring to it as
the "Chatsworth scandal." In 1979, nine "abandoned, thawed, and decomposed

'patients'" kept by the Cryonics Society of California were discovered.[16] This discovery began when the father of a seven-year-cryopreserved girl checked to ensure that his child was being properly cared for. After some investigation, the father of the girl discovered that her body had seriously decomposed, and he had her buried in a more conventional manner. The failings of the Cryonics Society became public knowledge later that year when a reporter broke into the tomb where the bodies were supposed to have been kept. The report's description is worth quoting at length, as it conveys not just the perceived horror of the place but also the complete absence of anything like due care, even when one considers the fact that at that point in time there was no sense of what the standards for proper cryonic preservation would be like. The report recounts the following:

> The accessible parts of the crypt were in total disrepair, obviously abandoned and unattended. . . . The stench near the crypt is disarming, strips away all defense, spins the stomach into a thousand dizzying somersaults. The single lightweight lock barring access to the crypt dangles uselessly from a broken clasp. . . . Inside, a 12-foot aluminum ladder leans against a chipped and cracked white-washed concrete floor. . . . Directly under the ladder are two 10-foot-long white capsules, a blowtorch may have been used to sear open a gaping hole in the top capsule, exposing its innards. The body, if there was one, had been removed. Aluminum foil—sometimes used to wrap cryonic suspenders—is littered inside the capsule along with a pair of surgical gloves, discarded in apparent haste. A pair of filthy rubber gloves are on a ledge. The bottom of the capsule is coated with a thick murky slime. The bottom capsule sits intact, warm to the touch. In the middle of the room a stainless-steel capsule stands eight feet tall, its three-foot-long insulating end cap thrown carelessly in a corner. A black body bag lies in the corner, soaked and rotting. . . . Tubes and pipes, attached to nothing, wrapped in thin insulating material, lead into an adjoining room, its freezer-like-door locked from the outside. Warm air gently flowed between the cracks as the door is pulled ajar. Dials and gauges designed to measure liquid nitrogen pressure register zero. A thermometer indicates the temperatures was in the 50s, the high 50s.[17]

Even a year later, a visitor to the site could still smell "the offensive intermingling of rot and DMSO" (DMSO being an ice crystal inhibitor then often used in the cryonics processes).[18] Less grisly, but in some ways equally damning for the cryonics movement, was the fact that a considerable amount of funds had also disappeared. Money in the tens of thousands of dollars had boiled off like the liquid nitrogen from those ill-maintained cryonic caskets. This event has come to be remembered by both the community and by academic commentators as a catastrophic failure.[19]

There is a tendency to drop all the blame for Chatsworth at the feet of a single individual: Robert F. Nelson, the founder of the organization. He certainly had some culpability, having moved from Chatsworth to Hawaii in 1974, apparently making no arrangement for maintenance or provisioning of his patients. There are even claims that he had abandoned care for some of the bodies as early as 1971.[20] The courts certainly saw him as a guilty party, finding Nelson civilly liable for both fraud and intentional infliction of emotional distress. But this singling out of Robert Nelson is perhaps unfair.[21] Most attempts at cryonic preservation conducted in the 1970s came to a bad end. A postmortem of early cryonic failures conducted by ALCOR noted that "of seventeen documented freezings through 1973, all but one failed, while maybe five or six later cases, some of them privately maintained, were later terminated (or were continued under questionable circumstances, such as attempted permafrost interment). In most of these cases, finances were a factor."[22] If cryonics was to persist, attention would have to be paid not just to preservation techniques but to infrastructure, including social, institutional, and fiscal infrastructure.

§65

Legal infrastructure would have to be tended to as well. The ALCOR report on early cryonic failures noted that there was "one notable exception" to the rule that finances scuttled preservation attempts. This was also the only ALCOR-associated case discussed in that report. The report discussed a "woman frozen in 1990 at Alcor (name withheld), whose will, it was later discovered, stated she did not want to be frozen. Her cryonicist husband fought the case through the (California) courts, arguing that the will, which survived only in photocopy, had been revoked, but the decision went against him, and her body was committed to burial under court order in 1994."[23] All this was despite the fact that two ALCOR officers had given testimony that by the time of the preservation, Mrs. [name withheld] had consented to be cryonically preserved. As was noted in some of the ALCOR literature commenting on the legal proceedings, "[i]f there was any good news here, it is that you can expect the judiciary of the State of California to uphold your direction regarding disposition of your remains after legal death. The caveat is that we had better be damned swift in making those directives. Even if full suspension membership had not been completed (for whatever reason), I believe that a signature on any two of the three core documents which comprise Alcor's core paperwork package would be sufficient to produce a different outcome in circumstances similar to those described above."[24]

The word *believe* is telling. That is because, at that time, how the state would react to a disputed or problematic attempt at preservation was still an unknown. This lack of certitude was aggravated by the disaggregated nature of local, state, and federal political organs; in addition to creating multiple potential regulatory and law-code-derived hurdles, this jumbled quilt of different authorities also proliferated the number of actors that could intervene in the cryonic process, and who could place cryonic advocates in peril.

The potential perils include being charged with murder.

Dora Kent was the mother of ALCOR board member and longtime cryonics advocate Saul Kent. She was also not very well. For four years, she had suffered from dementia, a disease that was in effect a slow erosion of not just her mind, but of the physical integrity of her brain. Notes from one ALCOR volunteer at the time described her brain as being "slowly decomposed—an outrage we may realistically hope to halt through freezing."[25] The brain, after all, is the organ where personhood is situated for most cryonicists, and therefore the organ that above all others must be kept intact, making both this case and her continuing life almost tragic. Fed through nursing tubes, plied with oxygen, effectively vegetative, there seemed little quality of life. This condition was not necessarily hopeless from the point of view of cryonic preservation, but success was not guaranteed. ALCOR notes from that time mention a conversation "discussing the possibility of simulating a reconstructed personality in a computer to see if it is psychologically viable before committing oneself to a flesh-and-blood resuscitation of a patient." These notes also have a parenthetical comment that "this issue will hopefully be well thrashed out by the time any of us are brought back from suspension."[26]

Ever since Saul Kent had helped found the Cryonics Society of New York in 1965, he had been speaking with his mother about cryonic preservation, and she had made her wishes to be preserved quite clear. Given Mrs. Kent's state and her unquestioned intent, when she contracted what was diagnosed as a likely fatal case of pneumonia, Saul Kent decided to move her from her nursing house. She was instead placed in the actual physical ALCOR facility, then located in Riverside, California. The staff at ALCOR put her on a gurney and wheeled her into one of the offices there, and an all-volunteer suspension team was kept on alert. They shaved her head in preparation for the procedure. For three days, her condition continued to worsen, and finally, late one evening, she went into respiratory arrest. The notes report that "[t]here was a horrible obstruction in the throat, mucus or something, that produced a rattling sound as the breath was exhaled" prompting "mouth to mouth resuscitation and other heroic measures . . . her color pinked and at length she started breathing again, shallowly."[27] This resuscitation was not done to prevent her from dying; her staying alive was not in the cards, regardless of whether or not she was cryonically preserved. Rather, it was to ensure proper oxygenation and to prevent damage to vital tissue. Even after she was again taking in oxygen, her end, or at least a long pause in her life, was near, and the preservation procedure was at its cusp.

The notes kept by one of the ALCOR volunteers report that following this episode,

> [t]his breathing went on awhile then stopped. The heart rate slowed, weakened, and finally Mike and Jerry separately checked with a stethoscope and reported it had stopped. This was about 12:25 a.m. and very shortly afterward the protocol for freezing was started. The chest was cut open and the stilled heart was observed. Heart massage was begun and oxygenation of the tissues was maintained. At no time after that did I see the color darken as it had after the first respiratory arrest. From there the suspension went smoothly, with about the only complications being (1) one instance of ineffective mixing that led to a higher concentration of glycerol than intended, and (2) the advanced age of the patient causing brittleness of the vascular system. Mainly I filled an ice chest and did a few other incidental chores, then sacked out, to be ready in the morning for the cool-down to dry ice temperature. The decapitation of this patient occurred about 7 a.m., while I was asleep.[28]

Dora Kent had died, at least in the conventional sense of the term, on December 11, 1987.

The preservation process for the dead continued, with people shuttling dry ice back and forth until at least one person's fingers caked. The ice was not just for the woman's head. While her head was being carefully and gradually cooled to (hopefully) minimize ice crystal formation, the rest of her body (excluding her hands) was wrapped in "heavy plastic sheeting," and placed in dry ice (her hands were kept to study the profusion of the cryopreservant). The full preservation process took two more days, after which Doris Kent's head was placed in a dewar. In the middle of a quick meal at Pizza Hut that occurred during the process, one member of the response team joked that they should shred the diary in which the notes on the Kent case was kept.

The diary, left intact, would become an important document. Evidence of small irregularities in the preservation began to pile up. There had been a clerical error during the freezing process, during which the dry ice was not changed every two hours as it was supposed to be; this did no perceptible

damage, but it was worrying. Worse, an attending physician who had volunteered to be present for the preservation had not been contacted when Mrs.
Kent passed. Since there was no attending physician at her time of death, this
became a coroner's case, and it quickly became apparent that the coroner's
office was in no mood to blithely issue a death certificate. Coroner's office
personnel came to ALCOR facilities and took the body away, leaving the cryonically preserved head behind, at least for the time. After a while, a consensus
started to emerge at the facility on precisely what was at stake with the corner's
judgment: this was ALCOR's "greatest crisis," and the "decision of the coroner
could determine when Alcor stands or falls."[29]

The days-long wait for the coroner's decision felt interminable. At one point
there was even loose talk about absconding with the head to protect it, despite
the fact the initial gross autopsy the authorities conducted suggested that there
was no foul play, and also despite the coroner promising that the cranial elements of Mrs. Kent would only be autopsied if it was impossible to determine
the cause of death by investigating the decapitated body.[30] Ten days after Mrs.
Kent's procedure, the coroner's office came back with the determination that
the cause of death was pneumonia, which was what the ALCOR personnel
had claimed from the very beginning.

This finding, though, was not the endpoint that it seemed to be. The press
had gotten word of the event and turned its attention to ALCOR. The coverage
was not kind. Subsequently, the Riverside County coroner announced that he
had (re)opened a murder investigation. The county was now claiming that this
was not death by pneumonia, but rather death by barbiturate overdose, and
therefore a case of homicide.[31] Several raids of the ALCOR facility were conducted, one by SWAT officers. During these raids, everything from computer
records to the medicines used during preservation was seized. Several ALCOR
personnel were taken into custody and interrogated as well, though none were
placed under arrest. What was not taken into custody, however, was Mrs. Kent's
head, which the coroner's office could not find (it would be missing for well
over a year).[32] Eventually, the Riverside coroner's office dropped the charges,
though it ended up being a several-year ordeal. During this prolonged investigation, it took a court injunction to prevent the coroner's office from seizing
the other cryopreserved bodies on the ALCOR premises.

This was the last hostile brush with the law that ALCOR would go through,
though there was always the threat of some public relations setback, a danger
exacerbated by the morbid fascination that some segments of the population
held for cryonics. Years later, a former security worker at ALCOR would allege
personnel there played an improvised game of baseball with the frozen head
of professional athlete Ted Williams. He followed up by cowriting a book, bearing

the not-at-all-purple title *Frozen: A True Story My Journey into the World of Cryonics, Deception, and Death*. The book further detailed this supposed incident, as well as presenting a serious of other irregularities and at least one instance of "death hastening," though the author would later back down from all of these claims as part of the settlement of a defamation suit ALCOR initiated against him.[33]

Such incidents have not been a barrier to cryonics eventually achieving a certain degree of acceptance. As noted by the anthropologist Abou Farman, cryonics was initially dismissed as taboo science, with a great deal of the pushback coming from scientists engaging in boundary maintenance. Despite this, cryonics has become increasingly more mainstream, receiving serious attention from newspapers and periodicals such as *The New York Times*, the *MIT Technology Review*, and *The New Scientists*. Part of this has to do with new medical and scientific techniques, such as medical cooling procedures, or the first (halting) uses of nanotechnology—technologies that make the far more experimental practice of cryonics seem more plausible. Another part is linked to Silicon Valley's growing interest in life-extension technologies, which, while not necessarily related to cryonics, certainly makes cryonics speculation seem more grounded.[34] In short, the success of these other speculative technologies—speculative technologies that have long been imbricated with cryonics—was in the end, success for cryonics as well.[35] But such reconsideration is a recent turn. And while it may well give pleasure to cryonicists who were earlier regarded by the scientific community as beyond the pale, both institutional and personal memory of earlier threats remain. It is no accident that the ALCOR visitors-bay window is made of bulletproof glass.

§66

These technical, political, legal, and sociological challenges gave shape to cryonics as a practice and as a community. It also gave a certain shape to the cryonic imagination.

In the March 1990 issue of *Cryonics*, there was something unusual. *Cryonics* was for a long time one of the chief means through which ALCOR, as an institution, built community and awareness. Designed for the age of snail mail, *Cryonics* landed somewhere between a newsletter and a full-throated periodical in length and quality. It was ideally mailed four times a year, though there were inevitable gaps as the organization struggled to find both staff and some kind of permanent home for the institution as it moved from Indianapolis to Southern California to Arizona. (One of the earliest newsletters, back when the institutions that would later become ALCOR was operating under the name of *The Institute for Advanced Biological Sciences, Inc.*, started with a somewhat plaintive, somewhat defense title: "Yes, We're Still Here.")[36]

For a house organ, *Cryonics* contained a great deal of desperate, argumentative opinion pieces and (sometimes tendentious) essays from a variety of perspectives. In this, the magazine was reminiscent of other technophilic periodicals of the day like *Extropy: Journal of Transhumanist Thought*. In these periodicals, there was a mix of overt pose-taking about libertarianism (the then-dominant politico-economic sensibility in these circles) and free thinking, mixed with technological speculation that was leavened with various degrees of sobriety. But there was one chief difference between the magazine ALCOR published and these other journals. These other technophilic periodicals would jump vertiginously in subject matter, leaping from discussing space colonies to neuro-net-modeled computing to arguments against monogamy. However, the center of gravity in *Cryonics* was always, unsurprisingly, cryonics.

Despite the contentious nature of back-and-forth debates that they often contained, there was also a certain boosterism that suffused these periodicals. That was especially the case for *Cryonics*, which was dedicated to making sure that ALCOR could grow to the level needed to be able to live up to its promise of long-term preservation in the service of immortality. Which brings us back to that odd moment in *Cryonics*. This moment stood out because there was a break in tone from this generally positive attitude about ALCOR as an institution, that is, there was a moment of sharp self-criticism that made for an arresting read. Nestled between snippets with titles such as "ALCOR vs. The Necrocrats: An Update," "Cryonics as an Employment Benefit," and "My Dog Died," there was a description of something unusual: unhappy customers. Or rather, given the state of things, unhappy potential customers effectively turned away at the door. Oddly, this piece was in the third person, and the account laid the blame on ALCOR itself. This piece, however, was not penned by any churlish outside critic. The author was one of ALCOR's former presidents, and at the time, a research director there (he also had his dog preserved there, though he was not the author of the meditation on canine mortality contained in that issue). He complained that ALCOR had a problem with a very particular group: Christians. He recounted the visit of a local woman to ALCOR's site, "a lovely woman with skills of potentially great use to Alcor and a strong desire to participate in our program at every level." Every level meant, literally, every level—the author mentions that she was a licensed nurse, hinting that she could possibly have aided in the messy business of cryonically preserving bodies. She was, in short, "one of the most excited and seriously enthusiastic prospects we've seen recently."[37] She was also "a born again Christian and was wrestling with the issue of compatibility of her faith with cryonics."

This woman's Christianity created two problems. The first problem was ALCOR itself. "I have," the author stated, "in the past, witnessed several rather ugly incidents in which Alcor members were condescending, hostile, or intolerant of the religious beliefs of others. This behavior is inexcusable and it will not be tolerated at any Alcor function or event." But this bigotry was also a two-way street: the woman was warned by her fellow churchgoers that "cryonics was un-Christian and would result in her damnation!"

The article suggested the institution of a support group for Christian cryonicists. The need for a support group suggested that this was beyond the ken of those already associated with ALCOR leadership—this was an encounter with an alien mode of thought, and new networks had to be set up to make cryonics cognizable to Christianity, and Christianity cognizable to cryonics. There were other forms of nonreligious bigotry against cryonics, it was noted. (They complained about the cold shoulder that cryonics was given by the Society for Cryobiology, a professional organization that brought together physical scientists and medical practitioners and had traditionally been embarrassed by any association with cryonics.) But while similar at one level, the problem of Christian revulsion was also different. First of all, a support group could help with the skeptical Christian kin of potential cryonics patients, who, in addition to giving grief to the living candidate, "may also be hostile and non-cooperative in the event a suspension needs to be carried out." But there was also one other difference.

While groups like the Society for Cryobiology can make being a cryonicist feel like being caught in a metaphorical hell here on earth, the author notes, it is the Christians who can "threaten to send us to hell."

§67

ALCOR's Christian problem may seem unnecessary. As the existence of religious transhumanism proves, there are the intellectual and human resources to mount a positive Christian argument for cryonics. But then, there was also a previous awareness of the contours of the conflict that necessitated the incorporation of an alien, faithful apologetics, either. Previous to this *Cryonics* article, the ALCOR cryonics community had meditated on, or perhaps more precisely created, deathism through performative fantasy-projection. And because of the awareness/confabulation of deathism, there is some reason, from ALCOR's point of view, to believe that while it may not always take the contours of a fully articulated Christian jihad, this opposition could be an existential risk as grave as that posed by a county coroner or a hostile press.

We can see this anxiety about religion in the back-and-forth regarding one article, "Programming People for Immortality," that appeared in 1985. The author was not peripheral to ALCOR's history; he was a longtime cryonics activist, a founder of the Cryonics Society of New York, and he is still involved in the movement. He was, in fact, Saul Kent, son of the future cryonaut Dora Kent. Kent begins the piece (in less than classical *in media res*, but more like the literary equivalent of a television episode's cold-open) with a quote describing a Japanese fighter airplane hurtling down toward the USS Franklin. He describes the plane as engaged in what seems to be a dive-bombing run until the pilot continues past the point of safely pulling out and crashes into the deck of the ship. "The pilot, Rear Admiral Masafumi Arima, had dived to death deliberately," Kent writes, continuing, "The Japanese did not know what to call his action that day, any more than the Americans, but Arima was the first of the Kamikazes."[38]

The expected pivot to cryonics does not immediately occur. Instead, we are given the religious history of twentieth-century Japan in miniature. The article does this by providing a *précis* of a book by Edwin Hoyt, a popular nonfiction author who wrote on subjects ranging from American Zen guru Alan Watts to the Nixon clan to the Palmer Raids. Hoyt addressed these diverse topics when he was not engaging in his chief life work of writing copiously about war in the twentieth century. Relying on Hoyt, Kent describes the reasons behind the Japanese turn to kamikaze attacks (too few skilled pilots and far too many planes), about the way that kamikaze attacks were planned and executed, about how these tactics worked their way through the thinking of the rest of the Japanese military (there were also Japanese manned torpedoes and anti-tank soldiers who engaged in what a later age would call *suicide bombings*).

As the article continues documenting kamikaze strategy on up to the end of the Second World War, it seems that the topic of cryonics has vanished beyond any hope of reappearing. It is only in the very last pages of the piece that we finally get to both cryonics and the point. But well before that moment, some sly rhetorical preparation for the conclusion's argument has already begun. As it unfolds, this piece is redolent with religious-invoking terms: the author runs right against the edge of redundancy in breaking down the etymology of the word *kamikaze*, just to let us know how the terms for *God* and *wind* are run together to create the word *divine wind* (the name given to the unexpected typhoon of that wrecked Kublai Khan's Japan-bound invading flotilla). The term *sacrifice* is used repeatedly, becoming a kind of verbal tic (even the aircraft are called *sacrificial planes*). We are informed the pilots "attended their own funerals before their mission." But this is as much about continued life as the end of it. Pilots are told that they "would 'live' forever in the hearts of their Emperor and the Japanese people," and at the moment of combustion, a living torpedo pilot experiences "the 'glory' of this mission by being blown to bits."

This drive toward what Kent describes as *national suicide* is presented in a way wholly suffused with the language of religion.

This religious submission to the death drive is also suffused with a kind of automaticity. As a people, the Japanese "were programmed to die for their country and, in most cases, were perfectly willing to do so with no questions asked. Most Japanese were willing to continue to die even after it became clear that the only way to save their country was to end the war." Part of this *programming* is informed by a fear of what an unconditional surrender to the Americans might mean, but this was also due to the sway of a broader controlling imperative. "The Japanese were programmed for death," Kent says.

To Kent, the Japanese case is informative not because of strangeness, but because it is a limit case for *all* contemporary societies, all of which are also programmed for death, though in slightly less florid ways. The seeming inevitability of death as a part of a natural order, the opportunity to avoid the rigors of aging by instead opting for a moment of glory, and the "subjugation of the individual to the needs of Society," all work in conjunction to make opting for a purposeful death appear sensible. These three elements might be thought of as the *push* factors on the side of mortality. There are *pull* factors too, chiefly the (claimed) ubiquity of the religious idea of a postmortal afterlife, which is described as the "foundation of most religious thought." For Kent, this feature of an undying "exalted state" is responsible for both the popularity of religion and for people's "subservience to [religion's] tenets." The ubiquity of religious vocabulary in the description of kamikazes is no accident; while kamikazes died for the emperor and the state, the mechanism that made it all possible was religion.

And here is where we find the relevance for cryonicists. This religious thanotropic indoctrination means that the immortalist project of longer, fuller, and perhaps even unending life through technological achievements faces a problem: convincing those who are "programmed for death" to join them. As Kent puts it,

> The immortalist dilemma is that hardly anyone has had the courage
> and imagination to shed the death programming that characterizes
> human life. Although there are more than 4 billion people on Earth,
> fewer than 100 of them have actually chosen to engage in hand-to-
> hand combat with death by making preparations for cryonic suspen-
> sion. The evidence is solid and clearcut. Over the past 20 years, tens of
> millions of people in the United States alone have been exposed to the
> idea that it's possible for us to achieve physical immortality. During
> that period, millions of these people have died. A great many of them

could have been frozen, but none of them chose to do so. The truth is that only a handful of people have not been blinded by the deathist programming and propaganda that dominate this planet.

Kent has some suggestions on what can be done, now that "we are finally on the verge of being able to sell the idea to the public": establishing a permanent cryonics facility, opening up an insurance company dedicated solely to offering policies designed to make cryonics generally affordable, making it part of a larger life-extension regimen, earning scientific bona fides for cryonics so that it appears as "a science rather than a cult." But while he does not say it directly, there seems to be one negative plank that would have to be a part of a foundation. That is because, as he states in the pivot from deathist programming to immoralist counterprogramming, "as immortalists we have totally rejected the 'inevitability' of death as well as the 'spiritual afterlife' promised by religion." Immortalism is not just a positive choice for life, but also the denial of religion, the chief engine of the old deathist regime.

§68

A reader might question how complete the escape from religion has been. There is something evangelical about Kent's vision; his account of this reaction to learning about cryonics ("I was instantly enthralled and excited beyond anything I had ever experienced before") certainly sounds as if it was cut from the late American Protestant idea of being born again. (He regularly speaks what we might call a cryonic conversion; at another point, he says that his dedication upon hearing cryonics was like a "lock going into a key.")[39] This is a scientism capable of speaking to the ultimate concerns of religion, as observed by Farman, but also a scientism that expresses itself through a constellation of subjectivity, value, belief, and temporality that resonates with religion not merely in the way it reaches toward eternity, but also in the way it presents (perhaps with a sharpness heightened by hindsight) a singular biographical moment being tied to the discovery of an overarching vocation in the fullest sense of the term.

This expression of immortalism was familiar to the cryonics community, though it was not always considered a good thing, either as an actual structure of sentiments or as a concept invokable for rhetorical purposes. One 1998 essay complains that while some elements of the cryonics community are working hard to better cryonic technologies (the author references an endeavor with the somewhat grandiloquent name "Project Prometheus"), other cryonicists claim that future nanotechnology will make a society so flush with wealth and prowess that "we need take no special effort to improve our cryopreservation methods." The author of this critique, Thomas Donaldson, was a mathematician and cryonics advocate who had once preemptively (and unsuccessfully) sued the state of California in an attempt to win a constitutional right for assisted suicide committed in furtherance of his cryonic preservation.[40] When it came to this easy reliance on future technology, Donaldson felt that he had "heard these ideas before." As to where that was, Donaldson said:

> The answer is the Christian Apocalypse, all dressed up in new clothes. Rather than God, we have Nanotechnology, which will put us into Heaven. . . . This is Nanotechnology as religion. This simplifies lots of questions and problems . . . as long as you have faith.[41]

This reference to Christian apocalyptic faith is not intended as a positive assessment. As Donaldson states in the closing moments of his argument,

> neither God nor nanotechnology will save us. We must join together to save ourselves, and support research to the end, *now*. Doing so will take work of many kinds, as some cryonicists have already learned and all cryonicists should know.[42]

This essay was no one-time fancy. In a different *Cryonics* article published a full nine years earlier, Donaldson mocks those who anticipate nanotechnology creating "a new heaven and a new earth" or reaching "the Third Celestial Sphere."[43] This is not necessarily to reject the historical role that Christianity played. "Science ultimately stems from Christianity," the author observes, later adding that "it's not wrong to be Christian."[44] (Even this statement is controversial; a later *Cryonics* essay reacted to this part of Donaldson's argument by seething that while there are some Christian scientists, "Christianity [as well as other religions] has commonly been an enemy of science.")[45]

Even if our author says that being a Christian is not an automatic ethical error, though, it turns out that thinking crypto-Christian thoughts *is* a transgression. Donaldson concludes by stating that "it is dishonest to oneself and others to think that just renaming everything, and having a slightly different theory of how God works, frees one from Christianity."[46] In these arguments, we have an inversion of Saul Kent's reading of what the stakes are when religion and immortalist technology mingle. In this telling, when religion appears in the guise of science, it does not function as the enabling element of a culture of inescapable death. Rather, it acts as a disabling element that can endorse cryonics, but only at the cost of fostering a debilitating cryonicist delusion about the inevitability of technologically reinvigorated life. Even if it is not wrong, religious cognition is a threat. Religion either precludes immortalism through romanticizing death or it wears immortalism as a mask to neuter itself and eviscerate the likelihood of immortalism achieving what it yearns for.

§69

In addition to there being a scientism opposed through religion, or a crypto-religion expressed through scientism, there also is a scientism that is hand-in-hand with religion, where religion is the basis for and catalyst of scientism. Two issues of *Cryonics* after Saul Kent's identification of religion with deathism, there was a letter to the editor in response. Speaking on behalf of the author and his daughter, the letter writer wanted to make it clear that "Saul does not speak for us when he writes that we reject any spiritual afterlife. We are non-fanatical but staunch, hard-core committed Christians who know and love the Lord."[47]

This would not be the only time the writer, Michel Laprade, would appear in *Cryonics*. In addition to serving as the contact person for the Christian Cry-onicist Support Group called for earlier, a full thirteen years later he would pen a piece called "Can a Christian be a Cryonicist?" His answer was (unsurprisingly) yes, and not only is it possible to be a Christian cryonicist, it is theologically defensible. Laprade starts this piece by recounting a cryonics origin story not unlike that told by others (including Saul Kent): he recounts the moment he stumbled onto Ettinger's *The Prospect of Immortality*.[48] The same kind of lan-guage of religious conversation was used—"I was immediately sold on the irre-futable logic of the arguments presented"—and of a desire planted—"I just sought . . . to have an opportunity to live again sometime in the future."

Despite the formal similarities between religious and cryonic conversion, as Laprade tells it, his initial desire for cryonic suspension "had nothing at all to do with my being a Christian." But as he later came to realize, there were resonances with his religion; as Laprade saw it, cryonics was *pro-life*. As he puts it, "If cryonics is not the ultimate pro-life position, I don't know what is!" This claim allows the author to reframe cryonics in terms of contemporary medical interventions. For Laprade, cryonics is not a denial of some ultimate other-worldly reunion with God, but a tool, comparable to other currently existing therapeutic interventions.

> To me, there is a very short psychological step between a potential cry-onics suspension patient and an accident victim who is asked by the ambulance attendant if he wants 'extraordinary means' taken to allow his survival. . . . I don't think I am trying to circumvent God's plan by asking to be suspended anymore than I am by accepting conventional medicine.

Nor is cryonics too farfetched a proposition to be taken seriously, or at least it is not any more farfetched than other possibilities he sees as credible: "[F]rankly, I am surprised that Christians, who have the faith to believe in our Heavenly Father's spectacular eternal promises to us, lack the faith required to accept the possibilities of His granting us this amazing but (relatively speaking) *insignificant* achievement called cryonics."

Laprade's is a minority report, though it would be a mistake to see him as a minority of one (for example, a later President of ALCOR, Joe Waynick, apparently had a Seventh Day Adventist background). But when expressed in the pages of *Cryonics*, that minority position never goes unchallenged. With the kind of automaticity that seems an inevitable feature of any discussion of religion in the pages of *Cryonics*, there are protests. For example, a long-time ALCOR member's letter to the editor fought back against Laprade's testimony of faith in both God and cryonics. This writer—who, in an ALCOR member profile segment that appeared in cryonics over a decade later, is described as having gone through a teenage "'religious conversion' from Baptist to transhumanist"[49]—does not object to "heterodox Christians who want to change their theology to allow for cryonics," since "Orthodoxy theology is arbitrary to begin with, and anyone is free to invent a new theology." So, Laprade is welcome. Laprade's mode of reasoning, however, is not. "I do have a problem," the writer states, "with the attempt to paper over the difference between the scientific materialist assumptions behind cryonics and the 'supernatural' assumptions behind orthodox Christianity. The two worldviews are just not 'consilient' in any meaningful sense and frankly, theology has nothing practical to contribute to our quest for radical life extension."[50]

Laprade does not care. He has time to bear witness and change minds, and quite possibly a divine vocation to do so as well. "If I were suspended, revived, and lived a thousand years only to repay the process a hundred times, this would be but an instant in eternity's time frame. . . . I believe that the Lord breathed life into me not intending for me to 'check out' the first chance I got. I believe that my purpose is to be a witness for him; common sense dictates that to do this, I have to be *here*. I am saddened by the reported number of cryonicists who are atheist. Perhaps our Father is intending to use me to reach them by having us spend a thousand years together in a dewar."[51]

§70

It would be a mistake to take any of the positions in the pages of *Cryonics* as canonical (to use a perhaps inauspicious term) of who and what ALCOR was. Recall that these pieces were written in a time when the number of active cryonicists was measured in the hundreds, when there was an air of scandal and the macabre to the practice, and when there was no clarity as to whether any particular act of cryogenic preservation might be attacked by some coroner or district attorney. We might even say that this was a time when cryonicists were persecuted for their beliefs.

This was also a time when cryogenics was thin on the ground in the most literal of ways. There were a few specific geographic centers of cryogenic activity, such as Scottsdale (or Riverside before ALCOR's move), and the presence of a cryogenics facility in a city would sometimes draw individuals curious about volunteering in cryonic preservation to the region, anxious about being too distant from the only mechanisms that could turn death into merely the prelude to earthly eternity. These centers were also places where both these future immortals and the facility that was their gateway to forever could catalyze the interest of others to perhaps also participating in this new medical art. But for the most part, cryonicists were scattered and interacted with each other through a sort of twentieth-century semi-digitalized version of the republic of letters that was cobbled together out of periodicals like *Cryonics* and *Extropy*, dial-up computer bulletin board systems (BBSs), and occasional in-person conferences.

It is no surprise that a community this dispersed, this small, and this young would hold a multitude of theoretical positions. This is not to say that one could not identify regularity in these positions, while there were those who were willing to court Christians interested in cryonics, and while there were even cryonicist Christians, antipathy (and sometimes antagonism) toward them was thick and always on the ready. Given the air of rationalist free-thinking libertarianism that suffused the movement in those early days (libertarianism being a perfect position for a group then on the knife's edge of both legitimacy and legality), it should be no surprise that religion, understood as collectivist, conformist, and moralistic, could serve as some defining, oppositional other to them. To many immortalists, religion was an imagined space where cryonics values are completely inverted, allowing organized faith to serve as a mechanism of negative identification.

But even here, we have to acknowledge that we are dealing with a moving target. Many of these positions seem longstanding in the extreme; consider, for instance, that Thomas Donaldson's two articles accusing some cryonicists of being crypto-apocalyptic Christians appeared a decade apart, or that Michael Laprade's article-length apologia for Christian cryonicists appeared a full thirteen years after his first letter to the editor asserting that very same point. But as the cryonics movement matured, and as ALCOR solidified as an institution, other prominent cryonicists seem to have undergone a shift on the matter of religion.

The Oxford-educated Max More (originally christened Max O'Conner) exemplifies reconsideration of the issue of religion and cryonics. More—alongside the equally artfully named Tom Morrow—was one of the cofounders of the Extropy Institute, the publisher of *Extropy* magazine. One central historian of transhumanism also credits More as the person who "wrote the first definition of transhumanism in its modern sense."[52] Cryonics was not central to More's vision in the early days, or at least cryonics did not exhaust his imagination. It was not that he lacked a formative experience with cryonics. He was the founder of ALCOR-UK in the mid-eighties, which was an organization that was set up to provide "emergency services" as well as "sign ups, insurance, legal issues," and also to deal with the press.[53] Soon after this, he moved to the United States to earn a PhD in philosophy from the University of Southern California. Despite his early interest in cryonics, his writings at the time were more concerned with a broader human expansiveness, exemplified by principles such as self-transformation, dynamic optimism, intelligent technology, and spontaneous order.[54] To the degree that guiding imperatives this broad could be concretized, they included some of the transhumanist interests we might expect, such as consciousness-uploading, and some concerns regarding technical issues that have since become a standard part of mainstream AI research, such as neurocomputing and genetic algorithms. His early work also included material that seems farfetched even by transhumanist standards, such as a serious discussion of how one might engineer a science-fiction style time-travel device, drawing on special relativity, quantum mechanics, and an understanding of causal loops based on the logic of computer circuitry. But cryonics was a core element in his early work. And not just cryonics the technology, but *Cryonics* the periodical; discussions had a repeated tendency to spill over from More's periodical to ALCOR's, with *Extropy* publishing a rebuttal or commentary to something that originally appeared in *Cryonics*, though the sequence just as often ran the other way around.

In the early days, More's attitude toward religion might charitably be characterized as somewhat disapproving. One of More's earliest thought pieces in *Extropy* was called "In Praise of the Devil." True to the essay's title, it celebrated the Devil, even while denying that the Devil "exists in the sense that you and I exist."[55] This lack of existence did not mean a lack of importance, however. Rather, as a symbol, Lucifer was a "force for good" (with *good* defined "simply as that which I value"). Lucifer's value lies in how he challenges God, whom More describes as a "well-documented sadist." In More's pocket version of *Paradise Lost*, what in the Mormon religion is called *the war in heaven* is described thus:

Probably what really happened was that Lucifer came to hate God's kingdom, his sadism, his demand for slavish conformity and obedience, his psychotic rage at any display of independent thinking and behaviour. Lucifer realised that he could never fully think for himself and could certainly not act on his independent thinking so long as he was under God's control. Therefore he left Heaven, that terrible spiritual-State ruled by the cosmic sadist Jehovah, and was accompanied by some of the angels who had had enough courage to question God's authority and his value-perspective.

This narrative means that we have to understand the Devil as a champion of free thought and unshackled rational critique and discovery:

> Lucifer is the embodiment of reason, of intelligence, of critical thought. He stands against the dogma of God and all other dogmas. He stands for the exploration of new ideas and new perspectives in the pursuit of truth.

More also adds that these virtues come bundled with others: a love of pleasure, of selfishness (presented as the antithesis of a self—abnegating altruism), and of self-responsibility (More states that the idea of Jesus dying for one's sins is, in particular, an assault on this last virtue).

§71

An aside: There are many Mormon readings of the *war in heaven*, the battle between God and Satan that ended with the latter cast out.[56] But in almost all Mormon Church and folklore accounts of that war, the bones of the narrative are the same. Satan's plan was for all humanity to renounce our agency, and for Satan himself to take on the messianic role that was to be given to Christ; this plan would guarantee that no one would lose their spiritual way on Earth, but at the cost of the educative experience of working the will and testing righteousness that was the point of Earth in the first place. In the Mormon rendition, Satan's plan was rejected not merely by heavenly hosts but by us all. We were able to reject Satan's plan, and accede to the one put forward by Jesus, because we were present there as well, as premortal spirit children of God. As part of Jesus' plan's later execution, though, we had to erase our memories of premortal life. But such forgetting is worth it, as it allows a pure test of our character, and a full experience of depravity. If we are to return, thrive, and become Gods ourselves, then we must have both freedom and sorrow. As Brigham Young put it: "You cannot give any persons their exaltation unless they know what evil is, what sin, sorrow, and misery are, for no person could comprehend, appreciate and enjoy an exaltation upon any other principle."[57]

It would be a mistake to put too much weight on it, but Max More's Satan and the Mormon God seem more kindred than strangers.

§72

Max More's celebration of the Devil may read as a bit of libertarian grand-standing. Despite its tone, More advises us that when he is championing the epitome of evil, he is not *entirely* serious. The piece is, after all, a part of a grander pattern of More trying to incite a reaction. But it appears that, at least at that time, to More's mind this sharp turn against Christianity and Christian virtues was indissociable with *extropianism*, with transhumanism, and with cryonics. As previously mentioned, More is sometimes credited as the originator of modern transhumanism; scholars usually locate this mental parturition in an essay entitled, fitting enough, "Transhumanism: Towards a Futurist Philosophy" that appeared in its original form in a 1990 issue of *Extropy*.[58] And while transhumanism is the avowed subject, the essay's effective topic sentence suggests that transhumanism can only be understood through a juxtaposition with its other, religion.

This essay, seen as marking transhumanism's ultimate move into its contemporary form, starts out with the claim that, despite undergoing initial moments of "explosive expansion in knowledge, freedom, intelligence, lifespan, and control over experience," our species "persists in old conceptual structures which hold us back. One of the worst is religion." More promises that "in this essay I will show how religion acts as an entropic force, standing against our advancement into transhumanism and our future." And to do that involves first defining religion (providing definitions for already familiar folk-categories is one of the tics of early transhumanist, extropian, and cryonicist writing). For More, religion is a cross of Tylorean thinking, Weberian longings, and Durkheimian activities; he states that "religions hold that there is a god or gods which give our lives meaning by assigning us a role in a grand plan created and controlled by external supernatural forces. Our assigned function is to obey and praise these forces or entities. However, the essence of religion is faith and worship rather than any belief in a god."[59]

Essence, though, should not be confused with causal grounds, function, effects, or dangers. Religion has two origins. One is the neurological architecture of the brain. More mentions in passing Julian Jaynes, who hypothesized that humanity only integrated the two hemispheres of the brain three thousand years ago; prior to that, right-brain cognition was only received as voices and visions that seemed otherworldly, and thus often *divine*.[60] The other springboard is religion as a hypothesis, a way of accounting for phenomena that otherwise would go without any understanding—as More puts it, "in the absence of scientific explanation a religious or theistic explanation was almost inevitable." Religion-as-explanans was also present in the modern day, though More says that this aspect of religion "had given way over time to the superior resources of empirical science." To the degree that religion has any contemporary power, that power is therapeutic, reassurances of a divine plan in the face of setbacks or failure; in a similar vein, religion gives meaning and consolation, acting as a "spiritual band-aid." This consolation effect is especially present in promises of the hereafter, or as More succinctly states it, "So long as you obey the rules and believe you will be rewarded; you needn't be too concerned with being a loser." It can also act as a mode of creating and expressing moral categories and suturing the otherwise isolated subject to something grander:

> By 'letting in the holy spirit' or some other link to a divine being or force, one steps beyond the confines of one's self as it is and connects into a meaningful condition. This feature of religious belief is related to its explanatory role since the being or forces which provide the meaningful structure also have important effects—such as creating, sustaining, structuring, and destroying humanity, the planet or the universe.

Despite this capacity, though, religion still is a failing gambit: "As a strategy (generally unconscious) to create meaningfulness, religion is a failure." The only way in which religion seems to be working is as a control mechanism, as an "opium of the people" (to draw on Marx). And even here the success of this function is not guaranteed; More notes that "the radical and disruptive nature of some religious movements," and the "role that religions have sometimes played in undermine statist power" mean that no matter how much the hidden lever-pullers controlling society work to stabilize the situation, there is no guarantee that religion will not act as "a rival authority rather than a collaborative one."[61]

Contemporary religion, then, strikes out every time it's at bat. Not only is it failing, but it is being actively eaten away by the scientific education of the population (though More makes the typically libertarian complaints that scientific education "is extremely poor in our monopolized and primitive state schools").[62] Even what looks like religion's successes, such as the birth of new-age religiosity and American televangelists, is an illusion, a minor uptick on a curve that has been slouching downwards ever since the Middle Ages. For More, There is an "urgency of the need to replace religion with another form of meaning-fostering." One might wonder why More believes transhumanism should take religion seriously as an opponent at all. Part of it might have to do with a certain animus toward religion running back to his childhood. (Remember More's anti-papal explanation for why he dropped his natal surname of O'Conner.)[63] Part of it may be that religion still has a certain psychic hold, despite all his protestations; there is an odd moment in the essay when More takes a break from cataloging all the ways that religion fails and lacks any truth value to observe that "[e]ven if reality contained the entities and force claimed to exist, any remotely objective meaning would still be absent." He explains his hypothetical denial of God, even in the case He existed, thus: "Being a trivial element of a plan would not satisfy us," he says, adding that "we want to be near the center of the plan and to play an important and positive role."[64] This statement that even if religion is ontologically valid, it is still ethically void, could be read as the author looking over his shoulder while he speaks, just to make sure that ancestral Gods are not still somehow in the room. But it could also be seen as being perfectly in line with the thoughts of someone who was only *somewhat* joking when he praised the Devil: freedom from tyranny and an antitheist ontology could be two independent tracks.

For More, the driving reason that religion must be replaced is not that it is ethically empty, however: it is rather that religion is a force for entropy. This is because religion, all religion, is irrational: "This [irrationality] is true not only of traditional religions such as Christianity and Islam," More tells, but also of "their offshoots such as Mormonism."[65] This irrationality hobbles human capacity to build, improve, and explore; the resignation to the accepted order of things that More sees in religion seeps away the energy needed for these constructive projects. Transhumanism means going beyond the human, into unprecedented and perhaps unimagined ways of being. More calls for a *dynamic optimism* needed to *build* the possible unheralded paradises implicit in the transhumanist project, and this optimism is eaten away by the religious fantasy of passively *receiving* paradise, regardless of whether that paradise is located in this world or the next. He states that we must "dislodge virulent religious memes."[66]

There are elements in this piece that are particular to More as an individual thinker, such as its strong Nietzschean flavor; More is at pains to say that values are constructed, rather than found or received, and the transhumanist who has overcome his species-being is explicitly likened to the Nietzschean Overman, who has overcome his humanity. But in many ways, the antireligious thought in More's essay is not unlike other antireligious transhumanist thought, whether seen in the pages of *Cryonics* or decades later in Zoltan Istvan's quixotic presidential campaign.

This similitude in antireligious transhumanist thought may be for genetic reasons, if we credit this piece by More as the ground zero of contemporary transhumanism. Here, secular transhumanism's antipathy to religion would be a founder effect—an early decision by transhumanism's founders encoded into transhumanist logic. Origins and destinations, though, are not the same thing. In 2010, the executive director and president of ALCOR stepped down after a decade of service; the announcement states that she was leaving to pursue her law degree full time, something that she felt "[would] put her in a better position for ALCOR's long-term benefit."[67] And the incoming chief executive officer? Max More. Though, in at least one crucial way, it was certainly not the same Max More.

§73

In September of 2009, in the Tuscany region of Italy, the Diocese of Pistoia held its twenty-third annual theological week. Throughout each week, participants address one theological question related to contemporary existence. The discussion in 2009 was "the first study in Italy aimed at systematically addressing the revolutions scientifically derived from the combination of genetics, robotics and nanotechnologies." Specifically, the conference was concerned with what would happen after death was *exorcised* by the usual technologies anticipated by singularitarians. As one of the announcements for the event humbly stated it, the presenters would consist of "some of the most important Italian scholars," including representatives from the Italian Institute of Humans Sciences in Florence-Naples, the *Instituo Superiore di Science Religiose* in Florence, and even from the Pontifical Lateran University. This panel was billed as addressing "the idea of earthly immortality."[68] On the last day of the conference, the floor was opened up to electronic missives. One of those missives was penned by Max More. Though he is not listed as one of the speakers in the Italian language material promoting the conference, Max More had written an essay to be read out loud (albeit in an Italian translation). The document was titled "Why Catholics Should Support the Transhumanist Goals of Extended Life."

At first, it seems that More has not been mellowed by time. The essay starts with the antipathy toward religion that might be expected from someone who praised Lucifer as the model for rationality; after expressing the personal importance to him of intellectual honesty, More continues, "I am not religious. As the founder of modern transhumanism, I am a rationalist and do not see good reason to believe in the existence of a being who is omnipotent, omniscient, and perfectly good."[69] But More also builds off his rationalist self-presentation to present himself as someone who has "enormous respect for St.

Thomas Aquinas—undoubtedly the greatest of all Catholic theologians."
Aquinas, presumably like More, is a figure informed by virtue ethics (though
they most likely have a different sense of what constituted virtue). And it is
virtue ethics' interest in human flourishing, and a Catholic commitment to
the sacredness of life, that allows for More's transhumanist olive leaf regarding
the position of extended life.

Note the use of the term *extended life*, which More consciously and positively
contrasts with *physical immortality*. Instead of extropian, open-ended, creative
Nietzschean thriving for all eternity, the phrase *extended life* suggests we are
now merely kicking the can down the road for more time—though possibly
for a great deal more time. In this presentation of immortalism, the grim reaper
still waits, though not indefinitely. "Even if we succeed in fully understanding
and conquering the aging process—as I believe we probably will in the coming
decades," More writes, "our life spans will continue to be limited by factors
such as accident, murder, and wars." Even if these are avoided, all such success
means is that death's scythe must wait either for the closing moments in cosmic
life cycle or for the second law of thermodynamics. As More puts it, "[e]ven if
we live until the far-future day of implosion of the universe, that falls infinitely
short of forever. A trillion years is but an infinitesimal fraction of eternity . . .
literal physical immortality, then is probably not an option."

But this does not mean that this extra time is wasted. Extra time for More's
extropian transhumanists means more time for increasing transhuman capacities in numerous categories (including *transhuman spirituality*, a telling new
addition to the list). More suggests that for Catholics, extra time means a longer
time to practice "virtue and the duties of human beings to serve and glorify
God." Extending life through contemporary medical technology—an activity
that More portrays as being in no way problematic for Catholic moral reasoning—operates on the same measure as transhuman immortality longevity, even
if the amount of extra time won is different. It is just another instance of the
longstanding Catholic imperative to "stand behind efforts to alleviate the
suffering of disease and the maintenance and restoration of the healthy, flourishing physical being gifted to us by God." After all, More notes, Jesus was a
healer.

More's argument has a deft beauty to it. If life is a good, and if death is the
enemy, then more life is by definition better. It also treats the religious as being
susceptible to reason; instead of being caught in a failed strategy, programmed
for death, they can be entreated to realize that not only is the transhumanist
project not anathema, it is actually, in important ways, their own project as
well. And this shift in how cryonicists view religion seems to suggest a shift and
a democratization of cryonics itself. With this instantiation of transhumanism,

the Christian volunteer is welcome at ALCOR after all, and it seems as if Laprade would not, after all, need the full "thousand years in a dewar" to reach his fellow cryonauts.

The victory is not total, though, and there are a few seams in More's argument. Even as he begins by making sure he is not identified in any way with the people he is addressing, More is willing to address his Catholic interlocutors as reasonable in the same way that Aquinas was reasonable. But the contours of the project are warped in that immortality is taken off the table. More also views his audience as being susceptible to swallowing hooks hidden in some Biblical red meat. Biblical Gerontocrats like Noah, Jared, and Methuselah get name-checked, as do their purported lifespans (950, 962, and 969 years, respectively). "Whether we take those ages literally or metaphorically," More says, "the Bible seems to suggest that our current life spans are not as long as those of people clearly favored by God." This is not extropy recoded as virtue ethics, but as an appeal to authority (the exact sort of reasoning the Luciferian More rejected years earlier) predicted on a different logic than that which animates transhumanism. And More also hopes that the promise of a different immortality will make transhumanism more attractive, even if it means turning the posthuman paradise into a spiritual endurance test: "To the extent that the world remains imperfect—and far inferior to Heaven—a longer existence in the physical world might perhaps be regarded as a milder form of purgatory." This is not a world that justifies itself by its lack of limits, but a world that has trying limits, justified by some other realm. But even if secular, Nietzschean transhumanism and religion aspiration are always informed by different logics and different teloi, to the degree that each can see some value in the other, then on this side of either the apocalypse or the heat-death of the universe, they can be fellow travelers for a while.

§74

In 1992, the letters section of *Cryonics* had a particularly heavy back-and-forth about the just-announced plan to move to Scottsdale. It is safe to say that the tone of the conversation was a bit skeptical. There were questions as to whether Arizona was truly as seismically tame as the ALCOR board had stated, about whether California regulations were really as firmly against cryonics as some board members believed, about whether Arizona actually had a cheaper operational cost than California (the necessity of ice chests for buying groceries in Phoenix was mentioned, and the presence of poisonous insects in Arizona was also an issue). At that time, Arizona also lacked any doctors willing to cooperate with ALCOR, which also seemed important, given the trouble caused by the lack of an on-site physician in the case of Mrs. Kent.

Among the list of concerns, though, was an anxiety that government in the state of Arizona would be a poor partner for ALCOR. Government in Arizona was depicted in one letter as being "even more corrupt" than California. But there was also another alleged problem with the Arizona government. It was hinted that State government was inherently unreliable as far as ALCOR was concerned because not only was it crooked but also "more religiously oriented." The specific religious population that was seen as an imminent problem was not sunbelt evangelicals, despite all the anxiety about "Christians," or the more-than-four-century-long Roman Catholic presence in the American Southwest. The writer states that the government of Arizona could not be trusted, "given the large Mormon population."[70]

§75

It is the summer of 2015, and a new episode of a weekly podcast is released: the thirty-sixth episode of the (now discontinued) series "Back and Forth with Shawn and Larry King."[71] The Shawn in the title is Shawn Southwick, an "actress, singer and former homecoming queen at North Hollywood High School," as well as "a self-described 'good Mormon Girl.'"[72] Larry King is the crackle-voiced octogenarian of television and radio fame who has, since CNN broke off their relationship with him, been attempting to reinvent himself as an internet personality.[73] Shawn and Larry King are married; even a cursory listen to their back-and-forth suggests that their married life is, at best, troubled (they would eventually divorce in 2019).[74] In the introduction to the show, an announcer (in a pitch-perfect drive-time radio voice) promises that the podcast will "pull back the curtain on what goes on in their home life." In the episode at hand, this is a promise quickly fulfilled. Just minutes into this episode, Larry King announces in a matter-of-fact voice that "today's topic is when Larry passes." Shawn jumps in to add this additional bit of information about this topic: "By the way, gang, Larry has been so excited! He's like a little kid. All week he's been rubbing his hands together in anticipation and excitement."

There *is* an air of excitement in King's voice, but Southwick's voice is not without a shade of judgment. The reason for her degree of reserve quickly becomes evident. King starts to discuss more about the reason for this subject coming up now. "They're doing a documentary on cryonics," he states. "That's where they freeze you and—"

"Chop your head off!" Shawn interjects.

"They don't cut your head off anymore," Larry replies. "They can, but they don't in my case." Here his tone is certain. After mentioning a few centers, including one in Phoenix (a mistake—he means Scottsdale), he adds that cryonics is "for those of us who don't believe we're going anywhere, and I'm one of those."

"You're going somewhere. You're going to go melt," Shawn counters. At this point, the listener may become uncomfortable; it turns out that seeing what lies behind the curtain of Shawn and Larry's home life is pretty much having a front-row seat to a domestic dispute.

She starts out suggesting that cryonics is a mismanagement of resources, saying that "it is a total waste of time and money." Larry starts to refute the idea by saying that "the extravagant waste of time is these funerals and these extravagant—" but he is cut off again.

Her other talking point is to question the integrity of cryonics institutes: "Larry, that place in Phoenix? Bodies. Melting, all over the place."

"You're full of baloney," he snaps.

"Are you kidding me? There was an exposé." He asks for the exposé in a way that suggests that one must provide proof immediately of a point or acknowledge falsehood. But Southwick continues; she still has other talking points she wants to cover. The issue of memorialization comes up: "Where are we going to visit you? Where are your children going to go see daddy?"

"They'll go to visit my vault."

"Oh, great! And do we get to see you frozen? Do we get to see icicles dripping out of your nose? Or are you in like a box in a drawer somewhere? We're talking about it—let's talk about it."

"I'll find out," King says, attempting to shut down the issue of what physical place there will be where he can be mourned for by his kin.

Another point of contention involves religion. Larry King has already stated that he does not believe in heaven, and he observes that neither he nor his wife have been there. It turns out that this claim is also a matter of dispute. As Shawn says, "I haven't been there. But I've been there before. I believe that I've existed before this." In this conversation, heaven comes hand-in-hand with God, who, Shawn explains, gives a person life as a gift.

"I've never met him," says King, obviously referring to God.

Shawn corrects him. "Oh, yes, you did. You just don't remember." (This allusion to the Mormon doctrine of premortal life and the war in heaven seemingly goes right by King.)

On this issue of religion, Larry King is not content to merely play defense. He has planned a line of attack. At a different point in the conversation, he starts his move by asking the sort of establishing question a lawyer asks when setting up a witness in the courtroom: "You are a devout Mormon?"

"I am," she says.

That statement is Larry King's opening; there is a sudden edge to his voice that suggests that for a moment, he believes the trap is sprung and the prey has been captured. "I have a whole thing here," he says (misspeaking slightly). He immediately starts reading, "Mormons favor using medical technology to preserve and expand life—"

"Of course," Shawn interjects.

Larry King immediately replies by reading the statement, "the LDS Church has no position on cryonics."

"Exactly! But I do have a position!" Shawn Southwick is not going to give ground.

That does not stop Larry, however. His voice rises in volume as he says, "but it has an official position on euthanasia and on prolonging life, allowing for cryonics!"

This is too far for Shawn Southwick. She senses a snake in the grass. "Wait, wait, wait, back that up, that is, that is technically worded, it is so tricky. Who wrote that? Was it a Mormon?"

"They got this from the Church." Larry King sounds testy. It would appear that this line of discussion is not the magic bullet that he perhaps thought it might be.

King's statement also gives Shawn Southwick what she needs, as she starts to finally shut her husband down. "Ah, they got it! Then they reworded it so that they could—no, that's baloney." This rejoinder ends this particular religious leg of the conversation, though the more general debate on cryonics continues for quite a while longer. Even as the discussion draws on, though, it seems as if King's attempt to use Mormon doctrine against her in a moment of intended intellectual jujitsu still rankles. Minutes later, she will not let it go. "I didn't realize that you'd come equipped with, you know, printouts and data and all these things. What kind of crazy person are you?"

Southwick's last question goes unanswered, as does her question about the origin of the document King was reading from. That does not mean that the answer to the question is unknown, however. It turns out that while it was not from the Church, the document *was* written by a Mormon.

King had been (indirectly) reading from a blog post by Lincoln Cannon, the then-president of the Mormon Transhumanist Association.

§76

The email that was forwarded to Linclon Cannon originally came from Larry King's executive assistant and was sent out by ALCOR on the same day that the podcast was recorded. The urgency of the email was communicated not just by the terseness of the message, but by the time frame that Larry King's representative had to work with:

> Larry will be discussing this today on his podcast. Could you send me a page with arguments as to why it is ok according to the Mormon faith. If you could put it in bullet point form, that would be great. I need to get to him by 1 pm. Sorry for the rush.

When Max More forwarded the message to a small handful of MTA members that he knew from a prior speaking engagement at an MTA-organized conference, he added a bit of information about the relevant backgrounds of the major players, including an understated sketch of Southwick's position on the topic at hand: "Although Larry King is a secular Jew, his wife is a Mormon and isn't happy with his interest in cryonics."

It happened that not all the MTA members that More contacted were themselves Mormon, so the scope of people cc'd in the evolving conversation was expanded to include more LDS-adhering members of the MTA. Lincoln Cannon, one of the MTA members originally cc'd, responded with something properly enumerated, laying out the following four propositions:

1. The LDS Church has no official position on cryonics explicitly, but it has an official position on euthanasia and prolonging life that implicitly allows for cryonics.
2. Mormon scripture establishes a mandate for resuscitation.
3. Mormon scripture teaches that God wants us to use ordained means to participate in the work of God, which would include the mandate for resuscitation.
4. Mormon scripture teaches that science and technology are among the means ordained of God.

These points did not appear by themselves. Each point was accompanied by evidential chains. The first point was backed by a series of links to Church newsroom announcements and other official statements. Propositions two through four, though, grounded themselves in chapter and verse references to the Bible, the *Book of Mormon*, and the *Doctrines and Covenants*. (The *Doctrines and Covenants*, or DoC, is a collection of various revelations, primarily received by Joseph Smith, though it also includes revelations from later "Prophets, Seers, and Revelators" for the Church of Jesus Christ of Latter-day Saints.) While the actual text of one Church announcement is quoted in support of the first point (a passage that says that judgment about what constitutes reasonable life extension is "best made by family members after receiving wise and competent medical advice and seeking divine guidance through fasting and prayer"), the rest of the citations are represented as abstracted, declarative statements. They are planks in support of a syllogism, a series of axioms and inferences that lead to the mathematical statement that serves as the culmination of a proof. They are crystalline. And while More took those statements and sanded them slightly at the corners before forwarding them to the podcast's executive assistant, they were still in effect Cannon's arguments that constituted the "printouts and data and all these things" that King held in his hands when he was in the recording booth.

While the conflict was no doubt exacerbated by other unknown dynamics in their marriage, Southwick's primary objection was to the substance of King's arguments regarding cryonics; there can be little doubt about that. But in the moment of confrontation, it was also the form of the argument that Southwick objected to. What to her was "technically worded . . . [and] so tricky," what was "reworded" so that "they" could advance a "baloney" claim, was something else for Cannon. For him, it was a defense that religious claims could be mobilized in favor of transhumanist arguments. (Interestingly enough, the occasion of Cannon's original blog post was an online complaint by Zoltan Istvan that a "Canadian" law regulating cremation, interment, and funeral services was "anti-cryonic." Zoltan Istvan blames this bill on "religious opposition" to cryonics, though he acknowledges that the origins of this bill are "murky.")[75] But it was also a reasoned claim stating a truth that stood fast regardless of a particular audience or a specific inciting event. The blog post that Cannon drew on when formulating these talking points was originally put forward on his website and was also syndicated in *h+ Magazine*—a website operated by h+, the largest transhumanist umbrella organization—as well as on the website run by The Institute for Ethics and Emerging Technologies, a nonprofit think tank that often takes up transhumanist issues.[76] The same argument was presented to convince a Mormon that transhumanism was religiously acceptable and to convince transhumanists that religion was at least potentially acceptable to transhumanism as well. Like the Roman god Janus, the argument faced both directions. But as shown by Southwick's rejection of cryonics and Istvan's continuing hostility to religion, the fact that the argument faced both directions could be read as indicating that it had no real center. And there was no guarantee that the argument's gaze in either of the two directions would be returned.

§77

Much like the early cryonicists, the Mormon Transhumanist Association maintained an online bulletin board. Over time, the importance of the bulletin board has faded as more of the conversation has taken place on social media. And while the bulletin board still dutifully reports content such as updates to the website, the MTA's weekly newsletter, and various news announcements about conferences and other association events, it has become just another faded digital medium, sitting in the same cybernetic scrapyard as the MTA's former presence on the *Second Life* platform. Even so, the bulletin board stands as a testament to the group's history. Looking at back entries, one finds what one would expect of a future-interested, technology-friendly bulletin board: early postings of new stories about bitcoin (there are many early adopters of crypto-currency in the MTA); discussions of the reputation economy of the internet; people sharing the word on new books about transhumanism, and so on.

Cryonics floats to the top of the listserv every so often. One post asks if any LDS members have arrangements for cryopreservation. At that time, no one reports that they do. It is mentioned in the back-and-forth postings that at one moment around 2011 or so, the board of the MTA debated whether they could add a premium tier of membership where annual fees would be used to pur-chase a pooled insurance policy to cover members' cryonics expenses at ALCOR. It would not cover all the costs, and it would have to be limited to one particular stratum of membership, but it would, in theory, allow for a sub-stantial discount. (This proposed membership perk did not come to pass.) In bulletin board discussions, one of the organization's board members laments that the MTA probably could not provide sufficient numbers of top-tier paying members for the plan to work. The same person also notes that he does not "see a ton of incentives for providers to offer group discounts." Like many threads, the conversation withers eventually. The thread ends with someone asking whether "a piece of hair is sufficient for reanimating," explicitly refer-encing the end of Steven Spielberg's 2001 movie, *A.I. Artificial Intelligence*. It is hard to know if that last post is facetious. No one replies to it.

Other posts are speculations about the validity of cryonics, sometimes in the form of somewhat breathless reports of advances in the field. Other times there are complaints about the expense; one member complains that cryonics is too costly for him to seriously consider and is poorly regulated to boot. (Years later, he would bite the bullet, and purchase a family membership at ALCOR). In the same thread, someone mentions that even though cryonics comes up with some regularity in MTA discussions, there seems to be much less discus-sion of such suspension than in other transhumanist communities. Unlike the back-and-forth that led to the MTA's stillborn cryonics initiative, these conver-sations float in a sort of abstract plane, never descending to address any event or a turn in a specific institution.

There are exceptions. Someone reports a death in the family of one friend of the group and suggests that some MTA members come over to the next of kin to inform them of the cryonic option. The person delivering the news suggests that the visit should be made by at least two members, perhaps a faint echo of the Mormon missionary imperative to proselytize in pairs, or perhaps just a sensible tactic to make a seemingly outrageous idea less bizarre by show-ing that while rare, cryonic preservation is at least a shared concern.[77] Another member posts a reminder on the bulletin board that there are two companies that offer biostasis, naming ALCOR and the Cryonics Institute, located in Michigan. The conversation drifts off into the technological subjunctive again with posts of news articles about cryotherapy from technology new sites, all of which have a slightly speculative cast. Like so many other bulletin-board threads, this one closes before the discussion is exhausted, giving a sense of a conversation in a party that fades away as the huddle of guests breaks apart.

§78

Some bulletin board posts, though, are different.

One, posted around 2015, starts with the line "My Mom just died, [s]o apol-ogies if I shouldn't be trying to communicate my current feelings in this way to all I love and trust so fully."

This is a potent opening line, all the more potent when a little context is added. The title of the post is "What will my Mormon Seminary Teacher Sister Think about my Dead Mom." The term *seminary*, as used here, does not mean what it does in other contexts. In Protestant, Catholic, and some Eastern Or-thodox contexts, *seminary* refers to either a school or to a program within a school that awards postgraduate degrees in topics concerning religion, usually informed by the views of a particular religious denomination, tradition, or sensibility. This education is almost always designed with the intent of prepar-ing the student for a position as professional clergy. In the Church of Jesus Christ of Latter-day Saints—a religion that is, in essence, a priesthood of all male believers, and built around volunteers instead of religious career profes-sionals—seminary is something else. Here, it is the term used for coed religious education programs for teenage Mormons, designed to supplement to their secular education. Seminary courses are often scheduled early before school, sometimes after, and sometimes in areas with high concentrations of Mormons, classes are scheduled during release time, with students leaving their school during breaks in the day to attend religious education. While there *are* other Church and private programs more like the traditional seminary training offered for more advanced students, these classes are instead taught by the rough equivalent of uncredentialed high school teachers (and sometimes unpaid, uncredentialed instructors). They are unlike traditional non-Mormon seminaries in that the instructors are not sharing technical points of theology, the grammar of Koine Greek or Aramaic, techniques of public speaking, or the organizational knowledge needed to operate a small to mid-sized institution. The seminary instructors are instead sharing core scriptures through daily supervised devotional readings intended to help the students (in the words of the Church itself) to "recognize and follow the Holy Ghost's promptings" and "learn and feel the truthfulness of the gospel and associate with others who uplift and encourage them."[78]

This is a form of education that is about piety instead of scholarship, prayer instead of argumentation, and those that teach it are best served by faith and devotion. Seminary teachers are also well served by their orthodoxy and their orthopraxy. They are not served by an openness to novelty, such as cheating death through unorthodox technological innovations.

The bulletin board posting is disjointed, but its disjointedness does not come from confusion or inability to articulate thoughts. It is the kind of disjointedness that comes with too much grief, where the author leaps from point to point, reusing the same phrases, propelled by anguish rather than by expository logic. The issue at hand in this post is not death, or at least not simply death. Rather, it is whether a death can still be averted, even though it has already occurred. It is a family dispute about cryonics, though the word *cryonics* does not appear once in the initial email. We know this is about cryonics because the post's author states that he would fund her "$80k" preservation at ALCOR himself, a decision he presents as an extremely expense sacrifice that he is still willing to undergo, a sacrifice that he feels that his father and brothers might be okay with if it wasn't for his youngest sister.

He is tormented. "I'm a Transhumanist," he writes, "so it feels to me like we've decided to rot my Mom in hell. I'm having troubles sleeping now at 3:00 AM, thinking of how my Mom is, right now, rotting in that mortuary cooler, beginning what feels to me like will be more than a thousand years of lonely hell for the living." His other family members, though, are presented not as dreading dissolution, but instead as being buoyed by an illusionary hope. "All I can think about them now being able to sleep, naively thinking my Mom is now in a happy place, separated from us, the living. Her now being with dead people, Santa Claus, the tooth fairy, Jesus and God." (He leans far more toward atheism than many of the other members of the Association). He goes on to recall the horror his mother experienced while dying, a horror the family tried to wash away with morphine, a tactic not unlike the analgesic balm of belief in an afterlife. "Yeah, if you feel happy, or don't know what you're missing, everything is OK, and you're not in a worse hell, right?"

Our author has not taken his family's rejection of cryonic preservation lightly. He speaks of "storming out of the room, slamming the door, weeping and swearing about throwing Mom away and rotting Her in hell for more than a thousand years." Nor has his family taken his resistance lightly, either. "[T]hey've censored me, the first born son, from speaking at my Mom's funeral, hoping it will be good enough to give the dedicatory prayer to God over the grave." This request for him to pray at the funeral is something that they have made of him again and again, even though this cultural Mormon is a "staunch atheist."

Despite his being an atheist, though, there is a prayer inside of them— just not the prayer they might want. He wonders what his "Mormon Seminary Teacher Sister, and the many, many other temple worker friends of my Mom will think about IF I prayed something like the following over the grave:

> Heavenly Father,
> We come before thee, this day, to dedicate this grave, for the purpose of preserving my Mom's Body. It is my feeling that nobody will ever give up till that glorious day of resurrection, when her body, all of Her memories, and all that Mom is, is finally reunited with this body we now lay in this grave.
> For those of us that currently feel like throwing her away and rotting her in hell like this, is a grave sin of omission, if we are mistaken in our thinking, please guide and inspire those that know better, so they may be able to find the words to successfully communicate to those of us who are still mistaken, so that we may find some kind of comfort in this what now feels like a grave sin of omission worse than when ignorant slave owners murdered slaves.
> And for those that think God will do everything to resurrect us, while we do nothing, if they are sinning, help those that have the moral capability to realize this, to be able to find the words, to better communicate, so this kind of grave sinning of omission may cease, sooner. For it is our prayer that more people will soon be better preserved, so that more people may be able to be resurrected during the morning of the first resurrection, rather than rotting in the hell, in graves like this one, possibly for more than 1000 lonely years, if not forever.
> We so dedicate this grave, for the preservation of what is left of Mom's body, in the name of Jesus Christ, Amen.

He anguishes over his position—what can he say while being honest? He fears there *is nothing to say*. He closes his post with the line, "I must just be the lonely crazy Son, right?"

§79

Even when Mormons opt for cryonics, the Church is cold to the choice. One blog post by Lincoln Cannon mentions an LDS friend whose father had opted for cryonic preservation, a wish fulfilled when the father passed away and the body was forwarded to ALCOR. The difficulty in this instance is not familial resistance but silencing by the Church. Cannon reports what happened when the son (whom he gives the pseudonym "Joseph") tried to make arrangements for the commemoration of the father (whom he calls "Alvin"):

> Joseph and the rest of Alvin's surviving family members are now planning a funeral for Alvin. Because most of them are members of The Church of Jesus Christ of Latter-day Saints, they would like to hold the service in the church building of the ward that Alvin attended. And as part of the service, Joseph would like to talk about cryonics and Mormon Transhumanism.
>
> However, the bishop of Alvin's ward has told Joseph that he may not talk about those subjects from the pulpit at any funeral service in the church building. That is, of course, the bishop's prerogative. It may reflect an unfamiliarity with the compatibility of cryonics and Mormonism. And surely it reflects what the bishop sincerely feels is best for his congregation. However, as you might imagine, this has been heartbreaking for Joseph.[79]

Cannon then pivots to his own funeral (an event, it should be noted, that he speaks of without any of the grammatical markings indicating contingency). He says that he "would like Mormon Transhumanism, and the practical faith it advocates, to be a primary topic at my service." He wishes for the topic to be presented by whoever is then the head of the MTA (Cannon having already stepped down at the time this blog post was written). He says that while it may be most convenient to have the remembrance carried out in a church building, "if the bishop or other steward of a building disapproves Mormon Transhumanism as a primary topic at my funeral service, my funeral service should be held elsewhere."[80]

Jesus had "commanded his disciples to raise the dead," Cannon notes as he turns to the close of his essay. He writes that he trusts that the children of God will carry out that command, and also that he trusts that "God has prepared the means for us to accomplish that command."[81]

But he also notes that "[i]n the meanwhile, it's a beautiful day to be alive."[82]

§80

In the Church of Jesus Christ of the Latter-day Saints, there are rules for funerals. This is not unusual; there is probably no community that does not have explicit or implicit norms regarding how to deal with human remains. In the case of Mormons, though, these rules are helpfully written down in a guidebook meant to help bishops see to the Saints that populate their wards. They are contained (primarily) in sections 18.6.1 to 18.6.8 of *Handbook 2: Administering the Church* (originally titled *Instructions to Presidents of Stakes, Bishops of Wards and Stake Tithing Clerks* when the first edition of the book was issued in 1889).[83]

The initial instructional passages in the Handbook are somewhat anodyne, offering the sort of cosmic perspective ("Death is an essential part of Heavenly Father's plan of salvation") and human comfort ("As disciples of Jesus Christ, Church leaders and members mourn with those that mourn . . . and comfort those that stand in need of comfort' [Mosiah 18:9]") that one would expect from someone with pastoral responsibilities.[84] But things then slide from the homiletic to the concrete and the particular:

> "Church leaders should not incorporate rituals of other religions or grounds into Church services for deceased members."[85]
>
> "Church leaders" should be aware of the "requirements" of "governments . . . that regulate what occurs when a person dies."[86]
>
> "The bishop notifies the Melchizedek Priesthood leader who is responsible for the family so he and other brethren, including home teachers, can assist the bereaved family."[87]
>
> "If a funeral for a member is held in a Church building, the bishop conducts it. If a funeral is held in a home, at a mortuary, or at the graveside, the family may ask the bishop to conduct it. A bishop's counselor may conduct if the bishop is unable to attend."[88]
>
> Senior church authorities, if present, should be "extended the opportunity to offer closing remarked if he desires."[89]
>
> "Video recordings and computer or other electronic presentations should not be used as part of a funeral service." (Similarly, these services are not to be "broadcast" over the internet, either.)[90]

And perhaps more importantly for some, "[f]unerals should start on time."[91]

Among these instructions (calling them *regulations* is perhaps too far, though the word does contain an air of the authority these directions carry), there are some specific rules about the burial itself. The deceased should be buried in *temple clothing*, meaning not just the white garments continually worn under clothing but also the (often diaphanous and flowing) white vestments worn during rituals inside temple spaces, along with the green, leaf-patterned cere- monial apron received after the temple endowment ordinance. If, for some reason, this is "inappropriate or difficult" due to "cultural traditions or burial practices," then "the clothing may be folded and placed next to the body in the casket."[92]

These are dry words, at least on the page. But what they are trying to regu- late is quicksilver. There are, for instance, instructions about how the body is to be dressed in temple garments, and about who is to do the dressing; leader- ship is obliged to "ensure that this assignment is given to a person who will not find it objectionable."[93] But the dressing, often by family members or friends who are a part of the ward, has all the depths of emotion that one would sus- pect, running from tears to smiles, joyous farewells to lamentations. Other directives touch on topics that also are emotionally fraught, though the fact that they are charged subjects may not be apparent to an outside observer reading the text without any context.

One such object is cremation. The section of the handbook on cremation is short. It reads as follows:

> The Church does not normally encourage cremation. However, if the
> body of an endowed member is being cremated, it should be dressed
> in temple clothing if possible. For information about dedicating the
> place where the ashes are kept, see 20.9.

This minimalist statement is perhaps an underselling of the horror, anger, and disgust that some Mormons associate with cremation. I have heard Saints report that simply an offhand mentioning of one's desire to be cremated after death can trigger a family argument. The taboo nature of cremation is probably lessening over time; the number of cremations per annum in Utah, for exam- ple, seems to be increasing, though this could be the result of demographic changes, as slightly ever-increasing numbers of non-Mormons continue to choose to live in Salt Lake.[94] It is certain that the vehemence with which cre- mation is disfavored is lessening over time, at least in the official proclamations of the Church.

We can see that by looking at the language used beforehand to discuss cremation. Bruce Redd McConkie was a member of the Quorum of Twelve apostles until his death in 1985; during that time, he published a book entitled

Mormon Doctrine. The book was never formally approved by the Church; in fact, the first edition of the book had its publication restricted, in large part because of tonal issues caused by McConkie's sometimes pugilistic prose. That first edition has a reputation for being uncompromising in its positions and for showing little deference for the feelings of those who might hold different views. Despite that, the book was influential in the creation of other reference books on Mormonism and is still used as a guide in some of the more traditional Mormon circles.

As might be expected from the book's reputation, the language on cremation in *Mormon Doctrine* is not precisely nuanced. It is not celebratory language, either. It reads

> Cremation of the dead is no part of the gospel; it is a practice which has been avoided by the saints in all ages. The Church today counsels its members not to cremate their dead. Such a procedure would find gospel acceptance only under the most extraordinary and unusual circumstances. Wherever possible the dead should be consigned to the earth, and nothing should be done that is destructive of the body; that should be left to nature, "for dust thou art, and unto dust shalt thou return." (Gen. 3:19)

This language, even given the book's reputation, is stringent. It is also consistent, appearing verbatim not just in the first edition, but also in the chastised, marginally more generous-of-spirit second edition; while other entries were qualified and softened, the language on cremation was not.

§81

Why is cremation an issue?

Recall the temple garments, either used to dress the corpse before burial or, if forced by circumstance, folded and laid alongside the body. These garments are placed there not just for their symbolic value; they are there for use on the day of the resurrection. When the dead rise from their graves, they will need something to wear that is fitting for the occasion. Clothing is (at least in part) necessitated by the fact that this will be a *material* resurrection. Considering the Mormon refusal to consider spirit as anything other than a form of matter (albeit matter with properties much different from the gross materiality we are confronted with on a day-to-day basis), the resurrection could be nothing else. Many forms of Christianity imagine the resurrection similarly, with physical bodies, though these bodies are often made perfect. There are even some Christian intellectuals—a minority, to be sure—who are so comfortable with the materiality of the resurrection that they are able to question the existence of the soul.[95]

Mormons, though, have committed themselves to the materiality of resurrection in a different way, in part because they have committed themselves to materialism in a different way. Not only is spirit matter, but God has a material physical body as well. Mormon teaching even denies the ubiquitous Christian concept of *ex nihilo* creation. Given this commitment to materiality, it makes sense that when bodies are resurrected, they are (nominally) the same bodies as those that die. This continuity may seem hard to square with a Mormon belief in resurrected bodies made perfect, in risen bodies that do not hunger and do not tire. It is a common Mormon tenet, backed institutionally by the Church, that this continuity is made possible because when resurrected, bodies have spirit-matter, rather than blood, in their veins; depending on the account, blood is either replaced with spirit, or is *quickened* into spirit. That blood should, in the resurrection, be *replaced* or *quickened*, rather than, say, *recreated* with spirit, shows that the Mormon concept of resurrection also entails a physical continuity of some kind between the body that dies and the body that is risen. One is not given a new body that is like one's earlier earthly body, with continuity of self, guaranteed by spirit-matter, one is given the same body, crafted from the exact, specific, and same material that constituted the body one had before.[96]

This concern with divine ability to summon up a body after desiccation— or even a worry regarding postresurrection material continuity of the matter that consisted the body—can seem odd even to people who have a background in the Church; one friend brought up Mormon jokingly referred to it as "[t]he unfathomable problem of which atoms are whose." This issue was shrugged aside in some comments by Brigham Young, when he described the resurrection:

> When the angel who holds the keys of the resurrection shall sound his trumpet, then the peculiar fundamental particles that organized our bodies here, if we do honor to them, though they be deposited in the depths of the sea, and though one particle is in the north, another in the south, another in the east, and another in the west, will be brought together again in the twinkling of an eye, and our spirits will take possession of them.[97]

But at the same time, we should remember that statements like Young's suggests can be read as much as a palliative to alleviate anxiety as it can be read as a proclamation shutting down a problem.[98] And such a problem can only seem to be a cliché to people who have grown tired of hearing it explored by others.

§82

Concerns about burial are not only about proper *commemoration* but also proper *preservation*, and even if the divine hand is capable of wonders, it makes sense that the disintegrating and possibly dispersive practice of cremation might make the more literally minded believers nervous. This anxiety about cremation may seem to have nothing to do with the furor over desires for cryonic preservation that we see from some believing Mormons. After all, when compared to cremation, the degree of preserved bodily integrity is far greater. Arguably, what cryonics promises is a degree of preservation and concentration of remains that, presuming all works well, far surpasses that of the grave. (This is perhaps a less compelling argument in the case of neuro-preservation, and as seen in Chadsworth, there historically also have been cases where bodily preservation was not in practice in what one imagines was initially hoped). And if there were still lingering doubts that the underlying cryonic process could work, to mainstream Mormon imagination that should be no worry, since a divine hand can summon the deceased back from the subzero grave on resurrection day just as quickly as it can desiccated, plowed-over bones.

It would seem that other Mormon objections to cryonics should in theory have even less traction. As we have seen, skeptics can have strong feelings about what might seem to be the waste of resources that comes with cryonic preservation. But in the two cases discussed earlier, we are dealing with people who are either incredibly wealthy (the case of Mr. King) or with an individual family member who is volunteering to take on the burden alone (the mourning brother of the seminary instructor). Preservation may also seem to challenge the commemoration process that is an inevitable part of burial, disrupting the capacity for the living to remember the dead even as they move on without them. But as we have seen in the discussion of familial dewar-side visits to the ALCOR facility, situated remembrances proximate to the material remains (or in a cryonic vocabulary, to the *patient*) is not impossible. The interstitial nature of the deceased may be disturbing, as we also saw in our discussion of commemoration in ALCOR. Still, for those skeptical of cryonics in the first place, framing such preservation as a state of extreme and incredibly prolonged medical emergency is unlikely to be taken seriously. And regardless, if there is faith in the day of the resurrection, then just like the cryonically preserved, the kind of temporal disturbance inherent in mourning the passed is potentially haunted by the thought of a return as well.

Given all this, even though a certain kind of faithful Mormon—a seminary instructor, perhaps—may doubt the cryonic claims about both possible future results and the present status of the preserved, in some ways Mormon death and cryonic preservation are more similar than different. First, they are *both* passing states; in the best-case scenarios, according to both understandings, death is something that can be waited out, in the sense that reanimation is in the cards. And in the details of this reanimation, there are still other ways that these two procedures rhyme. The trading of blood for cryopreservatives, for instance, could be seen as an uncanny mirror for the spirit-filled veins of the resurrected. And as we saw from Max More's anticipation of his post-thaw state as that of a superior, more youthful body attached to his original head or brain, cryopreservation and the day of resurrection also seem to offer perfected bodies. Even the silver encasement of the dewar and the white-robed covering of the faithful dead seem to share some imaginary contours, with both garments forming bright reflecting shrouds. It is certain that vision of the joyous reunion of the resurrected with their family shares some of the same emotional resonances found in the future reunion of the two-year-old A-2789 with her family, as imagined by the ALCOR press release announcing her preservation.

This should not be taken to mean that the elements of an orthodox understanding of Mormon burial and the features of secular transhumanist understanding of cryonics can be mapped onto one another one-for-one. These are resonances between structures, and not isomorphisms. In Mormon thought, the scope of the resurrection is broader: all of humanity will live again, even if the judgment meted out to all of them will not be the same. In cryonics, only those who are preserved will see the future. Furthermore, there is more of a catch in the throat when the cryonicist speaks of further life that is to come. As we saw with ALCOR's anxieties about everything from tectonic disturbance to a need for bullet-proof windows, there is a fear that intervening events could undo all their labors to preserve the past for the future. Cryonics places much greater trust in the agency of humans, both alive and unborn, much as hope in the resurrection places greater trust in the divine (though, as we will see, even the day of the resurrection has dependencies on human agency). Secular cryonics and faithful interment in the hope of the resurrection may rhyme, but they are certainly not conceptual synonyms for one another.

This conjoined similarity and difference between cryonics and burial may explain why it is that with some exceptions, people seem to hold to either one or the other of these two methods but seldom both. And as just discussed, this either/or nature of the cryonics/resurrection opposition cannot be based entirely on a sort of game-theory logic. Neither option precludes the other. But only one or the other seems to be a "live" choice for any particular subject. And opting for cryonics appears to be a denial of the resurrection; think again of the almost villainous understanding of Christianity common within early ALCOR. Even when discussing cases involving the MTA, the brother of the seminary instructor spoke in the voice of an unalloyed atheist. It is as if one *cannot* think of both cryonics and divine resurrection at once. In the cognitive structures we have at hand, only one account of resurrection can be placed in the structural thought afforded for rebirth in a singular cognitive imaginary.

We should also notice that these particular practices are quickly laminated to larger totalities. In cryonics, religion (or deathism) is not just an objection to a specific technical practice but an existential threat. Not just a technological mode of preservation but the entire transhumanist project is, by necessity, both opposed to religious thought and vulnerable to it as well. And conversely, not just cryonics as a practice but even the discussion of cryonics are so beyond the pale to an orthodox Mormon sensibility that they have no place at either the pulpit or the graveside; and this is a topical foreclosing in religious tradition which may discipline and police members, but which also holds itself open to public meetinghouse testimonials and talks that are uncontaminated by prior restraint.

It may be objected that there are exceptions to these cases. This objection is important. Advocates can imagine that other persuasions could be open to adopting the logic they militate for; we've seen this in moments of attempted outreach such as More's argument that the Aquinas-infused logic of Catholic theology demands a Catholic acceptance of cryonic preservation, or in King's somewhat aborted talking points on how Mormonism's doctrine could be read as being simpatico with cryonic preservation. And there are even people who imagine a kind of ideological superposition of technological and theological resurrection. Despite the hostility to Christianity that seemed to supersaturate ALCOR, we still have a few figures, such as Laprade, joyfully anticipating spending a few Evangelical millennia in dewars with nonbelievers. And then we have the MTA itself. The full-throated atheism displayed by the seminary instructor's brother is not so rare in the MTA that he can be considered an absolute outlier. Still, such atheism does not exhaust the religious positions found in the MTA. Furthermore, this atheism cannot be considered paradigmatic of the MTA position regarding God. Many Mormon Transhumanists in good standing with the Church of Jesus Christ of Latter-day Saints now have cryonics accounts.

Outside of some first glimmers in Mormon Transhumanism, there may be something like a Pauli exclusion rule for the imagination of embodied life after death. Under this rule, either technological or supernatural accounts of resurrection can be endorsed and relied upon as the grounding presumption that informs actions. But for cognitive-structural reasons, both concepts cannot be held as simultaneously present. There is no mediating position. Further, the ritual nature of any of these options also constrains possibilities.[99] While the question of life after death touches on cosmology and the mythic, these rituals also have to be physically enacted by a set of people acting in concert. And the material entailments and technological constraints can collapse possibilities as well.

But if so, what sense do we make of those cases that appear to break with this rule? How is it that some can hold religious and scientific-technological framings at the same time? Perhaps this simultaneous openness to both accounts can only occur through a reconstituting of the structure such that elements can no longer fill the structural position usually accorded to them. Perhaps this is done through a reconfiguration of human agency or divine identity. And perhaps through scientific speculation regarding bringing back the dead, which would look less like cryonic triage and more like the Mormon day of the resurrection. What is (now) impossible at the level of the material, the technical, and the collective may well be possible in the mythic imagination; and creating and circulating these could well be the first step to making them something more palpable.

Seventh Series: "as if awakening from a night's sleep"

§83

It was a beautiful, two-story house with an expansive yard and a splendid view of the mountains. Of course, almost every residence in Provo, Utah, has a splendid view of the mountains. The house was close to Brigham Young University. But then, little in Provo isn't close to BYU, given the size and presence of the University and the way that the town at times feels as if it could fit in the palm of the hand. The house was also relatively close to several technological startups, many of which were just a hop or two up I-15. But again, this was no accident, given that the Silicon Slopes has been fed by a slipstream of disciplined talent emanating from around the university. The house, by the way, is being spoken of in the past tense not because of any theoretical allergic reaction to, or virtuous stand against, the convention of the ethnographic present tense.[1] The past tense is being used because on this bright day the owner of the house, his family, and a few neighbors gathered together to watch as a mining excavator tore the building apart.

The event had a festive, party-like atmosphere. At one point early in the process, a boy asks if he can break one of the house's windows (the answer is no). Near the end of the roughly half-hour procedure, a girl shouts directly into a camera "I'M WATCHING MY HOUSE BE DESTROYED! IT'S AWESOME! BUT ALSO SAD !" Loose and playful as the event is, this demolition was the fruit of years of contemplation. For a long while, the owner wanted to make what Germans call a passive house: a building engineered so that it would use available thermal resources so efficiently that it effectively used no outside power for heating or cooling, vastly reducing its ecological footprint.

The owner, a software engineer who, like a handful of other Mormon Transhumanists, had recently, and unexpectedly, become relatively well-off due to some very minor investments in cryptocurrency, was now in a position to commission an architect to design and oversee the production of a passive house on his property. And he was also able to see that his current house was destroyed.

The impetus behind this act goes back even further than an interest in home energy efficiency. Even before he had heard about passive houses, he had for years considered the possibility of living off the grid, even looking into the use of greywater (repurposing used house-water for other purposes) and solar energy. These earlier investigations may seem to be just a prior moment of ecological consciousness, which was undoubtedly an element of his desire. The contemplated locale of this off-the-grid house in rural Wyoming, however, suggests that this isn't all there is to the story. As he acknowledged in conversation, what he euphemistically referred to as the Church's teachings was a significant element in this original desire.

§84

Why would an off-the-grid house in rural environs be an aspect of the Church's teachings? A Mormon tradition of preparedness helps answer this. The Church encourages emergency preparedness at both family and the ward/stake level. Families are encouraged to store food. (Current recommendations are now for at least three months of stored food, with having a year's supply being the ideal goal to work towards; at times, the Church recommends purchasing types of food that can be stored for as long as thirty years.) Mormon bookstores often carry various "food supply" buckets and boxes with enough provisions for periods ranging from seven days to three months, alongside self-help books with titles such as *Not Your Mother's Food Storage*. The Church even runs a dedicated website entitled *Provident Living*.[2] The tastefully minimalistic website features wholesome looking photographs of families that could be mistaken for clipart if these photos did not also involve bare metal shelving holding large cardboard boxes and outsized plastic tubs of food.

It could be a mistake to associate food storage too closely with Mormonism. There are plenty of non-Mormons whose imagination is caught by the idea of shortage and catastrophe, and these nonbelievers read or follow many of the same prepper books, newsletters, websites, and Instagram accounts that Mormon preppers do. And there are plenty of Mormons whose degree of orthodoxy, orthopraxy, and church involvement may not be the highest but who were still deeply invested in food storage. More pointedly, the Church's advice regarding food storage is now honored as much in the breach as in the observance, especially in younger generations of educated members; for many middle-class, professional Mormons, the days of bunker culture are over. Food storage is no longer a topic addressed by authorities in general conference talks, and some Mormon believers report sitting in the pews for years on end without hearing any mention of it. But preparedness and all that it ethically entails still colors the Mormon experience. For a long spell, having a year's worth of stored food was considered a measure of one's faithfulness. One Mormon friend recalls the closest ties to her Mormon heritage in her youth not being irregular, sporadic Family Home Evenings or Sunday Church attendance but the boxes of stored food placed under her parent's bed, which were used in lieu of a box spring mattress.

What is this preparation for? Part of the impetus for emergency preparedness lies in a deeply inculcated ethos of self-reliance, a corollary to the Mormon prizing of agency and responsibility. Preparedness is also a backdoor opening to charity, as aging foodstuffs can be gifted to needy Mormon families as a form of community welfare. And part of this food storage practice is certainly preparation for disasters situated in a more naturalist register, such as earthquakes or tornadoes. Undoubtedly institutional memories of anti-Mormon persecution during the nineteenth century are also an element in this. But food storage-related practices and material also are preparation for the apocalypse and the birth-pangs of the millennium. This is the church, after all, of the *Latter-day* Saints. Though it would vary in its intensity depending on the level of perceived danger to the faith, Mormon thought has been apocalyptic from its inception. This should be no surprise. The Book of Mormon itself, charting the impressive growth and complete annihilation of a Jewish settler civilization in the New World, is, in its own way, an apocalyptic narrative of an end-times that, for the people who were involved, had already occurred.[3]

Despite the apocalyptic rearview mirror clarity bequeathed by the fates of the Nephites in The Book of Mormon, the Mormon apocalyptic imaginary is at once overdeveloped and undetermined. The dispensationalist apocalyptic narratives common among Christian Fundamentalists and Pentecostals are, by comparison to Mormon accounts, detached hermeneutic exercises since almost all of these more familiar Protestant end-times scenarios combined exercises in imaginative literalist exegesis of allegorical apocalyptic biblical text with an assumption that true Christians will be exempted from these horrors. But Mormon apocalyptic thought is different. Born at the same time as the Millerite movement that ended in the Great Disappointment, Mormon thought has elements of the Protestant apocalyptic roadmap, drawn from the same slipstream of nineteenth-century eschatological imagination. For instance, like many other premillennialism dispensationalists, they foresee a Jewish regathering in Jerusalem as an integral part of the eschaton's arrival. But unlike most Protestant premillennialism, there is no assumption of a rapture as an escape hatch. The faithful will experience the apocalypse rather than safely view it from a supernatural remove.

These apocalyptic fantasies are hermeneutically limited as well, with much of the usual Protestant literalist tendencies undercut by the possibility of continued, though vague, revelation from Church leadership. The "prophets, seers, and revelators" who run the institutional Church are careful to only allude to apocalyptic scenarios while still emphasizing the importance of these future events, speaking in a way that at once excites anticipation but occludes any prognostic clarity. A typical example was when Russell Nelson, the President of Quorum of the Twelve Apostles, in a speech before a youth conference stated that he had "spiritual impression" that the term millennial was "perfect" for this generation because of their eschatological mission during what was "the eleventh hour." As he put it, "[a] True Millennial is a man or woman whom God trusted enough to send to earth during the most compelling dispensation in the history of this world. A True Millennial is a man or woman who lives now to help prepare the people of this world for the Second Coming of Jesus Christ and His millennial reign."[4] The end times are invoked and, in a way, made near tangible through the reliance on contemporary prophetic authority, but they are articulated in a way that makes it impossible to pull any hard, prophetic details from the narrative.

Given how open, unfocused, and yet pressing the Mormon apocalypse is, there should be no surprise that outside of the institutional church, there are three chief sources. One source is a nineteenth-century tradition, forged during the most active days of Mormon persecution; tales of sealed sacred books rediscovered on the eve of the apocalypse; the anticipation of the end-times figure of "The One Mighty and Strong"; the return of Jesus to Independence, Missouri, a spot which also happens to be the original location of the Garden of Eden. These more doctrinally grounded pronouncements combine with a cloud of Mormon folklore, made up of narratives such as the "three Nephites" (immortal prophets who wander the Earth until Jesus' return), or the most likely apocryphal "Constitution" or "White Horse" prophecy (a vision of the Church of Jesus Christ of Latter-day Saint's coming in to save the United States Constitution when it is hanging from "a single thread"). This material is introjected into a broader American collage of war films, disaster movies, science fiction, urban fantasy novels, and right-wing conspiracy-oriented nationalist pamphlets. Russian helicopters, disguised U.N. ground forces, relocation camps, and Roman Empire–style religious persecution are just a few of the constituent shards continually being pieced together in websites, speaking tours, and small-press-published books.

As can be seen from the distributed nature of the conversation, this is a project of collective, decentralized imagining. The institutional Church is not entirely comfortable with these apocalyptic fantasias; the carefully controlled allusions by Church leadership to the end times can be understood as a way to clamp down on the more ornate scenarios that continually bubble up from apocalyptic entrepreneurs. But despite this pushback, the exact nature of the Latter-days is a topic too fascinating not to obsess on.[5] As one Mormon friend said to me, Mormon apocalypticism is one way for believers to participate in authoring a deeply emotional investment with their religion, something outside the rigors and conformist pressures of sacrament meetings. These are scenarios that seem wild but are grounded enough in the here-and-now of church institutions, existing geopolitical entities, and current or emerging technologies to appear at once mad and plausible. Mormon apocalypticism is, in its way, another form of speculative religion, but it is one that is not so much concerned with exploring the future or charting the past but with extinguishing the present.

§85

These apocalyptic presumptions may sound a bit to secular ears like some kind of fever dream. And these end-times narratives certainly don't sound like the Teutonic engineering dream of a passive house. Touring the completed home years later, it looks nothing like a bunker. Its decorative stained-glass windows and art deco wooden wall panels makes it appear like an enormous craftsman bungalow, more arts and crafts than survivalist brutalism. One could imagine that the proper narrative regarding this act of house destruction and home construction is one where a secular concern for utilitarian ecological self-management over time uprooted irrational Mormon apocalyptic anxieties, and where yearnings for something off the grid in rural Wyoming are replaced by desires for something incredibly efficient in suburban Provo. That was not the case—or better put, that was not only the case. While those framings are not entirely wrong, this was also understandable as a shift to a different eschatological vision and not a reduction to any immanent frame or the abandonment of an apocalypticism for a technorationality.

Specifically, the passive house is a moment of a Mormon vision of carefully calibrated, continually advancing perfection. Part of this is religious millennialism (the thousand years of peace that follows Christ's return), but this vision of continual uplift cannot be reduced to that. There is a tradition of Mormon tinkering and innovation in the furtherance of efficiency. Joseph Smith was obsessed with design, laying out in 1833 a protomodernist plat for the anticipated city of Zion. After the Mormon flight to Utah (where Smith's urban planning vision was eventually realized in Salt Lake City), Brigham Young (head of the main body of the Mormon Church after Smith's lynching at Nauvoo) attempted to propagate a Mormon-invented phonetic script called the Desert Alphabet. It was comprised mostly of new characters, though a few letters made it in, such as O, W, L, and in a mirror image form, N. Used for a protracted period in the late nineteenth century, the alphabet was intended to simplify and rationalize the English language. It was taught in schools, used on street signs, and also utilized in documents and books before the project was abandoned. This list of projects could go on; there is the attempt to institute a Mormon system of currency, nineteenth-century church experiments with broad-ranging socialism, or even the tradition of Mormon inventors. In their way, all these projects go back to Joseph Smith's Plat of Zion. Zion was not planned as a heavenly paradise but rather as an actual physical locale, first intended to be constructed at Kirkland, Ohio, then Jackson County, Missouri, then Nauvoo, Illinois, and finally at Salt Lake City—the exact location kept shifting as protracted government-sponsored campaigns of violence triggered a series of Mormon exoduses. The plan was for an ingathering of Saints, along with converts to this new American Religion, to leave their home in Europe or the Americas and relocate to a singular point. There, they would coconstruct a Utopian community as the rest of the world slid into chaos. Since the early twentieth century, the Church no longer encourages converts to leave their homelands, advising them instead to do the equivalent of shelter in place. Still, the imperative to build Zion remains strong, and it is not uncommon for converts to find themselves eventually moving to Salt Lake City and environs.

And this is the engine behind a dual voiced view of the future. Because the eschaton was spatialized, apocalyptic nightmares and millennial hopes could be laid out alongside each other as Zion was built up and the rest of the world collapsed. Rather than absenting faithful Christians through mechanisms such as the rapture, the Saints indwelling and perfecting Zion creates two simultaneous human trajectories, one of abject destruction and one of evolving perfectionism. This patterning has important, though subtle, effects. This is not to suggest that the former-and-future homeowner either drew up the passive house blueprints or tore down his old domicile in a religious fugue. But in the narrative of how this dream house developed over time, one can see how apocalyptic sensibilities may have unconsciously torqued the imagination.

§86

While it is almost prismatic in the variety of views that it contains, one can also see similar broad and contrasting apocalyptic and millennial sensibilities informing secular transhumanism. For secular transhumanism, both of these trends are informed by the same conceit: the inexorable increase of intelligence in the universe. Many transhumanists claim that Moore's law is just the latest expression of an ever-continuing exponential increase in complexity and order (read here as a synonym for intelligence) that goes back to the Big Bang. The pace of this change accelerates as new tiers or epochs are reached, with evolution working much faster than bare chemistry in the production of novel forms, and technology working faster than evolution.

Recall that many transhumanists and immortalists believe that a new tier is coming, that we are approaching a moment when intelligence, complexity, or order (and these categories are used with a certain degree of fungibility) will advance faster than human consciousness can keep track. This would seem to indicate a hard limit to advancement and to identify the point at which the process collapses. To transhumanists, this identifies not the end of this process but its real beginning; the expectation is that either augmented humans, or nonbiological intelligences created by humans, will then take over the intellectual work of producing complexity, order, and novelty. Put differently, they anticipate the singularity. Recall the difference between the mathematical sense of the term; and the sense used here is that in mathematics the incapacity to plot is considered to be objective. But in the technological singularity, this "unplottability" is perspectival. This intensification of intelligence will continue, and eventually all matter on the earth, in the solar system, and in the universe will become smart matter, engaging in this ceaseless ramping-up. However, it will be impossible for nonaugmented humans to either foresee or participate in this process. Despite this limit to any calculability regarding how it will unfold once it begins in earnest, it is still very much anticipated. For transhumanists, both secular and Mormon, the singularity is both real and imminent. One hears predictions placing it anywhere from the end of the twenty-first century to the looming 2040s; the singulitarians' interest in radical life-extension regimens and cryogenics is so that they will be able to participate in the singularity once it occurs. Again, death is a technical problem.

There are also tensions between those who see this upcoming transition as in essence beneficial and those who see it as also posing existential threats to the species. However, this portrait is still enough for us to have a rough sense of not just the centrality of the new for transhumanism but its contours as well. It is invention, as much as that they hold this to be a process that must be engaged in, at the same time, it is the discovery of an existing cosmological telos centered on increasing intelligence. While this teleological edge allows some of the features of the new to be foreseen, this is only up to a specific, and relatively close, temporal horizon. Some, like Ray Kurzweil, see this anticipated state as a utopia, where death will be finally conquered. Others, like computer scientist Eliezer Yudkowsky or the philosopher Nick Bostrum, believe, to varying degrees of credence, that these developments rather auger humanity's extinction as the species is replaced by its own creations. Interestingly, this is a binary sort of identity—most favoring one side or the other, and with few either acknowledging both results are possibilities or that there might be some intermediary resolution between extinction and theosis.

Not so much with Mormon Transhumanists. In the invocation of the 2017 annual conference of the Mormon Transhumanist Association, the incoming president and cofounder started by, without any preliminary discussion, sternly reading from the podium the "Little Apocalypse" from the Gospel of Matthew, which anticipates earthquakes, famine, and war as the world comes to its conclusion. It was unnerving, unmooring, a moment of seeming fundamentalist rancor and doomsaying. For a second, I wondered if, despite all the years spent talking with Mormon Transhumanist, there was some dark, literalist streak in the movement I had somehow overlooked. But this was merely my being caught up in an aesthetic trap that the speaker intended. Immediately after finishing this reading from Matthew, he shifted tone with a theatrical, intentionally comedic *or*, and he read Isaiah 11:6, the Hebrew Bible verse which describes the wolf and the lamb, the calf and the child, lying down peacefully together. These are not sequential events but alternate paths. "What makes the difference between these contrasting visions of the future? I think a lot of it depends on what we, collectively, do." End times pessimism and the optimism of building up Zion allowed the Mormon Transhumanists to consider both the apocalypse and the millennium at once.

§87

Millennial thinking also means thinking about the resurrection.

Recall again the dutiful transhumanist son, requested to pray over the gravesite of his mother, the son who frets over what to pray, over whether to pray, who feels that he is just lonely, crazy. He is less alone than he thinks, at least if he steps outside of the family; this community of transhumanist Mormons and fellow travelers, years old at this point, reaches out to console and to share, to give what one listserv respondent calls "solidarity and love." On that listserv, some write to advise about the ostensible issue behind that email, that is, what sort of prayer, if any, should he give at the graveside; if he cannot pray in good faith, one friend suggests, perhaps he should decline the request so that he does not carry even greater grief. There are other imaginations of prayer offered by some friends. These approaches see these sorts of supplications as an existential act instead of as a statement to God, suggesting that perhaps they can be prayed after all without hypocrisy. Many write to share the pain at the passing of their own parents, both recently and long deceased, who were not cryonically preserved. One man frames the issue not by saying that he lost his parents to death but rather that because of his parents' explicit rejection of cryonics during their lives, he lost his parents "to burial."

Some see another issue here. One long-term transhumanist mixes his condolences with the statement that he disagrees "that religion (Mormon or otherwise) is necessarily doing harm by encouraging faith in an afterlife. . . . there are huge deficiencies in any belief system that leaves people bereft of hope in the possibility of an eventual reconciliation." After all, he states, "people without hope do bad things." There must be consolation fitting to the circumstances; "[p]eople," he notes, "are different and come from different backgrounds, educational levels and awareness levels." A moment later, he writes by way of apology once more, "this is not the time for this discussion." Instead, what he wants to stress is "that we care about you." Listservs are made for conflict, but here succor rather than argumentation is the ultimate intent.

There are two responses that offer a different form of consolation. Both of these responses acknowledge the pain of loss, with each writer remembering the painful passing of a parent. But each author looks forward, imagining that faith in the resurrection may not be as empty as our grieving atheist presumes.

§88

One is from an Italian transhumanist; he is not a Mormon, but he is a fellow traveler of his Mormon Transhumanist friends and a full member of their association. He's a physicist and a computer scientist who used to work on large research projects, but after early retirement now makes his living through a mix of writing, consultation, and advocacy. Over the years, he has grown to see transhumanism as something *akin* to religion. Like religion, it gives life meaning (in other writings he uses the phrase *transcendence*), and it also offers the possibility of a second life; however, because he understands transhumanism as being based in science, he views this more as an "unreligion," a vision of the world that could take on religious trappings and make their illusionary hopes real.[6] In his message to the dutiful transhumanist son, he remembers sitting bedside while his unconscious mother passes. He only brings one book as his companion, Frank Tipler's *Physics of Immorality*. Written by a professor of physics at Tulane University, this book posits that, according to the laws of physics, in the fullness of time—and he means the fullness in its ultimate sense, thinking in terms of tens and hundreds of billion years—the universe will form an Omega Point, an (effectively) infinite concentration of intelligence that will have an (effectively) infinite amount of time before it. This Omega Point, Tipler argues, is indistinguishable from the Judeo-Christian God in its power, but also in its benevolence, since it resurrects us as perfect simulations.

In his message to the grieving son and brother, this letter writer doesn't close by imparting the message of Tipler's book; this text is well known in religious transhumanist circles, and a mere allusion to it is enough to carry the argument. Our respondent does take the time to explicitly hope that future "heavenly Persons with God-like powers achieved by means of science, will choose to resurrect my Mom, and me, and all our loved ones who passed away, for us all to be together again." It does not matter that it won't occur for "a thousand years, or more." This is because "[s]ubjectively not a moment will pass. I hope I will close my eyes one last time, and then open my eyes again, and see my Mom."

§89

Our other author is Lincoln Cannon, then President of the Mormon Trans-humanist Association, though he is not writing in that official capacity here. Like the other responses, he is giving his posting to the grieving transhumanist primarily as a friend. Therefore, it is not surprising that, in many ways, his post is like the others. He recalls the death of his father, who passed when Cannon was a teenager; at other times, he had pointed to his father's death as the turn-ing point of his life. In this message here, he does more than try to offer em-pathy. He also attempts to comfort by observing that what was lost when the sister rejected cryonics was not all that dependable in the first place. Cryonics, at least in its current state, is no guarantee of proper preservation (Cannon describes it as, at best, a "high cost low chance proposition for most of us"). But Cannon also offers a third form of comfort. He links it to an essay he wrote in memory of his father, "with the intention of consoling others in similar situations."

Linking to an essay may seem like an odd choice and perhaps even a bit tone-deaf; an essay would seem to lack the immediacy and the intimacy that characterizes these message board exchanges, especially when discussing a topic as raw as this. But this essay is a death-haunted piece of writing with its own strange closeness. A brief, italicized prologue sets the scene by referring to a funeral oration given by Joseph Smith—a sermon that occurred "less than three months" before Joseph Smith's death, the prologue notes.[7] The prologue to Cannon's essay states that the essay is both "a tribute to and in reverence of Joseph's words." Referring to Smith by his first name is a rather traditional Mormon convention. In this context, speaking of Smith in this way makes him close to us. This air of belonging continues in the first line of the piece by saying, "[f]riends, I'll call the attention of this group while I address the subject of our dead."

Like the King Follett sermon that the essay references, it is at once about the cosmic but also about those present at the virtual time of the essay. It is not "the dead," but, "our dead," it is not for "the reader" but for "this group." The intimacy grows, even as the author brings himself to the fore; Cannon tells us that the "anniversary" of the death of Layne Cannon, "my father," is what "has led me" to the topic. We are given another temporal point of orientation and are brought into a kind of familial possessiveness by way of the first-person possessive pronoun. But that does not mean that this pain of loss is Cannon's alone. "[M]ost, if not all in this group, have also lost family or friends," which serves as a warrant for Cannon to "feel to speak on the subject of our dead in general, accordingly whatever wisdom and inspiration I might have, and offer you my ideas for consolation." But this intimacy with "our" dead quickly becomes cosmological. "To understand the subject of our dead, we just start with an understanding of God." Or rather of Gods, as we are told Cannon wishes (to quote an old Mormon hymn), "to go back to when Gods without beginning find themselves making worlds without end."[8]

The original King Follett sermon contained the startling revelation that God was once like us. Lincoln Cannon shares this faith. He starts stringing together pearls in the way of scripture quotations, telling us that "the Son can only do what he sees his Father doing" (John 5:26), that eternal life is to know the only true God (John 17:3), that we're in this God (1 John 5:20-21), that we will become heirs of God and joint-heirs in Christ (Romans 8:17). And, slipped dreamily between these Bible quotes, is Cannon's claim that "in these times of accelerating change, we should take this more seriously." This time, though, the citation is not to scripture, but to Ray Kurzweil.[9]

The tone starts drifting in a new direction: the language of the posthuman and analytic philosophy begins to stand in the place of the language of scripture and earnest religion. "I suppose I shouldn't speculate excessively beyond what experience and logic warrant," Cannon says. "If I do, the skeptics will ridicule me." He says instead that he will "comment on" and "analyze" the "arguments of philosophers." This is also the moment when we encounter a syllogistic inversion of the King Follett experience. If we are doomed, then it is likely that other intelligences before us were doomed as well; but if we can survive, then there is no doubt that others before us have also lived on, increasing in power and capacity. "Skepticism, reason with me. If God does not exist, then we will not become God." If we are to have hope for our future, then it stands that others have existed and overcome, too, and have earned for themselves powers that are more than human. Others have won world-making powers. Perhaps even cosmos-making powers. And though Cannon does not have to lay this out as explicitly—the claim here is backed by an unconscious but unyielding logic in a Mormon context, but not necessarily in other frames— the idea is that which is more than human is (to many) the very object of religion.[10]

We then have another shift and a return to scriptural language. We have now traced out a lineage of Gods. "The soul, the spirit, or the mind: where did it come from?" It was not created *ex nihil*, we are told. "That idea lessens humanity." Instead, we are information and intelligences are capacities, facts, forces that have always existed. We are eternal, and we remember that eternity. "We are the memory of eternity, cascading through space and time as cause and effect, moving and monitoring, forming and feeling." Even after we pass, we are present, telling us that "information persists":

> So, when we mourn, what have we lost? Our friends are only without their bodies for a short season. Their minds have faded only for a little moment, as it were. Yet they still exist in a place where they interrelate like us. They're not far from us, our thoughts, feelings and emotions. Indeed, all who have bodies have power over those who haven't. We're inhabited by them. They're right here. Remember our dead.

Here, we have a parenthetical citation to *Teachings of the Prophet Joseph Smith*; these are the passages that tell us that "Brother Adam has gone to open up a more effectual door for the dead. The spirits are just exalted to a greater and more glorious work, here they are blessed in their departure to the world of sprits. Enveloped in claiming fire, they are not far from us, and know and understand our thoughts, and are often pained therewith." These are the passages that tell us that "[a]ll beings who have bodies have power over those who have not," that tell us to "hasten our temple work."[11]

The incorporation-by-reference of the temple into the argument marks a point where the logic and language shift yet again. We have a responsibility, "an awful responsibility" to "our" dead, Cannon says. We are obliged to remember them. To resurrect them. Referring to the visions for John the Revelator, "[t]he book of life is the record of eternity, and the other books are our records, history and genealogy." Now, he tells us that we use this genealogy for the baptizing of the dead. And we use technology in service of this genealogy. But we are only at the edge of what genealogy can do.

"Imagine a posthuman child," Cannon says, musing on what this descendent will do with the genealogical knowledge afforded to this future more-than-human entity.

> Using the tools of quantum archeology, she traces backwards through time and space from effects to causes. Sampling a sufficiently large portion of her present, she rediscovers you. Attaining a desired probabilistic precision for a portion of her past, she recreates you. The future you is distinguishable from the present you, but only as the today you is distinguishable from the yesterday you. As if awaking from a night's sleep, you are resurrected, and you learn to do the same for your parents.

And so the chain of resurrection begins, working back through history.

> And, hence, the consolation:
> Although a mourner, I've reason for consolation. My father, Layne Cannon, is gone only until the resurrection of our dead, until the transfiguration of those who remain. In the resurrection, my father will rise and reunite with my family and his friends in love and light, far beyond expectations and hopes that we can now conceive. The sublime esthetic moves me against fear to boldness: I will meet my father in the morning of the resurrection.
> It is here that we return, not leaving the cosmological behind, but rather marrying it to the intimate, inclusive, and possessive which marked our point of departure.
> We have parents, siblings, children and friends who've gone, but they're only absent for a moment. They persist, and we'll soon meet again. Even if you or I depart, we'll yet greet our family and friends, and everyone else who's gone, beyond present notions of suffering and death, in eternity, not as a euphemism for death, but rather as eternal life that's real as light and warm as love.

But this is at once a contingent certainty, if such an oxymoron is not too much. This is a certainty that is hardwired into existence by the universe's creators, but only assuming that our species, and hence other prior divine species-creators, can survive. And it is also a contingent certainty because it requires of us an orientation, an attitude that (up to now) has only been partially realized. As Cannon asks in his closing, "[f]undamentalists, how are you going to save us with authority alone? . . . Atheists, how are you going to save us with science alone?" One stance would place all the labor on the side of God and never take up the necessary task to bring its vision into reality, and the other stance requires a purposeful direction (Cannon calls it "engineering") that it does not have. Both positions need to come to peace with their other and transform themselves in the process. As Cannon states in the closing passage, "Fundamentalists, atheists, and all humanity, we must change. Change! Be Christ! Trust in God, that we'll join posthumanity to the extent it exists, and that we'll make posthumanity to the extent it doesn't. Otherwise, neither science nor religion will save us or our dead from extinction or worse."

§90

Opposites, one anthropologist has said, are two things that are alike in every way but one.[12] Think of sugar and salt, both white, granular, crystalline, edible, but set apart and made each other's antithesis by the one feature of taste. By this light, what we have in the vision of resurrection in Tipler's *The Physics of Immortality* and Cannon's *The Consolation* are more than opposites; they are a single thought, growing at once in two different directions. Both documents are visions of some futuristic restoration of the departed that occurs not through a breach of nature's laws—but through their fulfillment. But this shared dream of resurrection is dressed up in such different intellectual garb in each of these two writings that seeing them both as only different articulations of a single stable common idea would be error. They are different expressions of a shared problem, in different idioms that index different modes of reasoning and different audiences. In other words, for whatever underlying similarity there might be to an abstract and decontextualized presentation of the claims being made, in their concrete context, each piece of writing instantiates different worlds.

But what form do these different expressions of the same take? The easy answer, perhaps the too-easy answer, is that what separates these two pieces can be articulated as a difference of opinion over whether religion or science should be given pride of place when dreaming of a resurrection that is in harmony with natural law. Take Tipler, for instance. It would be a mistake to say that Tipler's presentation is *not* theological. At the least, theology is not cartoonish for him. He intimates that he takes theology more seriously than do many of the people who study theology academically. In his book, Tipler recounts being shocked when, while attending a theology conference, he heard a room of "theologians and religious studies professors" laugh at early medieval wrestling with the question of what material the body will be reconstituted out of as on the day of resurrection: St. Thomas Aquinas struggling with the problem of reconstitution of bodies and the problems posed by resurrection of cannibals (who by definition are comprised of the bodies of others) was quaint to this set of believers. But it is not quaint to Tipler. While his theory of resurrection allows him to sidestep that issue, he considers this topic—the very question that vexed some Mormons when considering the appropriateness of cremation—to be a serious and technical problem.[13]

But for Tipler, the way to take theology seriously is to attend to it by using the language and reasoning of physics, mathematics, and computing. With little apology, his book quickly starts articulating everything through technical lexical filters. In his book, we have scientific notation, the use of c to designate the speed of light, discussions of universal Turing machines, descriptions of virtual Turing machines running only as emulations on yet other Turing machines, and the mathematics of self-replicating von Neumann space probes. There is even a chart for "Important Events in Future History," marking such milestones as the "sun expands to engulf the Earth," "Galaxies evaporate from clusters of galaxies," "Longest-lived stars use all their fuel and become either neutron stars or black holes," "Neutron stars cool to 100 degrees K."[14] And perhaps most striking is the book's reductionist definition of human beings as "finite state machines"—mechanisms that cannot exceed a limited set of experiences and reactions (albeit a finite-state machine that has 10 to the tenth to the fifteenth number of possible states).[15] The use of codes associated with empiricism, science, and engineering does not preclude his making axiomatic statements; for instance, at one point he declares that life will, in effect, reconstruct the geography of the cosmos, transforming everything into a "habitat for the expanding biosphere," rationalizing this claim by imploring us to "[r]emember that in the very long run life has no choice: it *must* take the natural structures apart if it is to survive. So I conclude that it will."[16] This almost *a priori* certitude seems at variance with the technical and scientific empiricism that the statement relies upon. The physics that Tippler bases his argument on have also not aged particularly well. The Omega Point, where there is effectively infinite intelligence with access to infinite energy for an infinite amount of time, is located at the end of the universe, when the cosmos recollapses into a singularity. And while there are still some physicists who feel that a contraction of the universe is not an impossibility, the consensus in the field is that the topology of the cosmos is flat and that the universe is most likely going to expand forever—a state of affairs that would preclude this omega-state singularity from forming.[17] But regardless of the consensus, Tipler's ultimate warrant and native language is still physics.

By comparison, as we saw in *The Consolation*, Cannon speaks piously. It is enough to recall that the essay was written "in reverence" and "as a tribute" to one of Joseph Smith's most well-known sermons, or to recall that stretches of the argument consisted of a long series of biblical references and snippets of

Mormon scripture, all quilted together to form a whole. The hard language of science is effectively absent, appearing only in vague references to technology and posthumanism that lack any of the technical argot that suffuses Tipler's argument. Putting aside its reverent tones, this in itself seems to be enough to suggest that this homiletic form and religious striving mark a contrast with Tipler.

However, there is another difference between the two pieces: where the authors locate agency. Tipler's book removes any agency, or it at least places agency out of the hands of both himself and his reader. It is "life" or "humanity" or "the Omega Point" that performs these works and wonders, and it does so with an automatic surety that makes these actions a part of the geometry of the cosmos. In *The Consolation*, agency is granular, and various figures are spoken about with a specificity, even if there is still the sort of openness about the exact identity of these figures that comes from the use of pronouns and other deictics. And where there is a call to agency, there is also the threat of failure that comes with the imperative mode. People have to do things. They are being exhorted to do things in furtherance of this vision. Consider Cannon's call to both the fundamentalist and the atheist; they must change their ways to help make this world. We see this imperative in other moments as well. In one passage, almost offhandedly, Cannon says that

> I'll make a few more remarks on ritual and engineering. Ritual without engineering is important, and engineering without ritual is meaningless. They are necessarily and inseparably connected. We must empower meaning to become God.

This passage starts with what appears to be a truism, a pair of short, almost aphoristic sentences about ritual and engineering. They are axiomatic, like Tipler's declaration that life persists. Those two sentences may be couched in a different code than that of Tipler, but as declaratives, they can be seen as sharing a certain formal didactic edge with Tipler's statement that "because life must, life will." But instead of thinking in terms of generic life and cosmic inevitability, here Cannon turns to what it is that *we* must do. And we should note that Tipler's "must" and Cannon's "must" are different. For Tipler, the must is the imperative that is the generative force behind life's expansion through the cosmos. But that expansion is never left to any particular mode or collection of life and is in doubt. With Cannon, we can see from his use of inclusive plural pronouns that we are the ones he is speaking to. This is not about what we are ontologically compelled to do but what we are morally obliged to do. And the simple fact that we must do it, in the ethical sense of the term, suggests that still it might not be done.

The difference between what *will* happen in Tipler and *must* happen in Cannon can also be seen in the agency of the one who accomplishes the resurrection. On one hand, it is the Omega Point, a culmination of space-time, that brings us back as it dreamily simulates every person that has ever lived. By contrast, here again is the moment of resurrection, as described by Cannon.

> Imagine a posthuman child. Using the tools of quantum archeology, she traces backwards through time and space from effects to cause. Sampling a sufficiently large portion of her present, she rediscovers you. Attaining a desired probabilistic precision for a portion of her past, she recreates you. The future-you is distinguishable from the present-you, but only as the today-you is distinguishable from the yesterday-you. As if awakening from a night's sleep, you are resurrected, and you learn to do the same for your parents.

Attend to the verbs that are used regarding our posthuman child: *using, traces, sampling, rediscovers, attaining, recreates.* This is the language of labors and efforts. Even the resurrected self, the passive recipient of this work, is by the end of the passage engaged in his or her own activity as newly risen persons *learn* to accomplish the same works as well. Recall the common secular transhumanist critique of religion as a passive, deathist resignation to the fact of mortality, an accusation found in places as diverse as the *Cryonics* letters column and Zoltan Istvan's campaigning. Whether Tipler is engaged with religion, he certainly absorbed passive acceptance. What Cannon presents, by contrast, is a call for *striving*. Whether what Cannon presents is also religion is a different question.

§91

Lincoln Cannon is not the only transhumanist that mourns for his father. He is certainly not the most famous who does so either. In 2012, as part of an interview with PBS News Hour correspondent Paul Solman, Ray Kurzweil was asked a question that did not make it into the broadcast, though it was included in the outtakes posted to YouTube. The question was this: "How much of your project was motivated by [your] dad, his early death, and your yearning for him?"

Context helps with this question. While Ray Kurzweil—as an inventor, as an author, as a businessman, and as Google's "Chief Futurist"—would have multiple projects, the question is asked in a way that suggests that Solman was not referring to some specific endeavor. Rather, we were referring to all these various projects and all of the different professional roles as if they were facets of a single effort. Even this is not quite right since it seems that what we have here is a single endeavor with multiple subsidiary stages, which explains why Kurzweil could be described in so many different ways.

Kurzweil's ultimate endeavor is to defeat death.

Defeating death, in this case, means not just escaping his own death, but also undoing the death of his father. Kurzweil's father died of cardiovascular disease when Kurzweil was young. Kurzweil himself developed type 2 diabetes while he was in his thirties, something that he took as a sign that he was susceptible to the medical conditions that took his father. Kurzweil has responded to this susceptibly by an extended health regimen that famously includes his ingesting "250 supplements, eight to ten glasses of alkaline water and ten cups of green tea."[18] The goal of this regimen is to live long enough to see firsthand the singularity, which Kurzweil dates as occurring roughly around 2045.[19] At that point, Kurzweil anticipates that, due to technological transformation of the species, and primarily due to nanotechnology and artificial intelligence, death will no longer be a necessity. As he stated in a book cowritten with the physician who helped him develop his medical and supplementary regimen, "[t]he knowledge exists, if aggressively applied, for you to slow aging and disease processes to such a degree that you can be in good health and good spirits when the more radical life-extending and life-enhancing technologies become available over the next couple of decades." Kurzweil and his coauthor are both comfortable with (and confident enough about) this to write what amounts to a roadmap for the process, penning a book that "is intended to serve as a guide to living long enough in good health and spirits—Bridge One—to take advantage of the full development of the biotechnology revolution—Bridge Two. This, in turn, will lead to the nanotechnology-AI (artificial intelligence) revolution—Bridge Three—which has the potential to allow us to live indefinitely."[20]

It is evident that the death of Kurzweil's father served as a cautionary tale regarding the costs of ill health; Kurzweil speaks of his father as someone who died before his genius could be either expressed or recognized. But the importance of Kurzweil's father's goes much further than as an object lesson in self-maintenance. Death is not just to be avoided but also possibly undone: Kurzweil wishes to make this denaturing of death retroactive by resurrecting his father. In an interview with Paul Solman, Kurzweil explains how he intends to accomplish this, saying that

> I've got hundreds of boxes of documents and recordings and movies and photographs, and I'm in the process of digitizing all that—uh, an actual good way to express all this documentation of my father who's a musician—I have all of his music, I have all of his letters, I have movies

and other artifacts—would be to create an avatar that an AI would create, that is as much like my father as possible, given the information we have about him, and including possibly his DNA, and, uh, that person would be very much like my father, you could argue that he would be more like my father than my father would be, had he been alive now, which would be, would be, close to a hundred, so it's possible he could be alive, but he wouldn't be similar to the way that he was fifty-eight and died over forty years ago. And, uh, he would pass a Fredrick Kurzweil Turing Test, with me as the judge, or anyone else who knew him.[21]

During this portion of the interview, Kurzweil speaks with a kind of bored, singsong automaticity; he is not ordinarily emotive in his interviews and talks, but there is a mechanical air to this part of the conversation that seems to contradict the wondrous nature of the claims and plans being made.

This automaticity is probably not because of a lack of some kind of psychic investment on the part of Kurzweil; rather, it may be a function of how routinized this discussion has become for him. In a *Huffington Post* interview three years later, he says the same thing regarding his father, in almost the same words. After discussing what he imagines will be the fruits of having swarms of nanites in the brain, he responds to a question about his theory of immortality in this way:

> I'm talking about this concept of a replicant, where we bring back someone who has passed away. It'll go through several different stages. First, we'll create an avatar based on emails, text messages, letters, video, audio and memories of the person. Let's say in 2025, it'll be somewhat realistic but not really the same. But some people do actually have an interest in bringing back an unrealistic replicant of someone they loved.
>
> By the 2030s, the AIs will be able to create avatars that will seem very close to a human who actually lived. We can take into consideration their DNA. In the 2030s, we will be able to send nanobots into living people's brains and extract memories of people who have passed away. Then you can really make them very realistic.
>
> I have collected and keep many boxes of information about my father. I have his letters, music, 8mm movies and my fading memories of him. It will be possible to create a very realistic avatar in a virtual environment or augmented reality. When you actually interact with an avatar physically, it will ultimately pass a Frederick Kurzweil Turing test—meaning he'll be indistinguishable from our memories of the actual Frederick Kurzweil.[22]

Kurzweil sketches the same picture in other network-news stories and feature-length documentaries about him as well.[23] The hardness to the account, the lack of apparent emotion during discussions, doesn't suggest an apathy but rather something foundational, that variations of tone and affect may wash over or around without changing the bedrock.

Despite its anticipation of some marvelous and almost miraculous technology, Kurzweil talks about both immortality and resurrection in very utilitarian terms, giving the project an air of pragmatism. While the exact content of specific steps is vague, which is understandable if one is talking about anticipated yet undeveloped technology, there is always a determinate number of steps (or, as we saw in the above quote about how one is to live long enough to live forever, "bridges"). And we can see an example of his pragmatism in his evasion or, put perhaps better yet, his shrugging off of the question as to whether or not the future simulacrum of his father *is* his father; he dismisses that question as being one of those "philosophical issues, that which you can't really resolve scientifically."[24] Rather than philosophize, his tendency instead is to see these problems as being, in essence, engineering problems that fall to the individual.

This individualism runs through the entire project. Kurzweil's cowritten instructional manual on how to live long enough to see the singularity reads like a self-help book; it certainly places the onus on gaining radical life extension on the person, and not on any wider system. Something similar can be found in discussions of advances in technology: they are either referred to as something without cause, an approach, device, or technique that just arrives one day on its own, or alternately they are the result of some endeavor that a particular individual (usually himself) undertakes.

Kurzweil could be understood as a man who sees himself as realizing a future; this would certainly explain many of his numerous projects. But there is another way to frame this. Kurzweil has an oft-stated belief that exponential technological improvement is wired into the very workings of the Universe, that it is a continuation of earlier exponential growth in human intelligence and toolmaking, and of even earlier exponential growth in the capacities of certain life forms through biological evolution. Given this, it might be better to describe Kurzweil as someone who is trying to position himself such that he has the greatest role in, and reaps the greatest benefits from, this self-realizing future. Hence a shuttling between the agentive-individual and world-historical stages in the form of discrete but revolutionary technologies.

What is missing is an intermediary between the individual and the universal-cosmological, where the responsibility is shared by more than a single person volunteered for the role. What is missing in Kurzweil is a sense of a first-person plural inclusive that indexes a group that seems at once open and also seems to have determinate bounds. What is missing in Kurzweil is a sense of a common, shared project, of capacities coming from the labor of others without resorting to bequests from an infinite Omega-Point-derived God. What is missing is the sacral, agentive, and communal language of Lincoln Cannon. What is missing is a moral collective. What is missing is a social component in the life of the dead.

§92

We are at one of the monthly MTA meetups, and a new format is being tried: a few opening readings, some sacred music played, a recitation of the Mormon Transhumanist affirmation. But after this, the real business begins: a wide-ranging conversation that is still going strong as the evening grows late. The conversation has wandered onto life-extension techniques, and one participant passingly acquainted first-hand with Kurzweil is discussing Kurzweil's supplement regimen. It was not being celebrated; Kurzweil, we are told, is cutting back on some of his supplements as they were making him ill. We are also told that Kurzweil lives in terror of dying before the singularity. Another (distant) acquaintance is spoken about, someone associated with Singularity University, an unaccredited institution that describes itself as "a global learning and innovation community" that uses "exponential technologies to tackle the world's biggest challenges and build a better future for all."[25] The person referenced, it is recounted, refuses to fly as ease of transportation is not worth the risk of dying in a plane crash before "immortality escape-velocity" is achieved. If he were to use air travel, we are told secondhand, he would wear a helmet to protect the brain, because if that organ is lost, one is extinguished beyond any recovery.

Both Kurzweil and this other singulitarian, the speaker comments, live in such fear that they "are not really alive." The circle of twenty-odd people muse on these instances, viewing both of them as, at some level, pitiable. A second-generation MTA member says that in contradistinction to these cases, one of the things that he likes about the association is its focus on resurrection. Focusing on resurrection means that, if something goes wrong and he dies before immortality becomes a technical achievement, there are people who have his back. Thanks to community, he does not have to carry the burden of protecting his potential immortality alone.

§93

What does a social life that includes the dead look like today?

Mormons have a name for the genre of speech we are about to discuss: Testimonial. Testimonials are on one hand attestations as to the proof of the

Bible, of the Book of Mormon, and of the "prophets, seers, and revelators" who constitute the leadership of the Church of Jesus Christ of Latter-day Saints. But as we saw in the discussion of fast and testimony sacrament meetings, testimonials are also a means of fashioning an identity and an ethic, of crafting a narrative for one's self that one can inhabit. They are a way of making sly editorial comments on the Church or fellow believers, a manner of expressing oneself aesthetically, and, when shared before gathered cobelievers in a Sunday Sacrament meeting, a method of performing vulnerability and building community.

But testimonials are, in the end, foremost a means of proselytizing. That is certainly the case with this particular testimonial we are concerned with here. Unlike most testimonials, which are usually short improvised talks, this one was practiced, videoed, edited, and posted at the Church's LDS.org website; the video is full of interstitial scenes and voiceovers that run on top of scenes shot outside the room where the testimonial was given. The video starts by showing a heavyset Southern Californian man wearing a generously sized Hawaiian shirt. He begins to talk, describing his childhood as a family that "was LDS, in a very good LDS ward." As he talks, we cut to an image of an awkward, decades-old smiling school photo, presumably of our narrator. We cut back to him as he continues his testimonial. After high school (jump to a grainy, oddly color-saturated picture of him as a smiling, confident young man dressed in what looks like high-school-prom white), he tells us that to raise money for his time as a missionary, he went off to work in an oil field. But rather than being a step to greater faith, this expedition ends up occasioning a break with the Church. On his return, he tells us that he stops taking his LDS commitment "too seriously." Excommunication follows, as does addiction: addictions to alcohol, to drugs, to pornography. He tells us that he gets "wilder and wilder." While he recounts this narrative, the video cuts to more still photographs of him as a young man, to contemporary shots of him driving (his fingers tap the steering wheel nervously), shots of him walking beachside while the waves crash on the sand. In the background, inspirational music swells, and he recounts shaking some of these addictions through a combination of scripture reading and participation in twelve-step programs. He is better, he tells us, but it is important to note—not all better. He tells us that there were still dependencies that he could not uproot: "coffee and cigs and pornography . . . and sex and blah, blah, blah." He tells us that he was "depressed and despondent." The background music turns slightly maudlin.

We already know this will happen due to the genre we are dealing with: he is saved, of course. He loses his remaining addictions and returns to church.

How he does so is unexpected. He recounts a phone conversation with a brother of his. At this point we cut to the brother. He has close-cropped hair and rimless glasses; he is wearing what looks like a funeral directors' suit and a lavender tie. The brother tells us that our lost protagonist was literally "begging the Lord" to help him "find a way back." The brother recalls that during this pivotal phone call, the brother mentioned in an offhand way to our prodigal protagonist that he had been indexing. As he says so, the brother leans forward ever so slightly in an almost subliminal moment of emphasis. There is a flicker of a smile.

We cut back to our testimonial narrator and hero. He mentions indexing as well, and tells us that on that very same evening, inspired by his brother, he "downloaded the program" and "got into it." At this point, we cut to a shot of him typing with the computer screen visible. As the video continues to play, we cut away too quickly to track what is on that computer screen, but if the video were to stop, what would we see? A gray interface with various icons on top. Underneath the icons, in the primary space of the window, we see the fields that he is in the process of filling out. It gives us "record type," which is being filled out as "deceased," "Deceased's Given Name" (Gina T), deceased surname (Gronseth), "Deceased's Gender" (F), Age (83), "Death Month" (Apr), "Death Day," (25), "Death Year" (1979), "Death Town or City" (Dorvay); "Death County" and "Death State or Country" are also listed fields, though they are not filled out. We cut back to our protagonist, now in profile as he inputs the data, while in a voiceover he tells us that "almost from the beginning, I could feel the Spirit. I could feel a difference." We cut back to the brother, who tells us that our protagonist called the next day to share that he had indexed his first hundred names and was staring on his next hundred.

And there, the transformation begins. Our protagonist realizes that he has gone the longest stretch he could recall not looking at pornography (this turn in his behavior is something that gets him excited). Although still excommunicated, he starts talking to people in his bishopric. All sorts of temptations drop away, including the internet and television. And when they do return, they can be pushed away in "1.2 seconds" by clicking on the "icon" on his computer screen that fires up the indexing program. He has found a way to conquer himself. He prays, reads scriptures, and dreams of being rebaptized. And sooner than he imagined he finds himself accepted by his local bishop and once again a full member of the Church. There is a still shot of him in (re)baptismal white, arm wrapped around his brother as he tells us that he now "just doesn't feel like doing anything wrong when I'm indexing . . . . it brings the spirits almost at will. Redeeming the dead has redeemed me. And it can work for anybody—anybody."[26]

§94

The link between baptism and indexing exceeds this single testimonial, much as the link between our testimonial's protagonist and indexing exceeds family ties. Let us start with that latter point and work back towards baptism. The indexing that both our prodigal son and his still-faithful brother were doing was not the creation of a genealogical database of their immediate family members. Other than the hypothetical kinship that comes from a common descent from Adam (a common, but not universal, Mormon understanding), those names that came in batches of a hundred for voluntary data entry were almost certainly not the names of any relatives of theirs. Rather, they were names of the deceased, pulled from various digitized records. Nor are these names being indexed by one person alone. Each name is indexed by two different people, in an attempt to ensure that they are being rerecorded correctly (instances of disagreement between indexes of the same individual are refereed by a third person).

The indexing here is in service of baptizing the dead, a ritual carried out in Mormon Temples worldwide. During the ritual, an individual dressed in white robes will be submerged in a large baptismal font held up by twelve columns, each in the form of an ox. The submerged individual is not being baptized in his or her name. Instead, he or she is serving as a proxy for a known, dead individual, a person who is believed to have not been baptized (at least during worldly life) in the Church of Jesus Christ of Latter-day Saints.

One usually doesn't serve as a proxy for such a baptism just once. This ritual is often repeated with an almost Fordist dedication to mass production. On a single day, someone may be baptized in the name of several individuals; in some temples, the names are shown on a teleprompter so that the person performing the baptism can say them aloud. Typically, the proxies are teenage LDS members, brought in as part of their LDS religious education, so there is often a line of young men or women, one stepping into the font when another leaves to keep things moving at a pace. It is not entirely unheard-of to be baptized in the name of an actual relative; the church does recommend that one have one's own kin baptized, covering the four previous generations. But except for these times when the proxy goes out of the way to arrange standing in for their kin, one is usually baptized for an unknown individual. Parenthetically, this is much like how one rarely indexes one's own family members; it is the wider project, not individual ties, that is being attended to. Since we are speaking parenthetically, we should also note that at least in Mormon understanding, this act does not coerce the dead to become Mormon. The usual account is that in the afterlife the departed must still willingly accede in some way to the baptism for it to take. But, hypothetically, if Mormon baptism of the dead is effective and the Mormon cosmological vision is accurate, it is hard to imagine under what circumstances the dead would demur.

§95

There are other ways of reaching out to one's ancestors.

If you google the corporation Ancestry.com and the term *Mormon*, one of the first hits, immediately after a link to the Wikipedia article on the corporation, is a "people also ask" box, which contains the following questions in a drop-down menu: "Is ancestry.com owned by the Mormon Church?" (No), and "Is ancestry.com free for LDS members?" (Yes).[27] These seem like reasonable questions, given the company's origins. Two BYU students founded Ancestry, and for its first few years, the company made money selling various Church publications on floppy disks. Later, it transitioned into an online information and indexing service, where for a small monthly fee, one could access a trove of literally billions of historical records selected for their genealogical value: draft cards, census material, newspaper pages, and the like. These roughly eleven billion records have been used to create "more than 50 million family trees containing more than five billion profiles."[28] Ancestry has not been resting on these genealogical laurels. In 2014, they released the Find a Grave iPhone app, which itself as the "world's largest collection of burial information." (Occasionally the Find a Grave website gets new data on burial grounds from Mormon missionaries doing service for their community.) More importantly, in 2012, Ancestry also began offering genetic testing that purported to show a person's origins. Based on information from customer DNA (extracted from a small sample of spit mailed to the corporation in a prepackaged vial), Ancestry creates online percentage breakdowns and maps, used to convey the business's estimate of the various constituting ethnicities from which the customer has descended. They also indicate DNA matches with other Ancestry customers (this latter feature can be foreclosed by selecting different privacy options regarding one's genetic information).

One could be skeptical about the value of this genetic data. These regional matches are based on the tested individual's DNA matching patterns of single nucleotide polymorphisms that the corporation has associated with specific regions of the world, based on proprietary technology. In the words of one biological anthropologist, this is a vision of things that presumes "a historical model of archaic pure races who lived at the margins of the Old World and subsequently commingled to produce modern peoples." That model, mapping specific genes onto specific regions, "was actually a nineteenth-century gloss on the fates of the biblical sons of Noah, who spread out to the corners of the earth and then became fruitful and multiplied—but that model intersects with reality at no point."[29] Of course, for those Mormons who are skeptical of evolution, a model of race ported directly from Genesis may be more feature than bug.

The questionable nature of the implicit racial theories informing the insights about one's progenitors does not mean that this information is of no value to other parties. In 2015, AncestryDNA, the name for the subsidiary that handles the direct-to-consumer genetic testing, announced a partnership with Calico, the Google spinoff responsible for researching life-extending technologies (or in the words of *Time* magazine, dedicated to defeating "death itself").[30] Under the agreement, Calico and AncestryDNA agreed to "investigate [the] human heredity of lifespan" through evaluating "anonymized data from millions of public family trees and a growing database of over one million genetic samples."[31] The results of the research were perhaps not what was hoped for: they found "that the true heritability of human longevity for birth cohorts across the 1800s and early 1900s was well below 10%, and that it has been generally overestimated due to the effect of assortative mating."[32] The partnership between the two entities was quietly dissolved.[33]

This happens in science, and to the degree that a study closes off false leads, the research that was the fruit of this collaboration can still be considered a success. But the true, lasting contribution to a sense of intimacy with one's progenitors that will be made by Ancestry's compendium of genetics samples and genealogical records may be of a different sort.

§96

Ancestry and the MTA are no strangers to one another, or at the least, Ancestry is not a stranger to the MTA. Both a past president of the MTA and the president of the organization at the time of this writing worked at different times for Ancestry. They were not alone; there have been times when as many as five MTA members were simultaneously working for the company, which is considerable given the relatively small size of the MTA. They were even at times a part of Ancestry's public face. Lincoln Cannon, for instance, wrote for the Ancestry blog, a website that features small stories about what sort of genealogical discoveries might be made, what resources Ancestry has, or how some of the technical aspects of the venture operate. Cannon's contribution is more along the lines of the latter, a discussion of the technological possibility conditions for Ancestry's existence; as he writes, adopting the voice of an Ancestry insider, "technology, and particularly computing, is essential to our mission to help everyone discover, preserve and share family history. Without it, we could still tell family stories to our children, but we certainly couldn't substantiate those stories from 12 billion historical records into 55 million family trees through the work of 2.7 million subscribers, as Ancestry.com does today across all its websites." Cannon addresses it, though, by referencing Moore's law— again, the observation that the ratio of computer capacity to cost of that capacity doubles every few years. It was only this ramping up in computing power that could make this ocean of records navigable. But Cannon goes furtherer. He also extrapolates Moore's law forward, predicting that, "[w]ithin a century, $150 could purchase more computing capacity than that of all human brains combined." This means that Cannon's meditation on the contemporary possibility conditions of Ancestry is a reflection on the future possibilities for the corporation as well. The future of Ancestry, though, is not simply greater in scope, but also different in kind. He suggests that through something like a melding of technologies similar to Google Earth, Oculus Rift, and Second Life, "[o]ne of the things we might do is tell stories about our family and ancestors at a much more massive scale and at a far deeper level." He even teases that such technologies would have "rather shocking philosophical ramifications" that he would address in his next blog post.[34]

But there never was a next Ancestry blog post from Lincoln Cannon.

§97

One morning, I was speaking with an MTA member and Ancestry employee, someone with a reasonable level of seniority in both organizations. His voice was gentle, his words carefully chosen. This trait allowed his speech to flow at a thoughtfully smooth pace, though he was prone at times to breaking that rhythm, racing ahead to excited, and ever-so-slightly dramatic, crescendos, when speaking about topics that particularly animated him. He was distinct in that he had a well-ordered beard with a sweeping and slightly waxed mustache; this was a break from the usual male Mormon Transhumanist style of facial hair (either clean-shaven, or a short office-casual beard).

It was a wandering, large-roaming discussion, which ran from this history of the Mormon internet (this was the man who coined the word *Bloggernacle*, which he originally presented as *The Bloggernacle Choir*) to where Mormon's speculative metaphysics is discussed and where it is not. Speculation about eternal progression and becoming gods will not be found in church periodicals like *Ensign* or sacrament meetings, he noted, but rather in internet conversations, private conversations, the back catalog of the Church-owned Deseret books, and temple ceremonies. During the conversation, I also asked about resonances between the MTA and Ancestry. He replied by mentioning in passing that Ancestry recently

> actually announced a really interesting experiment that they've done
> . . . by obtaining DNA samples from a fairly broad sampling of people
> who are related to each to other, we were able to do a partial genome
> reconstruction of one of their ancestors, a dead person whose DNA
> we do not have. This is really a fascinating step in that direction.

In the conversation, though, this virtual genomic recovery from the grave was presented as a first step, a mere proof of concept, as it were. This was in part because the DNA was not enough, but it was also because Ancestry had the capacity to recover so much more.

The same allure that Ancestry had for Calico, a way to match DNA samples and genealogical history in heretofore undreamt detail, could be the seed of even more audacious future projects:

> So, I think our expansion into the DNA space really opens up a host
> of additional possibilities that—the DNA by itself is not going to be
> sufficient. DNA as a repository of information is certainly critical, but
> the way that it ends up expressing itself is very dependent on environ-
> mental factors, the environment and all those sorts of things is really
> key. So, it's really a combination of DNA plus all kinds of other infor-
> mation that provide the richest idea of what these past people were
> like, who they were. And really, this is one of the reasons that I think
> Ancestry has become, it is the fastest growing consumer genomics
> database right not, and if its not already the largest—I believe it is—
> then it will be in very short order. And I think this is because people
> recognize that there is a very valuable combination between the histor-
> ical information that we have and the DNA information that allows
> you to discover and understand both yourself and your family, your
> ancestors in ways that aren't possible with just one or the other.

Part of the promise is that this library may make Ancestry something unique: the most complete single record of our species. "You know," he added, "I will say that, clearly, Ancestry does not articulate it, I think at the most senior levels [it] doesn't really envision itself yet, as this repository of the best information of every single human being who has ever lived, with a view to potentially reconstructing these people."

Anticipating a more powerful implementation of the same technologies Lincoln Cannon mentioned in his own blog post, this would mean a form of almost complete reconstruction. Speaking of how the various engineers and computer programmers imagine ancestor recovery, my interlocutor said that below the highest echelons of management,

> at the next level down, we definitely have conversations about this, quite explicitly we talk about an experience we imagine we could produce sometime in the future that would be something like the holodeck for *Star Trek*, where you could actually step into your ancestor's life, you could have a conversation with them, you could experience what their life was like, and this is something that at like my level, with my peers and my director and people at that level, we actually, have this, you know, forward-looking vision. That's not that far from the simulation argument, and that type of approach that the MTA is serious about.

This speculation was not dreamy-eyed. He noted, for example, that there were profound ethical challenges: what would it mean for our kin, or at least for entities modeled to be our kin, to also be corporate intellectual property? Still, this was set more in the key of a difficulty to be overcome rather than a fatal flaw that would make this project morally inconceivable.

This was a project that was not even at the stage of being vaporware.[35] It was, like every project or imagining spoken of in this series, an exercise in speculation. There is almost certainly no one in the Mormon Transhumanist Association who would say otherwise, and if that is the case with the MTA's more adventuresome souls, it seems unlikely that there was anyone at any level of Ancestry who would disagree with that claim, either. And yet there is a solidity to this last virtual ancestry project that gives it a different grain. It seems more agentive than waiting for cosmic expansion of intelligence and a collapsing universe's end of time, more extensive than a single set of files set aside to reanimate a single person. It is almost as if, in the din of corporate-owned server farms and cyclotrons, something else could be heard: the first stirring of the posthuman child from *The Consolation* waking to stretch her hand back in time to pull her ancestors out of the darkness of the grave.[36]

Eighth Series: Worlds without End

§97

Recall the Ancestry.com blogpost that wasn't there: Lincoln Cannon's promised follow-up to the idea of what a twenty-second-century genealogical corporation's interactive software might look like. This absent sequel promised some "rather shocking philosophical ramifications" resulting from the creation of an immersive world of simulated ancestors. The reason that this second piece was never published was that, as Cannon put it, Ancestry was "worried about the reaction of customers." It was, he said, "too much reality too fast."[1]

This may be an odd way of paraphrasing Ancestry's objection, because the post in question could be read as saying that there is no such thing as *reality* at all. The title of the piece was (or at least would have been), "Are we Living in a Family History Simulation?"[2] The post was another moment in the career of the simulation hypothesis: the idea that what we take to be the physical reality in which we are corporally embedded is actually a computer simulation, a program created by super-advanced beings. That virtual universe is usually spoken about as being most likely (but not certainly) an *ancestor simulation*, something not unlike what transhumanist-leaning engineers and programmers at Ancestry envision: a computer world peopled with sentient—but still virtual— entities consciously modeled after the progenitors of this simulation's authors. In the post, Cannon presents this idea by envisioning two people at some indeterminate future date using the latest, most immersive version of Ancestry-like software.[3] Taken in by the verisimilitude of this world, one of the two explorers turns to the other and asks whether it is possible that their own world, that is the ostensibly real world that houses the computers and programmers that made this near-perfect ancestor simulation, could, in turn, be an ancestor simulation for someone else—a rendering of the past of some even higher-level, actually corporeal reality.

§98

This thought is not unique to either Lincoln Cannon or to Mormon Trans-humanism. In 2003, Nick Bostrom, a professor of philosophy at the University of Oxford, published an academic paper with the then-odd title of "Are you living in a computer simulation?"[4] Bostrom's essay argues that if, in the future, humans are both capable of and interested in producing computer simulations of the past, and if these future humans ran multiple simulations (either con-currently or serially) then it follows that the number of humans and civilizations that exist only in such simulations would radically outnumber those that exist in the nonsimulated universe. Therefore, the chance of us being situated in reality as opposed to simulation is statistically quite small; perhaps so vanish-ingly small that it is effectively zero. It may be hard to imagine a world where such computing power is available. But Bostrom's expansive response to this problem includes envisioning worlds as possible computers; among the possible computing technologies that would be capable of running a simulation—which run from quantum computers to machines that use "nuclear matter or plasma"—he also includes the conversion of "planets and other astronomical resources into enormously powerful computers." Later in his essay, he states that a planet-sized computer, using only "far from optimal" contemporary nanotechnology design, would have a computing power of about 10^{42} operations per second. Computers of such a scale may be necessary for the project. While acknowl-edging that "it is not possible to get a very exact estimate of the cost of a realistic simulation of Human History," he suggests that we can use something ranging from 10^{33} to 10^{36} computer operations per second as a "rough estimate" of what would be needed to do so, noting that "[a] single such computer could simulate the entire mental history of humankind (call this an ancestor-simulation) by using less than one-millionth of its processing power for one second."

It would be an understatement to say that this paper has been well-received by the wider transhumanist community. Futurist icons such as Ray Kurzweil and Elon Musk have publicly toyed with the idea—backing it with varying degrees of certitude but backing it all the same. Some have even gone further than merely acknowledging the hypothesis and have used it as a foundational step in the construction of an ethos. The academic economist and transhumanist Robin Hanson, for instance, has noted that in our society, at least, simulations are prized for their "story-telling and entertainment value." Therefore, it behooves us to behave as entertainingly and unpredictably as possible, lest those running the simulation become bored and turn the simulation off. More darkly, he also suggests that "your motivation to save for retirement, or to help the poor in Ethiopia, might be muted by realizing that in your simulation, you will never retire and there is no Ethiopia." As Hansen distills his argument: "if you might be living in a simulation then all else equal it seems that you should care less about others, live more for today, make your world look likely to become eventually rich, expect to and try to participate in pivotal events, be entertaining and praiseworthy, and keep the famous people around you happy and interested in you."[5]

Bostrom also sees religious and ethical overtones in his idea:

> Although all the elements of such a system can be naturalistic, even physical, it is possible to draw some loose analogies with religious conceptions of the world. In some ways, the post-humans running a simulation are like gods in their relation to the people inhabiting the simulation: the post-humans created the world we see; they are of superior intelligence; they are 'omnipotent' in the sense that they can interfere in the workings of our world even in ways that violate its physical laws; they are 'omniscient' in the sense that they can monitor everything that happens. However, all the demigods except those at the fundamental level of reality are subject to sanctions by the more powerful gods living at lower levels.

This theology comes with an implied ethics, one that stands in sharp contrast to those Hansen imagines:

> Further rumination on these themes could climax in a *naturalistic theogony* that would study the structure of this hierarchy, and the constraints imposed on its inhabitants by the possibility that their own actions on their own level may affect the treatment they receive from the dwellers of the deeper levels. For example, if nobody can be sure that they are at the basement-level, then everybody would have to consider the possibility that their actions will be rewarded or punished, perhaps using moral criteria, by their simulators. An afterlife would be a real possibility. Because of this fundamental uncertainty, even the basement civilization may have a reason to behave ethically. The fact that it has such a reason for moral behaviour would of course add to everybody else's reasons for behaving morally, and so on in a truly virtuous circle. One might get a kind of universal ethical imperative,

which it would be in everybody's self-interest to obey, as it were, 'from nowhere.'

But Hansen and Bostrom are exceptional in the seriousness with which they take these discussions; for the most part, the simulation hypothesis is treated as idle speculation, something that, if true, has no real import for these transhumanists' current affairs.

§99

For many Mormon Transhumanists, however, the simulation hypothesis is the ethical and ontological scaffolding of the universe. I have heard MTA members publicly state and share in private conversation that they are certain they live in a computer simulation. And I have also heard first-hand from people who anticipate creating such simulated worlds themselves later on in their posthuman careers. One MTA member wondered if he had the strength of character to eventually create such a simulation; noting that "the world is a hard place," he worried that he might be too empathetic with his creations to placidly observe their suffering as he presumably tested and educated them for their own later creation of simulations within their strata of (virtual) reality. This concern echoes the Mormon tradition of "the God who Weeps," where God himself suffers with his children on Earth as they are tormented by the tests of the mortal realms. The only difference is, in this case, our potential God is a programmer.

This is not to say that we are dealing with the equivalent of simulation-hypothesis fundamentalists here. People show interest in other accounts of how this world might have been divinely crafted and how they might someday craft worlds after worlds themselves. Their language sometimes indexes different causal theories in the same conversation; the Mormon Transhumanist who was worried about the quality of his will for world-creation also jokingly referred to God as an alien several times in this and subsequent exchanges. And finally, most of the Mormon Transhumanists who still endorse and affiliate with the Church also had the ability to deploy the more standard and recognizably Mormon language of creation, such as one might hear during a sacrament-meeting talk or a fast and testimony-witnessing. But even if we are not simulated beings, that still does not mean that in a Mormon Transhumanist imagination, we are not just one moment in a recursive operation that stretches out to infinity on both ends. For Mormon Transhumanists, future trajectories and possible origin stories are the same thing, another moment in an endless, recursive cycle.

§100

In some ways, the simulation hypothesis is also part of the ethical and onto-logical scaffolding of the *institutional* Mormon Transhumanist Association as well. One of the first steps the MTA took was to author an article for *Sunstone*, a magazine that presents its mission as sponsoring "open forums of Mormon thought and experience." *Sunstone* is closely associated with the believing wing of progressive Mormonism, has a definite historical and intellectual bent to it, and is also known for regularly organizing conferences to facilitate the exchange of ideas by various Mormon public intellectuals. For these reasons, it is influ-ential in Mormon intellectual circles. It says something about both the indus-triousness of Mormon Transhumanists and about the novelty of their message that only a year after being founded, the association should be able to jointly author a cover story for a periodical of this stature in the Mormon intellectual community.

Their essay, though, does not read like other *Sunstone* pieces. Most Sunstone articles are single-author essays, presented in an academic or sometimes mem-oirist voice. This article, though, credited jointly to "Members of the Mormon Transhumanist Association" and bearing the title "Transfiguration: Parallels and Complements between Mormonism and Transhumanism," at first reads more like a piece of Mormon religious apologetics or even an artistic manifesto than like the usual literary, historical, or philosophical works that typically populate *Sunstone*'s pages. In fact, redolent with a kind of unquestioning surety about the truth of traditional Mormon religious claims, the article has an al-most fideistic edge. Consider, for instance, the two opening sentences:

> Mormon Tradition teaches that, throughout time, God has inspired
> and endowed humanity with knowledge and power in various dispen-
> sations, or epochal transitions in the relationship between divinity
> and humanity. In ours, the 'dispensation of the fullness of times,' God
> is restoring all the knowledge and power of past dispensations while
> continuing to inspire and endow us more rapidly than in the past, to
> prepare for a greater future.[6]

Note the use of the first-person plural in the last sentence, unquestioningly without qualification presenting the current age as *our* dispensation. Also note the use of *is* in that sentence, suggesting that this restoration is not hypothetical but a certain, ongoing process.

This piece may start in this serious, religious register, but it does not stay that way, though it takes its time to change its footing. It is only pages further when the tone of the article shifts in a (now) familiar Mormon Transhumanist direction and begins to not merely recount the various millennial and celestial promises found in the Bible and Mormon Scripture, but to also speculate on the mechanisms that have or will have brought these promises about. The specific pivot is on the believability of these promises; they claim that supernatural promises that can be rendered more technologically concrete can both give these processes a sense of reality and also provide a purchase for those who wish to be more agentive about making these promises come to fruition. As the article says,

> [M]any Mormons find it difficult to exercise faith in ideas and teachings like these except in abstract ways. Hence, some respond to these teachings by concluding that certain ideas are simply beyond our mortal capacity to understand. This response leads in turn to the question: What is the practical value of a belief in something one cannot understand? How can one possibly have faith in an idea if one does not understand it? This criticism is especially challenging to Mormonism, which emphasizes the importance of faith manifest in works. How can one work toward a future that is understood only vaguely, if at all?
>
> We, the contributors to this article, felt much this same way. During conversations across several years, we have observed that, although our faith was active in relation to many tenets of Mormonism, it was mostly passive in relation to the more concrete aspects of future salvation: transfiguration and resurrection to physical immortality, the paradisiacal glory of millennial Earth, the organization of new worlds, and so forth. We found our passivity toward these ideas troubling. What is the effect of faith that is not active? Is it even faith? If not, how can we change so that our faith is active in these ideas that we value? Beyond that, we wondered: Might active faith in these ideas be essential to realizing them?
>
> As we discussed ways to promote active faith in a Mormon view of the future, we observed that, in the broadest sense, science and technology are among the most obvious manifestations of active faith in the future: fighting disease and illness, improving communications, cleaning and beautifying environments, and extending life spans. This realization led us to ask how we could promote the application of science and technology to a Mormon view of the future, to bring these into dialogue with the plan of salvation, and specifically Mormon ideas about the exaltation of humanity to godhood.[7]

With this, the fideistic mood is broken and a new kind of religious empiricism and experimentalism called for. It is here that transhumanism comes sharply into focus. The usual instructional points are made (Moore's law, the importance of exponential growth, the possibility of a singularity). And various specific technologies are presented as allowing us to escape death and become something more akin to gods (genetics, nanotechnology, robotics). Again, what they say is at this point familiar to those already exposed to Mormon Transhumanist thought, but at the time this articulation was unprecedented. These phenomena are more than just balms for suffering and ways of overcoming ourselves; they give the reader ways to imagine divinization and eternal progression of knowledge as technical processes.

The list of conditions and innovations in the piece do not give ways of understanding world-making. This absence is surprising. World-making is essential; after all, "the discovery and creation of worlds without end" is listed by the authors of the piece as one of the objects that "scripture and prophetic commentary" make "replete" references to.[8] Given the importance of using transhumanist thought to give substance to Mormon cosmological anticipations, it is striking that the only real world-making technology the piece lists is computer simulation. But this is not just our simulating other worlds. It is also other worlds simulating ours. The only transhumanist account of the Mormon doctrinal claim that "others may have produced us, and we may yet produce others" is the observation that "nonhumans in another world may very well be simulating our world, and we may eventually simulate yet other worlds."[9] Looked at against the light of the other anticipated transhumanist innovations, it is the simulation hypothesis that allows for a technological account of the capstone of Mormon theosis. And hence it is also what allowed for Mormon Transhumanism to look like a complete vision rather than a mere loose alignment of slightly rhyming transhumanist and religious concepts. One may wonder if, absent the simulation hypothesis, there could have been an articulable Mormon Transhumanism at all.

§101

The simulation hypothesis stands apart formally from other transhumanist musings. Other singulitarian speculation focuses on technology that, in the future, we may be able to use to transform our nature. The simulation hypothesis, however, is not a claim about what may be possible in the future. Or rather, it is not just a claim about what may be possible in the future. This is because, in many ways, there is something infelicitous about using the simulation hypothesis as a supplement for Mormon Doctrine. The first difficulty is that as a word, *simulation* conveys a feeling of either falseness or hallucinatory unreality; in a sidebar to the MTA-penned *Sunstone* article, entitled "Is a simulated world 'real'?," the authors even quote the somewhat non-Mormon school-yard bromide "what have you been smoking?" as an example of the computer-generated-virtual dismissed as either addled or delusional. Their response to this objection is that living in a simulated universe does not erase either the gravity or the intensity of this world; living in a simulation may mean that we would have to adopt a different ontology, but it would leave our phenomenology the same. As the sidebar puts it, "whether we are living in a world computed by neohuman gods, our experience now remains the most real we know. We continue in pain and pleasure, misery and joy. Our memories neither decrease nor increase in poignancy."[10]

Perhaps more serious is the second issue: the obstacle implicit in Mormon thought regarding materiality. As mentioned earlier, one of the post-Copernican marks on the Church of Jesus Christ of Latter-day Saints—and the other churches, factions, and fundamentalist enclaves that see themselves as also following in Joseph Smith's wake—is to insist on the materiality of all forces in the world; *everything* is material. This included some of the entities and forces that are commonly more ephemeral or transcendent in the non-restoration Christian imagination. For the Mormons, God has a body of flesh and bone, and spirit is (a more refined form of) matter.

The difficulty that this presents the simulation hypothesis is apparent: depending on one's metaphysical interpretation, the computer-virtual is an effect of matter, but arguably it is not itself matter. A simulated reality seems to not only point to the existence of some immaterial entities and forces in the world, but in some ways, this claim could work to seep the very materiality out of everything, at least in this particular frame of existence. The usual way concerned MTA members respond to this objection is to conflate the reality of experience with the (presumably) material computing mechanisms generating this experience. It is still a material process, located in the far-future or post-singularity equivalent of processing chips, graphics cards, and RAM. It is just a far different material process than our senses, reason, and scientific investigations would lead us to believe. But just as often the response is not to address the problem at all, not so much as a form of evasion, but rather due to the way that the simulation hypothesis compels.

There are, after all, other approaches to the question of forging worlds after worlds.

§102

I am speaking with a Mormon Transhumanist, and he is unhappy. It is not the transhumanism that makes him unhappy, but the Mormonism, he wishes to be clear. He works as a science educator at a state university in a small Southern town and is part of a small but active and relatively accepting Mormon ward that is well-integrated within the wider community. Despite his warm feelings for the fellows in his ward, the Church often tries him. He speaks of the frustration he experienced when he expressed an early interest in evolution, back when he was attending high school on the Wasatch Front. His parents had no difficulty with his openness to the biological principle, but classmates saw evolution as something taboo, and the school required a signed parental permission slip to even attend the class sessions addressing Darwin's thought and legacy. Later, he supported the organization Ordain Women when they made a push for priesthood for women. He dislikes the Church's political conservatism. He grows his hair long and has a beard, a combination which is an incredible rarity among Mormons in good standing with the Church.

With all that, though, he wants us to know that this unhappiness does not equate with disbelief. He attends sacrament meetings with a healthy degree of regularity even though his wife (who now considers herself to be an atheist) doesn't attend. He is unshaken when he reads the CES Letter, an online document that is in effect a compendium of extant charges against Joseph Smith's character, the *Book of Mormon*'s historicity, the nineteenth-very-early-twentieth-century practice of polygamy, and the institutional Church's honesty. He recognizes that the Church has hurt many Saints, and also acknowledges many of the points made by the CES Letter, but in the end, he feels that most of the document's other points miss the mark, and that there is an unfair and unreasonable assumed standard of complete inerrancy that the CES Letter uses to measure the Church, the *Book of Mormon*, and Joseph Smith. On his own time, he has conducted stylometric analyses of the *Book of Mormon*; he reads the result of this research as suggesting that the various constituent books in the *Book of Mormon* were written by different authors, and perhaps even by the various prophets credited as penning these eponymous books, rather than by a single individual (which is what one might expect if the *Book of Mormon* was merely a fraud perpetrated by Smith). He even sees his work as a university educator as part of building Zion.

He credits the MTA as having worked to shore up his faith and to keep his unhappiness with the Church in perspective. The MTA has given him an intellectual community and new ways to think about his religion; it has allowed him to develop an attitude toward the Church that is, as he puts it, "good for my mental health." It has allowed him to embrace a hopeful millennialism and reject a hard and more traditional Mormon Apocalypticism.

But he doesn't believe in the simulation hypothesis.

When asked about the simulation hypothesis, he says:

I imagine actual creation of universes, rather than [a] simulation. And it just seems to me like—and I'm not sure this would be true—and I think they talk about it as having a basement or world, and then you simulate one upon that, and if I understand correctly, information theory says that every stage you're going up [you have] to lose a little bit of information, and so while you may be able to go on for a very long time with simulations and simulations and simulations, you'll eventually reach a point where there's too much information lost in the substrate that's being simulated. There's a limit to it. And so, I imagine— I don't discount the idea, but it's not what I imagine or aspire to. Even though I think it may be a step in the direction of—it may be one step in the process of learning how to create universes. Most of what we do now, we simulate before we try it. And, and maybe I think you probably could have real people, or people as real as we are, or indistinguishably different, or indistinguishably [not] less real inside a simulation, so I don't discount it, but it's not the goal I aspire to. I think of it more biologically, of biology and physics, and information theory.

I ask him about the time frame involved in physically creating a universe (as opposed to running a universe simulation); I am thinking that such a project must be eons away. He (politely) disagrees.

I see it as a problem that is already solved, and I think within Mormon theology we've participated in it already at some level, I think we've participated in creating the world we live in now. And so, I think we may already have a lot of the skills for doing it, that when memory returns, we will already have a lot of the skills for doing it. And I think it's the kind of thing you can teach, I *suspect* it's the kind of thing you can teach, just the same way you can write a computer program, give it to someone, and say "use this software, and you can modify things." I see it as a big project, I no longer view it as one God doing it all, unless God is a composite entity, which I also consider a viable definition, even though I tend to think of a personal God still, but I see it as a community of Gods, and I imagine a community of Gods doing all these things, so . . . and specialization, without everyone knowing everything, because I don't think anyone can know everything, even the Gods, and so . . . time frame? I'm not sure how much it matters because it's an already solved problem, it's one that we probably already have a bunch of knowledge about, and someone will be able to teach it to us, once we show that our character is trustworthy for having that much power.[11]

At other points in the conversation, he talks about black holes creating universes; he seems to be suggesting that this suspected phenomenon involving (gravitational) singularities is something that could be replicated in a lab. This is an account that is deeply Mormon, down to the idea of preexistence and the world as a means of educating and testing of moral character. But it is also strange, even among Mormon Transhumanists. The idea of a physical creation of universes through identifiable naturalistic processes is something that, at the time of our conversation, I have not heard before.

This conversation is enjoyable—he is pleasant, gracious, thoughtful, and motivated; like many of the other people I speak with during this project, he comes across as someone easy to be friends with. But while he is talking, I am wondering to myself if this is a case where an intelligent person has, over time, talked himself into an insane position. After all, what educated, twenty-first-century American—much less what university-trained scientist—could take seriously the possibility of a universe created through naturalistic means bearing messages from its creator?

§103

The name of the peer-reviewed academic journal is *Modern Physics Letters* A. The journal's contents "consists of research papers covering current research developments in Gravitation, Cosmology, Astrophysics, Nuclear Physics, Particles and Fields, Accelerator physics, and Quantum Information."[12] It has a respectable impact factor (1.308 for the year 2017, for instance), and more to its credit, it has published articles by researchers who were later awarded Nobel Prizes. Sometimes those awards were for research presented in the journal.[13]

In 2006, *Modern Physics Letters* A published an article, cowritten by Steve Hsu (then at the Institute of Theoretical Science, University of Oregon) and Anthony Zee (who was and still is with the Kavli Institute for Theoretical Physics, University of California, Santa Barbara). Hsu's faculty page describes him as working primarily in "applications of quantum field theory, particularly in relation to problems in quantum chromodynamics, dark energy, black holes, entropy bounds, and particle physics beyond the standard model," noting that he "has also done work in genomics and bioinformatics, the theory of modern finance, and in encryption and information security."[14] Anthony Zee's Wikipedia page states that he has "authored or co-authored more than 200 scientific publications and several books" on such topics as "particle physics, condensed matter physics, anomalies in physics, random matrix theory, superconductivity, the quantum Hall effect, and other topics in theoretical physics and evolutionary biology, as well as their various interrelations."[15]

The title of their joint article is "Message in the Sky."[16] It starts by asking this: "Suppose that the Ultimate Designer of the universe (assuming that there is one, an assumption that we do not address here) wanted to notify us that the universe was intentionally created. The question we would like to ask is: How would this Superior Being send us a message?"[17] Moving quickly to still any questions of the plausibility of a purposefully designed universe, the authors mention in an almost offhand way that "[w]ithin physics, several authors have speculated on the creation a universe in a lab using inflationary dynamics of false vacuum bubbles."[18] It then goes on to ask how this hypothesized lab worker could pass along a message. Species genomes and natural wonders are too narrow, too limited to one world: "The Ultimate Designer would clearly want all the advanced civilizations, not just in our galaxy, but in the entire universe, to know."[19]

The cosmic microwave background (or CMB) is something that would fit the bill. "We have convinced ourselves that the medium for the message is unique: it could only be the cosmic microwave background. The cosmic microwave background is in effect a giant billboard in the sky, visible to all technologically advanced civilizations. Since different regions of the sky are causally disconnected, only the Being 'present at the creation' could place a message there." Specifically, they are thinking of the probability distribution of the degree of variance in the CMB, since the view of the actual pattern made by these thermal fluctuations of the sky vary with location. We are also given some proposed possible content for the message ("we suggest that the coded message would simply be an announcement along the line 'Hey guys, the universe is governed by gauge theories, and the relevant algebras are such and such'") and even a hypothetical method through which that message could be inscribed ("suppose that before the Big Bang, the Creator fine-tuned the inflation potential with small deviations from flatness . . .").[20]

It is the conclusion, though that is most striking for present purposes. The authors offer, in essence, an areligious picture of the Temple-Mormonism speculative cosmology presented in a way that echoes many aspects of the Mormon Transhumanist imagination:

> Coming to the end of our short paper, we allow ourselves to indulge
> in a wild speculation. Even if a hidden message turns up in the CMB,
> nothing in this paper requires that the Ultimate Designer is theistic

rather than deistic in character. Perhaps the Ultimate Designer could himself, herself, or itself live in a theistic universe. There may be levels of universes, with Ultimate Designers all the way up. Even if our own universe, which we now understand to occupy the lowest level, may be merely deistic, some of these universes could well be theistic. We could easily persuade ourselves that the notion of a hierarchy of universes is more appealing than the notion of casually disconnected parallel universes.[21]

This is nothing more than "the discovery and creation of worlds without end."

§104

The official hymnal for the Church of Jesus Christ of Latter-day Saints is a mix of Protestant standards and Mormon history- and doctrine-oriented sacred songs of more recent vintage. "If You Could Hie to Kolob" is numbered among these latter hymns; the lyrics, first published in 1856, were written by William W. Phelps, an early Church leader noted both for serving as a scribe and ghostwriter for Joseph Smith, as well as for being excommunicated on three separate occasions (each time, he ended up rejoining the fold). Originally set to a tune commonly considered too difficult to sing, the hymnal started experiencing a wave of popularity in 1985 after the Church music committee set the lyrics to a more melodious English folk tune; in its present form, it is a favorite for choral arrangements, including The Tabernacle Choir at Temple Square (formerly known as the Mormon Tabernacle Choir).

The uptick of attention the song experienced in the twentieth and twenty-first centuries did nothing to take away from the nineteenth-century Mormon atmosphere conveyed by the lyrics, which work mightily to challenge the listener with the limitless nature of Mormon cosmology. This attempt to project a sense of grandeur and the sublime begins immediately with the first verse:

> If you could hie to Kolob
> In the twinkling of an eye,
> And then continue onward
> With that same speed to fly,
> Do you think that you could ever,
> Through all eternity,
> Find out the generation
> Where Gods began to be?

The hymn goes on to stress the limitless nature of space, worlds, matter, spirit, virtue, youth, priesthood, truth, and also of race (meaning here the human inhabitants of the myriad other worlds). But these other infinities are presented as logically and narratively subsequent to the initially addressed limitless divine genealogy.

One Mormon transhumanist has referred to this eternal production of divine entities as the "fractal lineage of Gods."[22] This choice of words stresses the recursive, self-similar nature of the process, where shifts in scale downward or upward do nothing to break the endlessly repeating pattern. His choice of phrasing also references the aesthetic quality found in both fractals and Mormon cosmology. These aesthetic qualities are important because the aesthetic edge to this concept is not merely that of well-patterned or curated beauty. Rather, the infinities contained in this concept suggest an idea that is sublime in the Kantian sense, as something that threatens to outrun the faculties of both reason and imagination. Absent the formalist, arid logic of Cantorian transinfinite mathematics, these cosmic infinities are only truly capable of being held in the mind through an implicit comparison with the singular, definite, and limited. And the recursive view of the universe, where future moments of creating new universes (simulated or otherwise) is a repetition of the inauguration of this universe, also only gets its imaginative charge and cognitive shape from the tension between this spiraling vision of different universes with separate but enchained temporalities, and that of a single, unidirectional time shared by all. In short, these are ideas only graspable by way of negating their cognitive antipodes.

In the previous two series, we saw other oppositions where Mormon, transhumanist, and Mormon transhumanist concepts that all rhymed were still differentiated by expressions built around abstractions that at once implicated one another and yet were also mutually irreconcilable, at least as presented. Differences appeared between the individual and the collective, the agentive and the passive, human technology and divine technology, religion and science. This was present in both the preservation of the dead and in the work of that dead's resurrection. With this concept of recursive creation, all of these rhyming movements imagine a fractal lineage of universes at least, if not of Gods. Here, all these groups are on the side of the multiple and the recursive, even if this means that they also implicitly reference the singular and solitary.

The secular technoscientific transhuman, the Mormon, and the Mormon transhuman, however, all actualize their orientation toward the multiple and the recursive with different degrees of centrality, certitude, concretization, and intensity. In short, each of the approaches to these infinities could be situated in different spots in a continuum that ran from the empirical to the purely imaginary.[23] The scientific and transhumanist imaginations see these recursive, multiple universes as possibilities, though not possibilities for them to enact. The emphasis is not so much on their own multiplication of daughter universes down the line but rather on the possible origin of this universe and implications of this synthetic origin for what is allowed and what is advisable, especially as the allowed and advisable speak to their chances of survival. It is not that there is no consideration of the technical aspects of this; recall Bostrom's estimation of 10^{33} to 10^{36} operations per second as a baseline of what would be needed for a credible ancestor simulation. But there is no real clear articulation of a desire or moral obligation to produce lower-level universes. It is not that there is no concern for ethics in Bostrom or Hanson, but it is more along the lines that ethics is a guide for escaping punishment from higher cosmic echelons. And this interest in continuation is present only to the degree that they show any interest at all; Hsu and Zee see this recursive universe as, at best, an intellectual problem and not one that touches on potential existential risk in the ways that Bostrom's and Hanson's formulations do.

The Mormon imagination heightens the certainty and centrality of this chain of being, at least compared with the more agnostic-leaning transhumanist thought on the simulation hypothesis or daughter universe creation. Yet even as they celebrate physical existence as a test of free will, the Mormon vision of this recursive, multiple universes leaches away any sense of individual agency in bringing it about, and Mormon thought offers little-to-no technical details on the mechanism of creation at all. All that is given is that for a certain select group of inhabitants of the postmortal Celestial sphere, world and universe creation is a possible aspect of theosis.

With Mormon transhumanism, we have foci on ethics, agency, and process. The focus on agency frames this recursive process as part of a test that they must undergo, inasmuch as there is a concern not only for being worthy of the opportunity to create daughter universes but in actually having the ethical fortitude to bring them about as well; recall the reluctance of one Mormon transhumanist to take on a burden akin to that of "the God who weeps." This understanding also highlights the mechanisms through which such creation could occur. This interest in the mechanism could be classified speculative-empirical, inasmuch as, while extrapolating wildly, these imaginings are doing so from concrete contemporary processes, such as computer engineering and laboratory physics. There are limits to this, of course—consider our materially minded Mormon university instructor's assumption that universe creation as a technical activity is "a problem that is already solved." But even here, careful thought had been given to the form of the medium (be it computer chips or high-energy physics) on which this off-the-shelf universe-creation kit would operate.

There is one way through which we could distinguish these three instanti-ations of stacked, infinite universes: the degree to which the linkages between universes are continuous or discontinuous. Secular-scientific and transhumanist thought stresses discontinuity. There is a gap so large between those in the bottle universe and the bottle universe's creators that it is speculated the only point of communication between them would be the moment when that bottle universe is created; they look to traces in the cosmic background radiation and not for contemporary messages to the living. And while the simulation hypothesis allows for more downward interaction with universes after their creation, the ontological disjunct between the simulated and the simulators seems a sharp divide. Mormon transhumanism, at least when favoring the simulation hypothesis, also seems to imagine disconnect, though its severity is eaten away by a hint of the continuous in that it imagines the universe's simulated entities are ancestors, making them virtual kin, and this virtual existence is just a way-station to a potential embodied resurrection.[24]

In Mormonism, though, the kinship element of the lineage of Gods is not a metaphor, and the privileging of the continual is not just hinted at, but openly endorsed. God, after all, was once a man, and men themselves can become Gods. And once that ascension takes place, the inhabitants of those worlds without end that those Gods create are their literal descendants. . . .

Ninth Series: Queer Polygamy

§105

Social ties and institutional forces anchor Mormon Transhumanism, setting the possibility conditions for both the association and for the larger intellectual project it fosters. This peculiar form of transhumanism is a way of navigating an at times treasured, at times problematic religious and cultural inheritance. It is also a way of immanentizing the eschaton, and of making a human's divinity imminent through their own agency. It does this through an unhalting endorsement of Mormon doctrines of the materiality of spirit, and by insisting that due to an inherent symmetry, the Latter-day doctrinal notion that miracles are ultimately technological also means that technology is miraculous. This takes the form of an exploration of the possibility-space in Mormon thought, and a mapping of Mormonism to transhumanism and transhumanism to Mormonism. Shaped by similar questions as Mormonism and transhumanism, and using ideational material borrowed from both domains (even if this borrowing results in almost alchemical changes to the shapes these concepts can take), Mormon Transhumanism highlights the way that these concepts often rhyme. At the same time, many elements of these intellectual complexes appear as structural variations or inversions of each other—conflicting visions of preservation, different imaginings of the presence (or absence) of agency and of the (a)sociality in the resurrection, of virtual or material modes of a new genesis.

As this speculation works outward from concrete problems (what do we do with the dead) to issues that are either only dimly practical or entirely notional, the harder aspects of collective social and material relations become less pressing. These ties and encumbrances are never abandoned, of course; all speculation is socially conditioned to the core. And this speculative thought, even as it escapes the material, is still dependent upon and shaped by the technological and material circumstances of the milieu where it is dreamt. But in these extended moments, speculation does not have to directly face the social friction of coming up against the real. There is a reason why sometimes Mormon Transhumanists refer to this thought as *myth*.

Inevitably, the social—in all of its hardened obstinance—returns, and the sometimes-rough grain of these formations decelerates speculation. Even in these moments, though, speculation sometimes has its own force, spilling outside of the intellectual boundaries of a conceptual space like Mormon Transhumanism, making way for new social formations that work to different ends.

§106

We are at one of the Mormon Transhumanist Association's annual conferences. There are three people on the dais. One is this year's version of the secular transhumanist keynote speakers that the association invites each conference. The other (ostensibly Mormon) keynote speaker is a former editor at a progressive Mormon periodical. With them, serving as a moderator, is Lincoln Cannon. It is late in the day, and the sunlight outside the auditorium windows is low. We are in the closing question-and-answer session, with the keynote speakers sitting alongside one another at the head of the conference room.

The back-and-forth between the panelists has drifted, with a slow certainty, to the subject of gender.

This comes up when the editor is asked what would keep her from being a Mormon Transhumanist. She says it is the same thing that makes her sometimes question her affiliation with the Church: the fact that it is a boys' club. She mentions she is happy to see more female faces in the audience; she parenthetically adds that before she arrived, she feared she would be the only woman there. Cannon, acting in his current role as master of ceremonies, explains that they are "working on that," and that the lack of female voices in the association is a problem. He also discusses the demographics of the group, not by way of excuse, but as explanation. It is a group of Mormon engineers, mostly, and most Mormon engineers are white and male. They are, though, reaching out. Still, the editor points out, it is as much notions as numbers when it comes to gender. During the conference, people spoke of God using male pronouns, she observes. She also notes there was no mention of Heavenly Mother, either. Changing the demographics will make problems like these visible, will make them articulable, but without some kind of working-through, those problems will still be there, she says. There is a round of applause from the audience.

Then, the next question. A sober-voiced engineer with a tightly trimmed beard is the next person at the microphone. He speaks patiently, clearly, carefully, without the small moments of disfluency that mark most people's public speaking voices. He has a question for the whole panel: "The family is a key part of the Moron aesthetics. And I was curious, not just from a Mormon Transhumanist perspective, but also from a strict transhumanist perspective, what are the opportunities or possibilities, and how it affects the family unit, that transhumanism would affect the family unit, and what are the risks?" The editor glances at the secular transhumanist keynote speaker on one side, and then at Cannon on the other, as if looking for assistance; the sense is that she feels the weight of answering the question falling on her. Instead, after a short pause, Cannon leans into the microphone, saying that one of the problems of "texts, whether ancient texts or unwritten texts," that speak on the family is that these texts also come with the inheritance of "many thousands of years of both constructive, and destructive, patriarchy." This makes it hard for us to sometimes "pull away from them and realize what some of us feel inside, that we have to learn and grow beyond parts of what those ancient texts, and modern texts, say." There are important things to learn about family, he adds, if we can attend to Joseph Smith's original intent: Joseph Smith did not just seal husbands to wives, and children to parents, but Smith sealed his friends to him as well. "And I would be ecstatic," he adds by way of closing, "if Mormonism were to reinstate such behavior, and to formalize friend relationships, and even consider other sorts of sealings that may be beneficial, to extending family in ways that we've not recently been accustomed to, but that Joseph Smith's behaviors challenge in quite positive ways."

There is a small moment of silence in the room, and then our secular transhumanist keynote speaker leans into the microphone. There is a touch of hesitation in his voice. "I don't know if I should say this," he intones, after which Cannon interjects, "please do." With that opening, the transhumanist keynote speaker goes on:

> I was talking about medical nanorobots, and nanomedicine, and things like that. At some point, we develop really good cosmetic surgery. I mean *really good*. And this concept that you are a man, and that's *it*, or you are a woman, and that's *it*, goes away. It's just *gone*. Because the mental patterns, the growth, the—who you are, that would be something that would be voluntary. But the physical structure is a physical structure. You can be anything you *want*. I don't

know what implications that has, but from a technical perspective, it's going to be possible.

His speaking style is not condescending, but there is a certain bluntness to the statement. The silence in the room seems to linger just a fraction longer than usual before the conversation starts up again.

§107

Recall our long-haired, science-teaching professor and Mormon Transhumanist who dreamed of, in the fullness of time, making a new material cosmos. When discussing the technology of universe-making as a technological problem already solved, he wavered when he spoke of God. As he pointed out in the conversation with a great deal of self-awareness, the problem that was confusing his speech was not the existence of God, but God's nature. When he speculated about the nature of this divine object, he spoke of God as a composite entity, as being a community, but then he also spoke of a community made of Gods. His language shifted and flickered, not despite but because of his self-knowledge. And even as he said he had something like a cognitive preference for a personal God, when it came to universe-making, his language about divinity became plural.

There is a long tradition of this kind of speculation about the multiplicity of God, going back to the Book of Genesis quoting the Ancient of Days as speaking to something like a committee or working group, saying, "Let Us make man in Our Image, according to Our likeness." Numerous other moments in the Hebrew Bible seem to suggest that it is not so much that there is only one God, but that among a wider divine council, there was only one God who had a covenant with Jews. Unlike many other Christianities, which have endeavored to exegete these Biblical problems away, these monotheistic-solvent aporias have been embraced by the Church of Jesus Christ of Latter-day Saints. An example of this is the Book of Abraham, a document that Joseph Smith presented as an interpretation of a fragment of funereal papyrus that came into his possession. The Book of Abraham presents the earth that we dwell on now as having been crafted by multiple Gods; key moments that Genesis presents as single speech acts, such as the proclamation "Let there be light," are depicted in the Book of Abraham as being chanted by a chorus of Gods.[1]

But even when Mormons speak of God as a singular entity, God is still not alone. In the orthodox Mormon doctrine, Heavenly Father is accompanied by and wedded to Heavenly Mother, who is the bearer of all spirit children. While Heavenly Mother frequently captures the imagination of Mormon feminists, for the most part, she is not spoken about much in the Church. There is not a total ban on her. While the first most famous Mormon poem is a single couplet crystalizing the King Follet discourse, arguably the second-most famous Mormon verse is a celebration of Heavenly Mother, written by Eliza R. Snow, a plural wife of both Joseph Smith and Brigham Young, and one of the few women to be commonly (and unofficially) considered to be a prophetess. (The text of the composition: "In the heavens are parents single?/No, the thought makes reason stare./Truth is reason: truth eternal, tells me I've a mother there./When I leave this frail existence,/When I lay this mortal by,/Father, Mother, may I meet you/in your royal courts on high?"). Heavenly Mother is also referenced in "The Family: A Proclamation to the World," the 1995 Church statement espousing the foundational role of the family in the cosmic structure. The "Proclamation" is no mere communique. It is a dispatch to which many believers are deeply attached. The document is commonly found framed and hung on the wall in Mormon homes, much like one would frame and mount the text of the Lord's Prayer, the Beatitudes, or the Declaration of Independence. Heavenly Mother also makes her appearance at general conference meetings. Different speakers have, from the podium, at times referred to her as a cocreator of worlds and as a codesigner of the plan of salvation.[2]

All the same, in attention paid and in implied eschatological importance, Heavenly Mother is eclipsed both by her consort and by her first son. There are various explanations as to why Heavenly Mother is not foregrounded in the Church of Jesus Christ of Latter-day Saints. One reason given for this is that since she is not part of the Godhead, worshipping her—and even attending to her mentally—is not a proper religious act. This is a rare opinion, though, and while it is hard to be certain about these things, thanks to the advent of something like a diffuse Mormon feminist sensibility, it is an opinion that seems to be increasingly becoming even more rare. It is sometimes suggested that the relative silence concerning Heavenly Mother is out of deference, and not disregard; the idea is that given the frequency with which the name of Heavenly Father is blasphemed, it is little surprise that he should endeavor to preserve her reputation. There is also the what-will-the-Protestants-think explanation, where some Mormons theorize that she is underemphasized because their American evangelical coreligionists might see her as heretical. Finally, it is sometimes said that when one is speaking of God, one is actually speaking of both Heavenly Mother and Heavenly Father collectively; thought of this way, there is an only a *seeming* lack of attention devoted to her. But this is a usage betrayed by pronoun choice; on Mormon lips, God is in effect always a *he*.

All this aside, there is another reason sometimes given as to why Heavenly Mother is not given her due. This reason is more whispered than stated—one hinted at in the complex marital status of Eliza R. Snow, the poetess known for exalting her. The reason is this: in the minds of some people, it is not entirely clear how many Heavenly Mothers there might be.

§108

I am in Provo, walking down the stairs to the basement of a house where one of the monthly MTA get-togethers is scheduled for that evening. Like many homes, the stairway walls are decorated with family photographs, typical framed pictures of family present and past. The color and the grain of the photographs tell a story of the age of the photos, not unlike the way that the different strata of rock and soil can date an object found buried beneath. The crispness of a photo points to something digital, then there is the grain and slight blur of a 1980s photograph, washed-out hues marking the seventies and sixties, and the black-and-white of photos taken before the easy three-color emulsion of Kodachrome.

There is a picture unlike all the others. The person who drove me here, one of the more senior MTA members, stops to bring the photograph to my attention. Even without his intervention, the image would have stood out. It is a large, sepia-toned photo, with an equally large and somewhat ornate faux-gold wooden frame. It shows seventeen gentlemen. Three of them are wearing what look like late-nineteenth-century business suits. The rest are wearing the broad horizontal black-and-white striped fabric associated with prisoners, though on closer inspection, it seems that many of these prisoner outfits are actually striped business suits—some of the apparent prisoners are wearing bowties. They are all posed in front of what looks like a substantial brick building with a thick, heavy wooden door, which is open. The same doorway also has what looks like an iron grate, which is shut. Everyone in the image is locked in the stiff positions of nineteenth-century photography, when longer exposure times necessitated an almost statue-like stillness.

Underneath the picture, affixed to the ornate picture frame, is a brass plaque. The plaque bears the following quote:

> It is a fact worthy of note that the shortest-lived nations of which we have record have been monogamies. Rome . . . was a monogamic nation and the numerous evils attending to that system early laid the foundation for that ruin that eventually overtook her.

The plaque credits the quote to a George Q. Cannon.

The person who brought the picture to my attention quickly sets things clear. The gentlemen in the prison suits are, in fact, actual prisoners; they were

all incarcerated by the federal government for the crime of polygamy. My friend
points to a particular individual in the picture, a man in the very center of the
photograph, sitting in a wooden chair as if it were a throne; he has a massive
silver goatee and is holding a small bouquet in one hand. He is the George Q.
Cannon of the quote, a former nonvoting territorial member of the United
States House of Representatives, an apostle in the Church, a member of the
First Presidency to the Church of Jesus Christ of Latter-day Saints, and an
ancestor (by one of his six wives) to the owners of this house.

My guide, looking at the photograph, says how impressive it is to be willing
to be jailed for your convictions. After a pause, he adds, in a way that suggests
that this point is obvious but should be explicitly stated anyway, that for him
it is the will to stand for one's beliefs, and not the underlying principle of polygamy, that is admirable. Given his tone—and having gotten to know both his
sincerity and his sensitivity over the years—I take him entirely at his word. We
walk down the rest of the stairs, into the basement and the meetup.

§109

In many of its self-narratives, but also its self-presentation to the wider world,
the end of polygamy in the Church of Jesus Christ of Latter-day Saints is pic-
tured as a sharp break. After decades of an extended antipolygamy campaign
from United States government agents, and also after various obstructions to
Utah being granted statehood, the Church gave way. Earlier in his tenor, President of the Church Wilford Woodruff felt convicted that the fires of the apocalypse would shortly come down to wipe away the United States and thus
protect polygamy.[3] Later, as pressure from the federal government mounted,
the fires of the apocalypse came to feel slightly more distant. Woodruff penned
a document that became known as *The Manifesto*. Citing revelation while
speaking on behalf of the Church, Woodruff stated that:

> [w]e are not teaching polygamy or plural marriage, nor permitting any
> person to enter into its practice, and I deny that either forty or any other
> number of plural marriages have during that period been solemnized
> in our Temples or in any other place in the Territory. . . . Inasmuch as
> laws have been enacted by Congress forbidding plural marriages, which
> laws have been pronounced constitutional by the court of last resort, I
> hereby declare my intention to submit to those laws, to use my influ-
> ence with the members of the Church over which I preside to have
> them do likewise.

There is nothing in my teachings to the Church or in those of my associates, during the time specified, which can be reasonably construed to inculcate or encourage polygamy; and when any Elder of the Church has used language which appeared to convey such teaching, he has been promptly reproved. And I now publicly declare that my advice to the Latter-day Saints is to refrain from contracting any marriage forbidden by the law of the land.[4]

Presented in the 1890 general conference, the Manifesto was reported by the Church as having been unanimously endorsed by those present. Other reports, though, suggest that there were some present in the general conference audience who voted against the measure, and many more who abstained. Many took the passage of the Manifesto as Woodruff himself would have done earlier: as a sign that the end times had begun.[5]

The Manifesto was not the end, though, of either history or of polygamy. There were already extant plural marriages, the vast majority of which were still respected by Saints after the Manifesto, with the honoring of these marriages often taking the form of continued cohabitation. But there were also plural marriages consecrated after the Manifesto; some believers understood the Manifesto as a ruse to trick gentiles and believed that they saw something like a wink in the metaphorical eye of the Church. For instance, in the fifteen years after the 1890 manifesto, "eleven General authorities . . . fathered seventy-six children by twenty-seven plural wives."[6] And it was an open question as to whether the Manifesto applied to Mormons who were outside of the United States, such as colonies in Mexico and Canada. More seriously, after the proclamation, several Saints continued to perform plural marriage, in the neighborhood of two hundred and fifty unions in those first fifteen years; sometimes, these marriages were performed under the authority of members of the twelve. Several were authorized, though not officiated, by George Q. Cannon.[7] A second manifesto, presented in the April 1904 conference by Joseph F. Smith, Woodruff's successor (and Joseph Smith's nephew), reiterated the message of the first Manifesto. What was different this time was a change in Church policy. This second manifesto was accompanied by a practice of only advancing monogamously married men to the general authority. People who officiated in plural marriages were even subject to church sanction, including excommunication. Among those sanctioned with excommunication for officiating in plural marriage were two apostles.

Even this was not the end. There were factions who felt that in issuing the manifestos and in turning its back on the principle of polygamy, the Church had fallen into precisely the sort of apostasy that they imagined had originally denatured Christianity millennia ago. These in time became known as the *Mormon Fundamentalists*, though the term is something of a misnomer. Unlike Protestant Fundamentalists, who prize a literalist hermeneutic and present the image of an unchanging Church, Mormon Fundamentalists are open to new revelations, produce complicated, allegoric readings, and are quick to experiment with unconventional forms of authority. But fundamentalists aside, polygamy was extinguished within those social and intellectual spaces that were controlled by the Church.

Or more properly, Earthly polygamy was extinguished. Consider this. Russell M. Nelson, the First President of the Church of Jesus Christ of Latter-day Saints as of the time of this writing, married Dantzel White in the Salt Lake Temple on August 31st, 1945. After nearly sixty years of marriage, nine daughters, and one son, Dantzel passed away on February 12, 2005.[8] On April 6th, 2006, Wendy L. Watson, a professor of marriage and family therapy in the School of Family Life at Brigham Young University, married Russell M. Nelson; despite her field of professional expertise, this was her first marriage. The couple was also married in the Salt Lake Temple.[9]

Note that both of these marriages were temple marriages, sealed and eternal, binding husband, wife, and issue together in a family for time and all eternity. On the day of the resurrection, or in the spirit world waiting for that day, which woman will he be married to? The answer is unclear. There is a tendency among some Saints to say with a sort of doctrinal shrug that "God will sort it all out." The more common answer is that President Nelson will be married to both.[10] This is an instance of *eternal polygamy*. This principle suggests that while polygamy is foreclosed in the here-and-now, it still exists in the afterlife as well as during the resurrection. The conundrums created by eternal polygamy are not hypothetical. Many Mormon women fear predeceasing their husbands for this very reason; after her passing, will he marry again, lashing her to a polygamous family for eternity? Even outliving the husband is not a guarantee, as there is the question of whether the husband has taken on additional wives on the other side of the veil, since many ordinances, including marriage, can also be executed in the afterlife. This danger of marriage is not necessarily understood as an exercise in concupiscence on the part of the husband; if marriage is a prerequisite to enter the Celestial Kingdom, then marriage is a way of allowing Mormon women to claim their full inheritance as faithful children of God. An unmarried woman is a woman potentially deprived of full divine glory and exaltation. Thought of this way, an additional marriage can be presented as an act of charity. Sometimes this is even used as a cudgel when women express anxieties about the prospect of eternal polygamy to, say, their bishop, or (when younger) to a seminary instructor: are you so selfish that you would deny a woman her access to the fullest heaven merely because of sexual or emotional jealousy?

There are other twists. A civil divorce from a husband does not break the eternal seal; the woman remains a spiritual wife even after the legal dissolution of the marriage by temporal powers unless she can get the seal canceled, a cumbersome procedure that involves submitting an application to the office of the President of the Church. There are women who are still sealed to abusive ex-husbands, just as there are women raising children with stepfathers to whom those children will not be sealed in the hereafter. A similar logic operates where the temporal bond of marriage is broken not by divorce, but by the husband's death. The woman will still be sealed to the first husband, and any children from a second marriage on this side of the veil will be considered children of the now-dead husband for eternity.

It is easy to overestimate the breadth and psychic costs of this problem; this is not a day-to-day burden for most Saints. But because it is a question that can silently haunt, gnawing on relationships, it is easy to underestimate its intensity and the depth of the damage that it does. This is because eternal polygamy can act as a synecdoche for all the other disjunctures and inequal-ities in gender relations and gender roles: exclusive male access to the priest-hood, an exclusively male hierarchy, the judgments and assumptions about kind and degree of paternal involvement in child-raising, the expectations about the necessity and qualities of a properly feminine and properly attractive presentation to the world. The Mormon poet and feminist Carol Lynn Pearson titled her book on this phenomenon *The Ghost of Eternal Polygamy*; her choice of the term *ghost* speaks to the way that this complex of doctrine, fear, and speculation haunts marriages, and also indexes the essential element of death and the afterlife. But this title also speaks to the spectral nature of this phenomenon.

Eternal polygamy is an issue that I have not ethnographically seen firsthand, but then it is one that I would be unlikely to see if it was at work. Regardless of its presence or absence among the particular Mormons I came to know, eternal polygamy is still important to attend to. It says something about the idea of sexual difference and polygamy in the contemporary Mormon imagi-nation. It tells us about the insistence of these concepts, and about their virtual mutability, the way that their internal logic and import changes with the wider circumstance that these ideas are embedded in. It is about a perduring stub-bornness, and a sometimes-cruel creativity.

§110

Polygamy is thus at once acknowledged and forbidden. Unsurprisingly, it is regulated as well, as are other forms of sexuality and love.

Before condensing the two volumes into a single text and putting them online in early 2020, there was limited access to the blue-covered first volume of the *Handbook of Instruction* for the Church of Jesus Christ of Latter-day Saints.[11] But that does not mean that its contents were occluded. This book was supposed to be read only by bishops, stake presidents, and other priesthood-holding Church officials who have some sort of oversight responsibility. But in addition to being known by those with past and then-contemporary callings, there were other instances where the contents of volume one erupt into public knowledge. One example of a handbook stricture that spilled out into public view was the November 2015 internet leak of new handbook rules regarding same-sex couples. Those in same-sex marriages were, according to these rules, apostates, which is an excommunicable offense. The policy also affected the children of same-sex couples, foreclosing their being baptized (which usually falls in a child's eighth year), or receiving the priesthood (which is given to boys when they are teenagers). Further, as the leaked handbook rules stated, the children of same-sex parents could not become missionaries. These strictures would hold until they turned eighteen, when, if they choose to, they could disavow same-sex unions, after which they could be approved by the office of the president, and thereby become eligible to be full members of the Church.

Despite the fact that it would be withdrawn in 2019, it is difficult to exaggerate the speed and the depth of the injury that the "November Policy" (as it came to be called) inflicted on the Mormons who were a part of the open conversations that exemplified the Bloggernacle. News of this new policy came sharply on the heels of what seemed to be a reappraisal of same-sex attraction by the leaders of the Church of Jesus Christ of Latter-day Saints—for example, the Church had played an important role facilitating the State of Utah's adoption of legislation that banned discrimination on the basis of either sexual orientation or gender identity.[12] There were other signs of change in the air. In some of the wards with more progressive reputations and less doctrinaire bishops, same-sex couples had started openly attending services, and they were often accepted by those present, even as their participation was limited due to their status. Together, these facts were read by many as a sign that the days of treating homosexuality as an abomination might be numbered.

This false hope meant that when news of the policy was leaked, there was a sort of whipsaw effect, a reaction to an apparent doubling down by the Church rather than to the anticipated stepping back. There were tremendous amounts of anxiety about how this would impact an already too-high suicide rate among LGBTQ Mormon teenagers. It was feared that these vulnerable youth would read this policy as saying that they would never be recognized by their families or the Church, pushing them into fatal despair. Then there was outrage regarding what was seen as punishing children for the innate sexual identity of their parents. Finally, there were those who had close relatives who identified as part of the LGBTQ community who perceived the policy as an attack on their *entire* family.

There were protests. There were mass resignations from the Church, sometimes as a part of large public demonstrations held across the street from Temple Square.[13] Many of those that had resigned were not active, ward-attending members of the Church. And there had been mass resignations from the Church in the past, going to the days of Joseph Smith.[14] Still, the *scale* of these resignations, reportedly in the tens of thousands, seemed epochal, as did the degree of anger that informed them.[15]

§111

Not all policies are known via recent leaks, however. There are policies that are widely known simply because they have always been the policy during living memory. Such as the strictures around polygamous unions and the children that spring from them. As set forward in the *Handbook*, polygamy *is* a form of apostasy, an excommunicable offense, and children of polygamists cannot go through the formal blessing and naming ceremony through which infants are welcomed into the Church. And the sons of active polygamists cannot become priests or serve on missions (or at least they cannot until they reach the age of majority). This does not mean that these children cannot attend church; they are welcome at sacrament meetings, as is anyone else. And every once in a while, some do attend; it is rare, but not unheard of, for a teenager who is a part of a polygamous family to attend LDS ward services, drawn by curiosity, by a desire for respectability and stability, or perhaps just by the idea of accessing a broader world (polygamist communities can sometimes feel claustrophobically insular). But the price for their full membership in the Church as adults is their denouncement of polygamy.

The echoes between these two sets of strictures are not lost on First Presidency and the Quorum of the Twelve Apostles, the collective body that, at different times, has mandated each rule. In fact, apostles publicly argued as justification for the November Policy that it paralleled or was analogous to policies regarding polygamous marriage.[16] This implicit comparison between LGBTQ Mormons and polygamists is not a syllogism that everyone agrees on.

And even when the comparison is treated as valid, it does not always result in the same conclusion. Some polygamists complain that the comparison between the two is unfair to polygamists, inasmuch as polygamy—as both a historical practice and a next-worldly possibility—is a part of the church's heritage. Some proponents of same-sex marriage or their allies find the comparison offensive, given that violence, patriarchal cruelty, incest, and the sexual exploitation of children are day-to-day parts of some of the most infamous Mormon polygamist communities.[17] This does not preclude some from using the parallels as an argument for their movement, nor does it mean that there are not moments of strategic identification and the filching of language between the two groups. Some polygamist activists, usually women, have adopted the rhetoric of *coming out* from the LGBTQ community, and see the social and juridical fate of these two communities as being linked, even as polygamists still tend to see homosexuality as sin and abomination.[18] But very few would conflate the two groups, seeing polygamy and same-sex marriage as not *essentially* the same thing but *actually* the same thing.

Very few, though, is not the same as *none*.

§112

We return to the Bloggernacle. The podcast we are listening to is called *My Year of Polygamy*, though it was not always called that, nor was this podcast always concerned with the question of celestial marriage and plural wives. In 2013, this podcast was associated with the progressive-leaning blog *Feminist Mormon Housewives*. One of the podcasters involved in this project was Lindsay Hansen Park. Later in her career as a Mormon public intellectual, Park would go on to become an assistant director of Sunstone Institute, an important node in intellectual progressive Mormon circles, thanks to its magazine and symposiums; the general sense is that in this position Park has brought a kind of openness to who speaks at Sunstone that the forum was previously lacking. But before this curatorial turn, as part of her work with the *Feminist Mormon Housewives* podcast in 2014, she took some posts on the history of polygamy she had written for the blog and began turning them into a series of episodes. These episodes started with "The Forgotten Women of Joseph Smith," and slowly expanded in scope to address not just historical plural marriage practices, but contemporary Mormon polygamy as well.[19] In time, the popularity of these episodes encouraged Park to leave the *Mormon Feminist Housewives* podcast and make this feature independent.[20]

The original desire was to present a history of polygamy through conversations with experts on the field (it was not unusual for an episode to append the equivalent of a bibliography of sources referenced or used in the episode). In time, though, Park began to include interviews with people who had escaped from polygamy, such as adult children of polygamist families, or wives who had endured them. Even after the series ended with its one-hundredth episode, Park continued the work and began to include the voices of not just polygamy survivors, but polygamy advocates and practitioners. Based on the broadcasts alone, it is hard to express precisely Park's attitude toward polygamy; the show presents the dangers of polygamy and denies that it is "a principle of God." At the same time, the show works against any kind of othering of fundamentalists by parochializing the demographically dominant Church of Jesus Christ of Latter-day Saints as just one of a number of historical and contemporary Mormonisms; the show also acknowledges that there are also happy Mormon polygamist unions and nonabusive plural marriages. While Park's connection with Utah and Mormonism gives her an unbreakable tie to the Church and its culture, her contact with Mormon Fundamentalists has put her in the same position as that of an anthropologist, leaving her with sympathy for, but not an identification with, the people she has documented. It is hard to speak in general terms of Park's various interlocutors. When talking about polygamy, there is a tendency for the imagination to turn to images of compound-dwelling women in prairie dresses. But that in no way exhausts polygamists, either in general or on Park's program. While such conservative fundamentalist groups do exist, there are also other polygamist groups such as the Apostolic United Brethren (or AUB), who are less set-apart from the wider world and are more open to engaging with popular culture (and at times even producing it, as with the American cable television reality show *Sister Wives*). Members of some of these other, more open fundamentalist groups are frequent guests, engaging in a respectful presentation of their understanding of what constitutes orthodox Mormon practice, belief, and church structure. And then there are those guests who can only be described as the survivors of polygamy, indistinguishable from anyone else one would meet on the streets of suburban Utah, with the critical exception of their psychological scars. But scars are not always present with people who have put their fundamentalist childhoods behind them; some of those guests who have left polygamous families remember them fondly, and simply feel that the principle is either merely doctrinally incorrect, or not a good fit for them personally.

Today's guest on the podcast is different. She is neither a practicing polygamist, nor a survivor, nor an academic steeped in Mormon polygamy's historical past or sociological present. Instead, she is a *theorist* of polygamy. Or rather, a theorist and a proponent—though not a proponent of a polygamy that has existed before or that currently exists. She is advocating for a polygamy that is to come. The person being interviewed is Blaire Ostler.[21] In introducing her bona fides to speak on the subject, Ostler lists her involvement with a series of LGBTQ-friendly Mormon organizations dedicated to supporting youths and adults who accept their sexuality in an often intensely hostile atmosphere shaped by religion: these groups have names like Mormons Building Bridges, Affirmation, and Encircle. She mentions her written contributions to *Rational Faiths*, a fairly well-read multiauthor website that concerns itself with progressive Mormon issues.[22] Then there is what Park and Ostler refer to, half-jokingly, as Ostler's "Mormon Credentials." Ostler calls herself, tongue-in-cheek, a "papered and pedigreed Mormon," a ninth-generation member (from the "Levi-Hancock line") with a polygamous family history. At another time, Ostler proves her Mormon credentials by invoking her ancestor, Amanda Barnes Smith, who had been courted as a potential plural wife by Joseph Smith—a courtship was cut short by Smith's death outside of the Carthage jail—who was posthumously sealed to Joseph Smith by none other than Brigham Young, who stood in as Smith's proxy during the ritual. "I'm too Mormon for my own good," Ostler jokes.

Mormon family history, however, is not the core focus of this episode. Rather, Ostler's purpose today is to help save polygamy from itself. Ostler begins by saying that "tonight, I'm really hoping we can clarify, maybe, some perceptions of polygamy that maybe don't necessarily have to be condemned right off the bat without introducing a little nuance and a little context to what the diversity of families can look like." Ostler does this by opening up what is contained in polygamy conceptually: noting that *Mormon polygamy* is the usual term, she points out that polygamy more generally contains several different potential categories. Polygamy itself is just "multiple marriages to multiple persons regardless of gender." The traditional form of Mormon polygamy is actually polygyny—multiple wives to a single man. Ostler notes that there is also the category of polyandry, multiple men to a single woman.[23] Ostler wants to supplement these categories with a term of her devising: queer polygamy. "It's breaking heteronormative boundaries and power dynamics that seem to be systemically rooted in views of gender."

As the show progresses, it becomes clear that the category of *queer* needs a lot of explication for a primarily Mormon audience. *Queer* is presented as an "academic term" (Park's words) for anyone who is "a member of the LGBT community" (Ostler's). Then there are other related senses of the word, such as its use in *queering the narrative*, which is described as an unconventional and perhaps even scandalous breaking with propriety that acts as the engine for the creation of something new. Mormonism, it is put forward, is an example of this queering. Given the dizzying transformations Joseph Smith outlined for kinship, sexuality, and divinization, Mormonism is "a queer theology in the context of its time," Park stresses in conversation. This queering includes the theology supporting polygamy, at least the theology of polygamy as Ostler understood it while growing up. As a child, Ostler was not haunted by the ghost of eternal polygamy. This was because her early reading of polygamy was that multiple sealings were allowed for both men and women. She recalls herself talking excitedly to her clearly unsettled mother and sister about polygamy: "I thought polygamy was that people could be sealed to who they wanted to be sealed to, and that's it. It's just consenting adults. . . ." She adds, "inside, little bisexual Blaire was screaming 'this is the best of both worlds!'"

And it is that early childhood understanding of polygamy that Ostler sees as the model for a future possibility: a polygamy without patriarchy, where all would be free to marry (or, as a necessary codicil, free not to marry) whomever they wish. It would be a polygamy without the coercion or violence or even the obligation that is polygamy's traditional traveling companion. It would be a polygamy without tears. Ostler is not suggesting that this egalitarian nest of cross-cutting bonds, indifferent to distinctions in gender, is what polygamy has been historically; she is quite open about the heartbreak that polygamy has brought in the past, though she is careful not to categorize past versions of poly-gamy as a mistake, or as entirely negative. Her rejection of past polygamy also should not be taken to mean that her vision is without Mormon precedent in scripture and practice, Ostler argues. As Ostler reads it, Zion is a community, built together, and this mutual building of Zion through ties formed by all sorts of love and affection is a part of how every person fulfills his, or her, or their, potential to become God. After all, Ostler argues, polygamy was not always just patriarchy: Smith himself sealed men together.[24] Letting people come together in all forms of family must be a part of this process.

But, Ostler suggests, this vision of polygamy will not come in a Church-sanctioned form anytime soon, in part because of Mormon anxieties about the LGBTQ community. In prior generations, Ostler says, Mormons were punished by the government for their involvement in polygamy, and this punishment fell not just on the patriarchs of polygamous families but on the wives in those families, who refused to testify against their spouses. The community at that time was, in short, persecuted for their queer practice. This, in turn, made it gun-shy in the modern day to adopt another form of queer love:

> We still haven't dealt with our polygamous roots—and by we, I mean the LDS community. We still haven't grappled with our history and come to terms with it, and figure out how we're going to move forward to address this. . . . In a lot of ways, we fear our own queerness, because

our queerness became a source of politicized shame. It's a classic case
of the abused becoming the abuser. And the [contemporary] queer
community is getting a beating.

Until there is a reckoning with polygamy, a reckoning with what family meant
to nineteenth-century Saints, the Mormon Church will be unable to deal with
the issue of LGBTQ Saints. But if that reckoning should ever come, what family
means could be more multifarious than even the most clear-eyed polygamous
Mormon, back when Utah was a territory and not a state, could ever have
imagined.[25]

Parenthetically: Blaire Ostler is a board member and the former CEO of
the Mormon Transhumanist Association.

§113

It would be wrong to describe Ostler's membership and relatively high profile
in the organization as a direct function of the years-long initiative on the part
of the Mormon Transhumanist Association to become more welcoming to
women. It is true that the leadership of the MTA has long been concerned
with increasing both the participation and the profile of female members. But
even if the association was originally "male, white, and nerdy" in its demographics (as more than one member has described it) women were always a
part of it. But more to the point, no amount of pandering would make someone
as dynamic as Ostler join a group with which she had fundamental differences.
And when Ostler first encountered the Mormon Transhumanist Association,
she *did* feel she had fundamental differences with it. Both Ostler and Lincoln
Cannon (who are related through the drunken spiderweb that is Mormonism's
extensive kinship network—again a part of the Wasatch Front Mormon polygamous inheritance) recall their initial discussions as constituted by Ostler
presenting blistering attacks on what she understood to be Cannon's patriarchal
image of God. It was only after Cannon insisted that his vision of God and
Zion was as a collective and emergent communities rather than any patriarchy
or male *singleton* (to use a common transhumanist phrase for an individual
locus of authority) that she began to consider Mormon Transhumanism as
something not necessarily beyond the pale.[26]

Even as she warmed to the concept, her understanding of Mormon Transhumanism, and of what transhumanism had to offer in general, developed a
different texture than that of most of the other (male) Mormon Transhumanist
imaginations. It leans into the pragmatic, millennial side of building Zion,
rather than the apocalyptic and transformative; when she speaks of technology,

it is near-future technology, or even of technology currently extant. Hers is a transhumanism based on an appreciation of technology, not on any claims of mastery. But it is also a transhumanism leavened by Ostler's interest in an embodied, maternal feminism.[27]

Again, feminism is in no way alien to Mormon Transhumanist sensibilities. Part of this is found in the social dynamics. I have heard more than one woman note that in meetups, organizing meetings, or official or unofficial conference discussions, they aren't talked over, ignored, or left out of conversational turn-taking, something that is characterized as rare among mix-gendered Mormon public gatherings. But part of this feminism is also at the level of more notional commitments. Many members of the Mormon Transhumanist Association supported Ordain Women when the movement was at its peak. And many still feel strongly that women should be acknowledged as having the capacity to perform priesthood blessings (a simple prayer for healing, comfort, and guidance, performed by putting one's hands on the recipient's head while praying aloud, usually in an improvised manner instead of recitation of a text; the blessing is redolent of familial intimacy and care, and thus often carries deep emotional attachments). Some have practiced religious feminism, instead of a merely conceptual feminism, which at times comes with social sanction. An example: when one bishop learned that women were giving priesthood blessings in one transhumanist family, he declared that the Holy Spirit "wasn't in that house." Finally, there are certainly strong feminists in the immediate circle of many members of the MTA, as well as part of the organization itself; one of the board members of the association is married to a best-selling author, housewife, and Mormon feminist who is well-known in online communities for her widely circulated argument that men's irresponsible orgasms and reluctance to wear condoms are "100% to blame" for unwanted pregnancies, and that therefore laws curtailing women's reproductive rights are not just misguided, but morally indefensible in that they erase male accountability.[28]

That being said, it sometimes seems that there is a sort of brittleness, or a sense of remove, to the feminism of some (male) members of the MTA; feminism is a notion, or a principle, that is sincerely held, but perhaps not given a driving primacy in their thinking. And more than one MTA member came to feminism by way of an interest in progressive Mormonism, rather than becoming a progressive Mormon because of preexisting feminist commitments. It also should not be forgotten that there are those members of the Mormon Transhumanist Association who do not see any motivated link between feminism and transhumanism. Some of these reject feminism and queer advocacy across the board. Others present themselves as sympathetic to these causes but understand them as unrelated to transhumanism, and thus feel that these discussions do not fit in the MTA's various forums, or that addressing these issues in a Mormon Transhumanist space give a false sense that it's necessary to endorse feminism and champion queer rights if one wishes to be a Mormon Transhumanist. These feelings ranging from indifference to antipathy do not surface much, and certainly do not represent the most active members of the Mormon Transhumanist communities. Still, those opinions are also present.

Blaire Ostler's embodied feminism stands out in sharp relation to these other sensibilities. Part of what makes it distinct is that Ostler takes her politics to the streets—literally. For several general conferences in a row, Ostler, alongside a few allies, has stood outside of the LDS conference center, holding a large posterboard sign reading "Hug A Queer Latter-day Saint." (She says that it is not unusual for conferencegoers to give her some words of support, or even take her up on the offer for a hug, as they stream into the monolithic building.) And she also has been quite visible in many Utah organizations concerned with LGBTQ Mormon adults or teenagers and their families.

But it is not just Ostler's other commitments that sometimes stand out in sharp relief against the feminism of Mormon Transhumanists. Within the confines of the MTA, what also stands out is her particular take on some of the core issues that inform Mormon Transhumanism, and the manner in which her thought developed. The best example of this development, ironically, is an instance not of growth but of repetition. In multiple talks and essays, Ostler has given an account of how a series of difficult childbirths and near-death experiences made her into a transhumanist. The thematic refrain of conception, childbirth, and death found in these pieces also gives us something more. It shows us not just a mind turned to transhumanism but also a mind rapidly intensifying the depths and breadths of a particularly feminist and nonheteronormative transhumanist imagination.

§114

The first of Ostler's accounts I want to address was only the third essay she had ever penned for an online resource and only her second piece for *The Transfigurist*, the Mormon Transhumanist Association's online journal. The other essay I want to lay alongside this one was delivered a mere five years later, during a talk to the 2019 Annual MTA conference. The time between those pieces is not much—a mere four years—but those years mark a great deal of change.

Unlike the version Ostler would present years later, the first account has a domestic edge, starting with a depiction of what seems to be a model, heteronormative Mormon family. Here, the credentials that she lays out do not refer to Mormon past, but a maternal present: "I am a stay-at-home mother with three beautiful children."[29] This is not just an opening gambit but a theme continuous throughout the piece. Shortly after, Ostler will go even further into what at first reads like Mormon mommy-blog territory, telling us that "[o]ur story begins on the campus of Brigham Young University." She recounts a young marriage and a short courtship, a marriage at age eighteen in the Portland Temple to a twenty-two-year-old man to whom she was engaged for only "several months." The marriage took. Striking a tone of female domesticity, she states that "I have not regretted marrying young, nor the man I married. If I were to ever become a mother and embark on the journey of parenthood, this man was going to be the father."

But these details regarding the marriage are to come a bit later on, following her self-introduction. And in that self-introduction, after giving us her status as mother and stay-at-home wife, Ostler quickly tells us that she is also a transhumanist. "It may seem like an unlikely pairing, but as you read you'll see it's quite natural. My journey towards Transhumanism started before I even realized it began." Her use of the word *natural* in communicating the arc of her transformation is important because what she presents is different from the heady transhumanism of simulated worlds and technological theosis. We are not setting things at a cosmological scale. Rather, this is a transhumanism that is drawn out of, not quite *everyday* experiences, but experiences that are grounded in common domestic events that have a very this-world feel. These are events that happen in identifiable, concrete places—in the marriage bed, hospital cots, doctor's offices, operation rooms, and intensive care units. It is not a vision of radical technological transformation, but of a short-term gratitude for, and extrapolation from, extant medical machine capacities.

The narrative begins in earnest after the marriage. She and her husband begin their "journey towards parenthood" via the "conventional means of procreation, sex." There are snags and complications, with Ostler's "mutated bicornate uterus that was tilted towards [her] spine" preventing any easy conception. In time, though, there is success, and we are treated to another pitch-perfect scene:

> I still remember the look on my husband's face when I showed him
> the positive pregnancy [test]. His expression of joy was evident through
> his tears. We held each other for what seemed to be a lifetime.

But this idyllic scene is immediately undercut. The act of conception, she writes, was only the beginning of our struggle. Pregnancy and delivery would prove to be even more taxing, complicated, and dangerous.

Rough waters lay ahead.

It is here that the tone changes, without losing the essay's maternal motif; what read a little like a Mormon mommy-blog posting transforms into a medical horror story. The pregnancy brings "painful complications," "erratic vomiting," and dehydration. Ostler's gastrointestinal tract, triggered by the spike in hormones that accompany pregnancy, refuses to digest food. Due to a freak medical condition, even her uterus swells, pushed outward with an excess of amniotic fluids to the point where Ostler describes her abdomen as having been "the size of a woman carrying twins or triplets." As the delivery day approaches, it turns out that things are even worse than imagined, and the baby is in a position that precludes any natural childbirth. Ostler is told by her physician that this will be a C-section.

The night before the procedure, she has something like an epiphany. But unlike other revelations that occur on the eve of a dangerous event, this realization runs not toward the divine, but rather away from the divine, and toward humans and human technological achievement:

> I was restless in bed that night, unable to sleep. I couldn't help but feel like nature had failed me. Perhaps God just didn't care. Perhaps my interpretations of God were simply wrong. One thing was evident, humanity cared. The combined efforts of people and technology brought me to this point—from the electricity that powered the machines that sustained my body's functions, to the car that rushed me to the hospital, to the combined knowledge of skilled physicians working with me, to the creators of the internet that connected me with critical medical information. It was evident humanity cared. Pondering these thoughts, I was grateful.

Ostler's Garden of Gethsemane moment happens in surrender not to God's will but rather to human capacities and ministrations.

The next day, the C-section goes well, and Ostler's child is born, carried away in his father's arms while physicians finish the procedure (though they do ask if they can bring in some staff members to show them her uterus, which is apparently unlike anything they have seen before). The doctor reassures Ostler that she should be able to conceive again, though only after mentioning her surprise that she could conceive with such an organ in the first instance. Ostler does conceive a year later, triggering yet another difficult, painful, dangerous pregnancy. Ostler endures. She writes that, "[w]ith the support of my husband, doctors, and more machines, we made it through." That pregnancy also triggered a moment of reflection on the eve of a C-section; Ostler comments that as she lay awake on the eve of the procedure, she thought that "[i]t seemed like there could have been a much more sophisticated way to create life that didn't involve so much risk."

This is a dream of further medical technology that could either lighten the load of carrying a child to term or allow that burden to be put down entirely as new life is created through other means. But that meditation on the risks of childbirth in its present form is also a bit of foreshadowing. During the delivery, the baby is not breathing. It takes immediate medical attention to start the infant's respiration. But that attention is not from humans alone. As Ostler lies on the surgeon's table, listening to the medical staff shouting directives, she hears "the clicking of more machines working to resurrect [her] son." Hours afterward, still suffering from the effects of the spinal block, she is wheeled out to see her son. Her description of the scene, though, transitions again into an almost devotional meditation on medical technology:

> I reached out to touch our baby's tiny hand. He was beautiful. He was hooked-up to machines that were teeming with life. The multiple wires around his body were accentuated by the bright electrical heating lamp above him stabilizing his body temperature. The sight must have been horrifying for other mothers, but not for me. The machines brought my son life. The machines brought me life. I had grown fond of them. They were the creations of humans doing what I could not and again, I was filled with gratitude. I welcomed the technologies and humanity. Embracing one without the other seemed dishonest.

This is not a pure technophilic moment; Ostler stresses the humanity that lies behind and works alongside machines. But still, these are machines that are spoken of as agents engaged in the work usually proximately attributed to human mothers, and ultimately attributed to God: the work of giving life.

She would have cause to appreciate the power of those mechanisms one more time. After roughly two years, Ostler tells us that she "couldn't stop thinking about having another child. My body longed for a baby. My arms begged for an infant. My husband agreed and we made preparations for the next year." She conceives one more time and suffers a pregnancy even more troubled and agonizing than the ones before. These complications take a heavy toll on her, and she states that her "body was falling apart. I did not doubt my will to give my daughter life, but my body wouldn't comply."

The C-section seems to go well, and the daughter is delivered safe and healthy. The surgeon once more asks if she can bring in medical staff to look at Ostler's uterus, and she complies. She continues to passively lie there as the doctors proceed to the next part of the operation: a tubal ligation agreed upon beforehand (the time for difficult pregnancies is over). As the work is being done, Ostler lies back, attending again to the devices that she has grown to have faith in: "I listened to the humming of my beloved machines that were intimately connected to my nude body." But the machines are doing different work now. "The gentle beeping of the machine monitoring my heartbeat began too slow," she recounts. Ostler begins to feel herself fade away. The operation room team scrambles to keep from losing her. Only the anesthesiologist's labor prevents Ostler from going under as the team works on her with greater urgency; he keeps her conscious by asking her the names of her children. Finally, her heart rate stabilizes and they close the incision.

The coda to this piece is an essay where she announces that via a distant family connection, "[a]bout a year later, I stumbled upon this strange group called the Mormon Transhumanist Association."

> They spoke of machines that could improve the human experience. They discussed ideas concerning reproductive technologies, genetic engineering, and nanotechnology. They had a curiously unique per- spective of God I had not considered. They discussed the possibilities of increasing our human intellect and physicality through emerging technologies. I couldn't help but be intrigued. Perhaps, I could spare my children the suffering my husband and I had experienced.
>
> Naturally, I joined. It was easy—like embracing an old friend.

This end has a tight symmetry with the beginning; it carries a tone of ma- ternal care and responsibility, as well as a rehearsal of the domestic dramatist personae, by referencing the husband and children. It even returns to the theme that an attraction to Mormon Transhumanism is *natural*, even as it involves advocacy for "machines that could improve the human experience." This is a transhumanism not of rupture and protean creation but of continuity and care.

§115

It is also a narrative that hews closely to the one that makes up the lion's share of a later presentation Ostler gives in a lecture nearly half a decade later. But *hew* here must be taken in both antithetical senses of the term, in that the second account is at once very close to and wildly divergent from the first ac- count.[30] The details of the two presentations are similar, with the second often borrowing language from the first, though this subsequent account is truncated in that it only recounts the second and third deliveries. These two births hold more dramatic weight in this recounting, as they are both presented not just as medical emergencies but as encounters where Ostler has met death. This account is simultaneously more expansive, however. It stands not as gentle explanations of a domestic, maternal, gradualist transhumanism oriented toward limiting human suffering, but instead as a full-throated radical transhumanism geared toward a final *overcoming* of human suffering that is as much political and religious as it is technological.

The religious is expressed in multiple ways, but the primary means is through engagement with the scriptures. The very title of the talk, "O Death, Where is Thy Sting?" is an allusion to the King James translation of Paul's first epistle to the Corinthians.[31] The passage itself appears in the body of the talk immediately after the account of the second child's C-section delivery and near stillbirth. Years later, Ostler is now in a position to make these children more than narrative elements; they are figures with their own voices. And the voice given to the middle child, her "now thriving nine-year-old son," is that of someone who has experienced resurrection. Though Ostler is careful here not to make her son's account of his resuscitation unnuanced. If you were to ask him to his face if he was resurrected, Ostler says,

> he likely won't answer you because he's shy, but I'll tell you what he told me. "If people don't believe in resurrection, they need to talk to me, because I know things about heaven and resurrection that other people just don't understand." In his mind, resurrection is not a matter of if, but a matter of when. Simply put, the longer someone has been dead, the harder it's going to be to resurrect them. We are resurrecting people every day, we just need to get better at it.

And here, as in standard Mormon theology, resurrection has eschatological import, though that eschatology is a grounded and immanent one:

> My son further expanded on his theory by explaining to me, "If I died and now I'm alive this is heaven right now. Because I died as a baby, then I was resurrected by the doctors here. So this has to be heaven— or at least one type of heaven."

In Ostler's mind, her son's religious speculation is one with firm scriptural support, both in documents particular to Mormonism, and in those that are part of a larger Christian or Judeo-Christian inheritance.

> For Mormons, this is not a new notion. In Doctrine and Covenants, we read the earth must be sanctified from all unrighteousness and prepared for celestial glory. And how do we sanctify the earth of

unrighteousness? We rid it of sin, and as the scriptures taught one sin
is death. In Corinthians we read "the sting of death is sin," and in
Moses we read the glory of God is to bring to pass the immortality of
humanity. Not only that, it is prophesied in our scripture that celestial
glory is right here on earth and celestial bodies will possess it forever
and ever. This is the promise if we choose it. But we have to choose it.
God doesn't override human agency, and we are not idle participants
waiting for God to do for us what we must do for ourselves.

According to Mormon scripture, my son is right. This is heaven and
we will not receive heaven's highest degree of celestial glory unless we
choose it according to both our faith and works.

And, as Ostler presents it, her son's "smiling face" is evidence that "doctors
and technologists" had decided that "death was a sin."

In this later account, Ostler's operating-room medical crisis during her final
delivery is again presented as an encounter with death. But this time, it is alluring rather than horrific, or at least, Ostler felt a kind of allure when she started
sinking into darkness on the operating-room table. Here, she presents death as
a passing temptation that contrasts with the enduring creative life-force of her
daughter. Ostler tells us of a time when her five-year-old daughter surprises her
by presenting a handmade aquarium constructed to hold all of the daughter's
toy *Finding Nemo* fish. When Ostler asks her daughter how she came up with
the idea, her daughter shoots back a reply in response to what she reads as her
mom's condescension. "Mom, don't you know I can create anything?"

Ostler sees her daughter's reply as showing us a capacity that we have supposedly lost in our adulthood:

At what age do we become so jaded or cynical that we lose our belief
in possibility? At what age did we stop believing that we can create
anything? At what age did we throw up our arms in defeat when our
visions became harder to achieve?

How could the impossible ever become possible if its possibility
cannot even be envisioned? Impossibility is a state of mind that favors
obedience to conventionalism over the courage to create. A child
doesn't waste energy imagining what could be impossible. She imagines what could be possible and how to make her visions a reality. She
truly believes, she can create anything.

This lost visionary and innovative capacity points at once to what Ostler
sees as her work, while this childhood power also points to the limitations of
her various critics. Those who criticize Ostler for focusing too much on social

issues and not enough on what they see as real transhumanist concerns fail to understand that her work is aimed at a total reinvention of life that has a quality exceeding the mere extinguishment of physical suffering alone. "What good is a healthy body if it's enslaved? What good is a healthy body if you are never granted social allowance to share with a lover of the same gender?" she asks. "What good is a healthy body, without love?" She sees a symmetrical failure in critics that see her transhumanism as either a distraction or a trap.

> [O]nce we've achieved gender liberation, racial equality, or sexual liberation, what's next? We all answer to the biggest oppressor of all, Death. This is an oppressor who doesn't care about your gender, your sexual orientation, your race, your likes, your dislikes, your religion, your family, your hopes, your dreams, your life. It doesn't love you, it doesn't hate you, it doesn't feel, it doesn't sleep, it doesn't eat, it doesn't get tired, or stop. Death exterminates, consumes, and negates life. Death comes for all of us. . . .
>
> We are all running from the same demonic monster, and while that monster consumes life at an unchallenged banquet table, you and I squabble over the scraps Death left lying on the floor. Death gave us so little time that we squabble like vultures over whose sexual orientation or gender identity is legitimate or not. We fight over nations and plots of land that don't belong to any of us. We draw imaginary borders on mother earth that somehow makes it okay to take a child from a migrant mother's arm. Neither death nor us should be the ones to take a baby from her mother. Death mocks us. We bicker with each other while Death feasts.

This attack on death itself, though, means a break with a sense of natural progression seen in the first account. In the *Transfigurist* article written almost a half-decade earlier, Ostler's turn to transhumanism was presented as *natural*, and hence not a break with either the established order or the implied heteronormative maternal family values that are the rhetorical backbone of her medically informed transhumanist origin story. In this later account, though, nature is the enemy—or at least certain conceptions of nature. Voicing the critiques of her imaginary interlocutors, Ostler says that "[n]aysayers will doubt that Death is an oppressor, let alone one we should take a stand against. They may claim, it's only natural." But if this is the case, Ostler points out, then according to nature, "I'm supposed to be dead."

> Do you understand what I'm saying? If death is something I'm supposed to just accept, then my children I and should be dead. I argue

that instead of building coffins for babies, let's build technologies that keep them breathing. We are already taking up arms against death every day. We are resurrecting those that would have been presumed dead and working towards a future where nobody is beyond resuscitation. As a species, we are fighting the oppressor and always have been. It's time we set aside our differences and take our efforts against Death seriously.

Nature is no longer the warrant for a maternal feminist transhumanism but instead is a queer transhumanism's enemy.

§116

When she reflects on how both her writings and the persona created by those writings have changed over the years, Ostler tells me in conversation, she sees part of this shift resulting from a willingness to be less guarded, and also from her developing a greater comfort with showing aspects of her personality that break with the atmospheric conservatism that characterizes Wasatch Front Mormonism. But that willingness to step out and take risks also comes hand-in-hand with the development of her thinking on these matters, development catalyzed by transhumanism. Transhumanism, Ostler says, "unleashed intellectual creativity within me."

Perhaps no aspect of Ostler's thought has been given greater freedom than her understanding of polygamy, though even here Ostler's earlier writing hints at the course her thoughts would take in due time. While Ostler first blogged at the MTA's *The Transfigurist* page, some of her early articles also appeared on *Feminist Mormon Housewives* (the same website that would also help birth the *Year of Polygamy* podcast). During the same week that saw the publication of Ostler's "How a Mother Became a Transhumanist," she also published a *Feminist Mormon Housewives* piece entitled "Raising My Daughter in the Church."[32] The post is a numbered list of shared concerns that Ostler and her husband have about what life in the Church of Jesus Christ of Latter-day Saints might mean for their daughter (two years old at the time).

The list is long, and touches on many familiar *FMH* concerns, such as the invisibility of Heavenly Mother and the question of women in the priesthood. Another concern found in the enumerated issues is that of modesty. Ostler's attitude toward this issue is complicated. There is a saying in Mormonism that is just as likely to be invoked ironically as sincerely: "modest is hottest." This mantra captures a wider aesthetic that simultaneously sets forward a standard of beauty for girls and young women while insinuating that something is disquieting, and perhaps objectionable, about their bodies. Ostler has no love for this adage. She worries that the sensibility that animates the saying promotes "fear surrounding women's bodies," and places the responsibility for male sexual thoughts on women's choices. Ostler does not argue for a rejection of modesty, though. Instead, she hopes for modesty's transformative reconfiguration:

> Might I suggest physical modesty being taught as an aesthetic that should exemplify a discipleship of Christ with a motive and love and respect for one's self? It could also be effective to broaden our understanding of modesty to include humility in regards to our actions, manners, and aesthetics, not simply hemlines and clothing that needs to be micromanaged by authority figures.

For Ostler, modesty is not an incorrect value, but its scope and import need a recalibration. It is about a care for the self, rather than a mere policing of the body for the sake of others.

This discussion of modesty is part of a broader pattern, where Mormon virtues and concepts that are problematic for religious feminists are not rejected out of hand but rather reimagined. Another example of this sensibility is Ostler's discussion of pornography. The Church routinely depicts pornographic consumption as a social and moral ill that not only threatens the ethical integrity of the person but eats at the family and the community, and they are not above using the language of communicable disease or addiction to describe the work accomplished by adult material. But the language of contagion is a mistake. As Ostler sees it, this equation creates something like a transitive property of shame, a daisy-chain of culpability where evil is associated with pornography, pornography is associated with sexual attractiveness, and sexual attractiveness is associated with the female body: therefore, young women come to conflate their femininity with evil. Ostler does not use the language of moral panic or plague, but she does not dismiss concerns about pornography either. Instead, she suggests that attempts to control pornography consumption through shaming female bodies are a misstep. She argues for addressing pornography in a way that presents it as damaging to all genders, while still allowing space for "individuals to forge a positive sexual identity with themselves." Again, this is not dismissive of Church practices or the conservative values of many of her coreligionists. It is instead a rearticulation of them in a way that undoes the sting of some aspects that can eat away at the self-esteem of young girls and women.

§117

Polygamy, though, seems harder to recuperate. Polygamy (number five on the list of concerns Ostler has about raising her daughter in the Church) is different from those other problematized aspects of the Church, in the kind of damage that it has wrought, in its deep ties to Mormonism's origins and founding Saints, and also in the weirdly liminal state it holds as a practice that is simultaneously banned in this life and celebrated in the next. Indeed, polygamy is such an expansive minefield that it takes more than one entry in Ostler's list: the byzantine and spectral aspects of the postmortal ghost of eternal polygamy causes Ostler's discussion of this doctrine to spill over to yet another bullet point (number six: "Inequitable Sealing Practices"). But even as the issue is apportioned between two different entries in the list, Ostler's discussion of polygamy's sprawling and convoluted nature makes it seem as if the subject is exceeding her grasp. This difficulty is something that she admits from the very start. She begins her discussion by foreswearing even a gesture toward a comprehensive discussion, stating that she "can't even begin to address the complex and sometimes horrible nature of polygamy." And she also writes about it as if it was a topic for which there is no therapeutic working-through. Even after Ostler believed she "had made amends with the practice," she shares that her equanimity was disrupted by a series of semiconfessional, semiexculpatory essays penned anonymously and published under the auspices of the Church. These are essays on polygamy and related subjects written by Latter-day Saint historians in an attempt to lance the boil of perceived Church secrecy by addressing the material used by internal and ex-Mormon skeptics to impeach the Church. It was, in short, a homeopathic tactic: a bit of controlled exposure to ward off the full brunt of truth in its greater form. But for Ostler, these essays only made the problem of polygamy live again.

Read quickly, this reaction to the essays sounds like another instance of someone shaken by Church deceit. But that is not the case. Ostler's description of polygamy as *sometimes* horrible suggests instead that something much more complicated is being expressed here. The problem, as Ostler lays it out, is not that the founder of Mormonism had multiple wives, but rather that the Church attempted to "subdue the voices of the wives of Joseph Smith." Ostler asks:

> Did they not give up their lives for the Church too? Did their work not matter? If a woman's work of child-rearing and housekeeping is so important then why not acknowledge their history as wives of the Prophet?

The problem is not that the existence of these polygamously married women automatically vitiates any claim that the Church has to legitimacy but rather that their absence in official Church material fails to honor them and does damage to the female descendants of these original polygamous Saints. Later generations of women are deprived of what would have been activist role models:

> Some of these women were amazing feminists working to secure the vote with the support of their husbands. Shouldn't my daughter hear their stories, too?[33]

Polygamy is not the problem, then. Silence about polygamy in the past is the problem, just as gendered inequalities in who can be sealed to whom is a problem in the present. But while silence and inequity are facets of polygamy's history, they are not (it is implied) a necessary part of polygamy itself.

This concern for the lost voices of polygamy also represents Ostler's concern for her own voice to be heard, since she sees her very being as inextricably linked to polygamy. As she puts it in a different moment:

> I descend from a polygamous heritage. Mormon polygamy is literally in my DNA. My foremothers and forefathers were polygamous— mortally and post-mortally. I am sealed to Joseph Smith via my polyg- amous foremother. My existence (and the existence of my husband and our children) is the product of an interconnected polygamous family. I will not denounce polygamy or my polygamous heritage which has made my existence possible.

Remembering polygamy, acknowledging her polygamous roots, in time transformed into advocacy for polygamy. The polygamy she argues for, though, is not what was endured by the founding women of the Church. It is instead that polygamy that is to come, one without "authoritarianism, sexism, racism, superstition, and coercion."[34] This new form is "a way to reconcile diverse de- sires for celestial marriage under a new model." It is a conception of polygamy that stands in opposition to the "sex-focused, androcentric, patriarchal, heter- onormative . . . Standard Model," the blame for which Ostler places at the feet of Brigham Young.[35] But as Ostler tells us, while the form and practice of this polygamy-to-come would be different, in its goals, a perfected, egalitarian polygamy would be the same as Young's version, just as it would be the same as the polygamy of Joseph Smith. The original doctrine of celestial marriage and adoption was intended to bind together all of Mormonism into one family; this new queer polygamy, with a capacity to acknowledge all kinds of relation- ships, also works to extend the family network, while strengthening the ties internal to this expansive family by multiplying them. It models this family, though, not as a hierarchical ladder but as a web of connections, forming "a grand and beautiful quilt that envelops us all," Ostler writes, adding that heaven "isn't heaven without all the people we love, and I trust God feels the same way. If not, heaven becomes hell."[36]

But doing the same work as other forms of polygamy and being of the same order of things do not necessarily come hand-in-hand. Putting aside queer polygamy's rhizomatic logic, which sets it apart from the arboreal logic of the Brigham Young practice, there are a few other questions we might ask of this twenty-first-century polygamy to distinguish it from the nineteenth century's standard model. We could question what the relationship is between trans-humanism and polygamy. Ostler presents these ideas as autonomous, independent of one another, but we might question whether this is so. We could also question the degree to which this queer polygamy is a merely conceptual practice, as opposed to something acted out in the world. And finally, we could ask about the conditions that brought this queer polygamy about. If ideas are often born of crisis, what might the crisis be here?

§118

There is such a thing as Mormon swinging.

Mormon swinging is challenging to discuss. First, there is its cloudy nature as an anthropological object. Swinging challenges what is for many ethnographers the very limits of participant observation, mandating the crossing of a line that many (myself included) would rather not; the literature on swinging, scant as it is, tends mostly to be based on interviews, surveys, or content analysis, rather than of the ethnographer throwing herself into the deep end of actual practice.[37] Then there is a thick fog of stereotypes and clichés that descends on any discussion of swinging: images of car keys in a fishbowl, of overly eager husbands and nervous wives mingling at a sex club.

This is also a topic that generates a lot of false positives. On this subject, the internet is even less reliable than usual: social media posts and blogs point to others wondering (in all senses of the word) at the existence of Mormon swinging, but if one takes on the drudgery of crawling from hyperlink to hyperlink, the reader is just as likely to find some fevered and suspiciously anonymous writings, imagined fantasy scenario or unsublimated projections, as he, she, or they are likely to find what reads as credible reports. Beyond the problem of fantasies presenting themselves as reportage, there is the clandestine nature of the practice itself. Even more than elsewhere in the United States, in Utah, to play at swinging is also to chance social sanction; therefore, people are more likely to whisper than to talk about the subject. And when people do whisper, the tale they share is often secondhand. People refer to places like Draper or South Sandy, green-lawned track-house bedroom communities strung midway between Provo and Salt Lake City. Here, supposedly, small parties take place, with everyone understanding in advance the *type* of gathering they will be attending. Different accounts suggest different dramatist personae attending these events. Sometimes stories are peopled with true believing Mormons (referred to as TBMs), or at least those appearing to be TBMs; these Saints are depicted as people who have become comfortable with hypocrisy, or at least with enduring some rather substantive cognitive dissonance. But often these stories are peopled with ex-Mormons, or those who are somewhere on the trajectory that leads to becoming an ex-Mormon. These narratives of ex-Mormon and near ex-Mormon sexual adventuring seem less fantasmic. Specific names are given, and sometimes the name volunteered is the interlocutor's own. One reason why discussions of this sort are different is that ex-Mormon swinging seems like a more motivated activity. First, it has been suggested that for some former Mormons, sexual experimentation acts as something like a psychoanalytic working-through, with long-inculcated anxieties about modesty and the danger of desire and the flesh confronted and traversed.[38] This kind of therapeutics is no doubt a part of this phenomenon. But these practices may not begin as a form of performative therapy for everyone who experiments in this fashion, and it certainly does not end up being the case for everyone, either.

The reason to doubt that this kind of sexual experimentation is a working-through in all cases is that it often seems to lack an element of control. A common refrain in the Church is that doctrine and morality are inseparable, and outside of religion, there can be no ethics. Absent the Church, there is no moral compass, and thus no reason to hold back. The statement also appears to inform how a few "apostates" govern their sexual lives after leaving the Church. They are fueled in part by a sense of making up for lost time or lost experience. Mormon sexuality is at once policed by ethical imperatives and constrained by social circumstance. In a world where premarital sex is frowned upon, there is an intense desire to marry young so that other desires can be exercised. The importance of eternal families centered around Church-sealed heterosexual marriages to the cosmic plan of salvation cannot be ignored, either; sexuality and salvation are fused, but only if sex is sacramental and carried out within marital bonds. Then there is the general sacrality of the body—the biblical quote comparing the body to a temple is taken quite seriously.[39] This results in pushes for modesty and purity, states which are themselves framed as sexually attractive: recall the refrain "modest is hottest." The absence of modesty and purity is not merely the lack of a positive aspect of the self but a state of pollution. During adolescence, boys and girls (particularly the latter) are told losing one's virginity or *virtue* outside of marriage turns a person into something like garbage. They are told that they will become refuse, a bit of waste that no one on the marital market could desire; a frequent analogy is a chewed-up piece of gum.

This should not be taken to mean that Mormons do not engage in premarital sex, though doing so is dangerous and the possible consequences onerous. Those under eighteen are supposed to report any immoral encounters, ranging from masturbation to full intercourse, to their bishop during *in camera* interviews in the bishop's office. It is not uncommon for the bishop to ask particularly detailed questions as part of these meetings. Authorities justify this degree of questioning as a necessary measure to discover the depths of the impropriety or to build up a dossier on perpetrators if they fall under Church authority. But the interview often comes across as not just intrusive, but abusive. (It should be noted that in the small world of pornography intended for Mormon and ex-Mormon audiences, the interview, alongside temple anointings and paired missionaries, is a common scenario.)[40] A history of violations of what is called the *law of chastity* can cause the Church to judge a mission candidate as unworthy and decline their applications to become a missionary. In educational institutions such as Brigham Young University, sexual activity is considered a breach of the honor code and can lead to expulsion; even sexual activity where the reporting individual is a victim of a rape can be the nub of an honor code complaint. Even adults can be disfellowshipped by their bishop for breaking the law of chastity.

Sexuality can thus feel heavily policed. And this is putting aside other social pressures, such as an imperative for women to present themselves in ways that are at once alluring and demure (recall the inordinately high number of plastic surgeons working in Utah). But when the Church can no longer exercise these forms of authority, and when the moral prestige of the Church has been worn away by doubt or unbelief, it may feel as if the shackles have been removed, and a season of sometimes quite intense sexual experimentation can begin. Many ex-Mormons use the phrase "I lost my body to the Church," and say that engaging in previously foreclosed forms of intimacy is one way of taking it back.[41] But the speed and dedication with which this lost time is made up gives it a frenetic edge. Sometimes, it seems like a pent-up desire rushes out all at once; there are anecdotes of people attending weekend seminars for Mormons transitioning away from the Church, merely out of curiosity, and then suddenly finding themselves having anonymous sex with strangers. I have heard Mormons and ex-Mormons observe that there are recent nonbelievers who seemingly feel obliged to act out the most scandalous predictions that the Church makes about those who leave the faith. An unconscious acting out of such an imperative makes a certain sort of sense; rather than merely being constrained, various modes of policing are known to do a great deal of work foregrounding sexuality by locking it away, and thus these strictures may ironically have done much of the work to produce what comes rushing out in their absence.[42]

Not all forms of sexual experimentation are animated by an appetite for the transgressive. Alongside things like conversion therapy, heterosexual marriage is one of the techniques that in the very recent past Church elders have suggested as an answer to what they refer to as *same-sex attraction*. Working off of anecdotal data, it seems that this is a therapeutic practice that only very rarely achieves its goals.[43] When the hold of the Church is loosened for some reason—whether it be by religious doubt or frustration with an unfulfilling marital bed—the marriage opens up to allow the queer partner to engage in the sort of intimacies that they often confess to having dreamt about for years, sometimes decades. A desire for equity will often push the heterosexual partner to open up the marriage on his or her end, too, though this is not always the case. These kinds of negotiated breaks with monogamy can sometimes work as a quick fix, and some families can turn it into a steady arrangement. But often juggling jealousies, attractions, and the psychic wounds that result from engaging in a marriage where one partner has to feign sexual attraction to the other long-term ends up being too much. The centrifugal forces tear these marriages apart.

People who have the MTA as part of their lives are not necessarily immune to this. The use of the phrase *part of their lives* is the more apt way of describing these relations, since the other individuals involved are almost always not all Mormon Transhumanists. The ties of marriage, love, and desire that are either created or tested by a falling away from Mormonism rarely involve whole sets of people within the association; it is rare to have both members of a marriage share equal levels of involvement, especially when that involvement is deep. And ties outside of marriage, whether forged by propinquity or dating app, almost always there are linkages to people beyond the MTA as well.

Most members of the MTA are not affected by this storm. Part of the work of the MTA is to change what Mormonism means such that those who might have fallen away can continue to be a part of the institution in some way; between reimagining what Mormonism is and being provided the support of a community of like-minded people, many members find a way to make peace with the Church, even if they cannot find a way to total agreement with it. But *many* is not *all*. Some find Mormon Transhumanism after they have lost their faith; and while it is not unheard of for someone to rejoin the Church after becoming a Mormon Transhumanist, led back by the power of this new manner of articulating the religion, others have no interest in finding a way back. Then there are those who feel hurt by the Church, or (even worse in some ways) believe that the Church has hurt those they love. Injuries of this sort corrode someone's ability to say that they "have a testimony that the Church is true." For these people, the Church itself is the enemy, a cruel institution with which one cannot make intellectual or emotional accommodations.

Some of those who feel most hurt by the Church are those in mixed-orientation marriages. These are relationships entered in good faith, either because neither partner knew, or even sometimes because they knew: the counsel from Church leaders to those struggling with "same sex attraction" often consisted of a heteronormative, marital "kneel and you will believe" type statement. And then as the years pass, and the sense of injury and frustration grows, something shifts in one of the partners. But there is also a desire to keep the family together; there are shared finances, shared memories, and most critical, shared genetic material and moral obligations in the form of children.

As an idea, queer polygamy helps these families. It suggests ways to decelerate and sacralize the threatened explosion of cross-cutting ties, to make the experiment additive rather than subtractive, and give these new relations a different patina, one that redeems a perfected past and populates a cosmic future. Now, these are people who are not practicing queer polygamy in the strict sense. There is (as of yet) no sealing between same-sex individuals, and it is unthinkable to have such ordinances performed outside of sacred walls of a temple. But the idea of queer polygamy opens up imaginable futures and gives cognitive tools for understanding the here-and-now. For some association members, it helps make readable various forms of intimacy that may otherwise feel strange. Many Mormon Transhumanists have the same pro-LGBTQ sympathies seen in Progressive Mormon circles, and there is an intense hunger among a good deal of them to be good allies. Here, certain nonstandard, nonheterosexual relationships are more immediately assimilated to Mormon sensibilities than others: gay marriage, presented as a permanent coupling between two individuals, and often seen as the first step toward the creation of a family and the construction of a respectable kind of household domesticity, is a form that can quickly be understood as analogous to the sort of idealized monogamous heterosexual families that make up most of the Mormon social landscape. More complex, overlapping nests of relations, though, tax the openness of some Mormon Transhumanists. Queer polygamy helps ease the way, here, though its work is done not through its acceptance as a platform to be endorsed, or even as an expectation of what will come to be in the future. But modeling these sorts of relations using a familiar language of Mormon piety and sacred familial celebrations makes it possible to imagine these relationships as something other than the hedonistic byproduct of ethical collapse.

Queer polygamy is not understood as a Mormon Transhumanist idea; when Ostler writes about it, or discusses it in a speech or a podcast, the only mention of transhumanism may occur when someone is presenting the one-paragraph biographical introductory sketch typical to these genres. As we have seen, Ostler herself sees a connection between the two: as she puts it, technological freedom from human limitation means nothing if one is constrained in whom a person can love, and in what form. It should be remembered at the same time, though, that some view Ostler's concerns as laudable but unrelated to transhumanism, and thus perhaps not an appropriate topic in social spaces, public events, and internet sites dedicated to discussing transhumanism. And then there are a few who, suspicious of models of family, love, and sexuality that differ from those promulgated by the Church, would even question the morality of what Ostler is promoting.

Given the autonomy of the concept from transhumanism, and the fact that as an idea, let alone as a practice, it is in no way universal to the MTA, to say that queer polygamy is a Mormon Transhumanist idea is to go too far. But queer polygamy is certainly an idea that rhymes with aspects of transhumanism. And viewing the way that each of these two ideas mirror each other's rhythms and structures suggests how the virtual capacities of religious transhumanist thought could catalyze the formation of the notion of queer polygamy. At a surface level, they are both solvents as far as gender goes. Queer polygamy dissolves gender by opening up the protocols that inform how and in what way gendered individuals can relate to one another. And transhumanism dissolves gender by imagining it as soon to be technologically mutable, giving it such a high degree of voluntary plasticity that speaking of gender roles borders on meaninglessness, and gender almost becomes a secondary and wholly aesthetic category. There is also a shared willingness between the two ideas to at once make use of the entire run of their Mormon inheritance while simultaneously feeling free not to be limited by what has been passed down to them. Rather than abandon nineteenth-century Mormon history, treating it as an embarrassment that threatens Mormonism's re-creation as a twenty-first-century pseudo-Protestantism, it recoups and reimagines this material. But this is not slavish fundamentalism; the ideas that it finds are redeployed in the present to solve problems far different from the challenges they were originally concocted to face. There is also a similarity to the sort of networks that queer polygamy and Mormon Transhumanism anticipate. Queer polygamy's cross-cutting ties are by nature antihierarchical, but then the same could be said of Mormon Transhumanism. Individuals cease being individuals to the degree that these networked actors draw upon computational capacities and technological prowess outside of themselves, and collaborate with others to bring into being even greater powers and possibilities. Even ascension to the cosmological realm does not undo this interbeing. While it is not impossible to imagine a divine single-ton in the MTA, the idea of complex chains of resurrections, of counsels of Gods and composite Gods, suggests that the networked self does not recede even as humanity is outgrown.

This anticipated technological and divine interbeing points to what might be the most foundational shared aspect between these two concepts, if we take *foundational* not in the sense of ground, but in an all-pervading patterning fostering possibility conditions for all other aspects of each concept. An almost protean sense of agency stands at the center of Mormon doctrine: spirit children came to earth and became embodied children so that they could be tested, and hence granting and respecting free will is, at least at the level of doctrine and self-justification, paramount. After all, the proposed plan of salvation that Satan presented in the council of heaven would constrain agency and hence ensure everyone's redemption; therefore, the imposition of any fetters, or any corresponding failure to own one's responsibility, is to break with the divine order and side with Devil.

But Mormon Transhumanism's dedication to human free agency goes even further than that. In traditional Mormon doctrine, humans are tested to see if they are worthy to participate in the process of divinization that God undertook eons earlier. But in most conceptions of Mormon Transhumanism, the human is not merely on trial for potential benefits after the resurrection. The human is also responsible for taking up the potential for self-divinization through the technology which God (however understood) seeded into this world as a po-tentiality, and for laboring to make this divinization actual. The day of resur-rection is not just something that happens; believing Saints have to bring it about through science and engineering, or at the least help pave the way for later Saints to fulfill the promise that one day "the trumpet shall sound, and the dead shall be raised incorruptible, and we shall be changed."[44] But this is an ethical challenge as well as a technical one. Since it is human labor that will bring about the day of resurrection, it is human ethics that will shape it. Technology must be made moral in large part because the fact that we will make the future means that we will be responsible for the tenor of the future. Queer polygamy is similar. It is not presented as something that will be handed down by the First Presidency. Rather, like the transhuman future, it is a form of attachment and loving that must be carved out in the here-and-now, through a mixture of freedom and mutual respect.

Series $\lim\limits_{n\to\infty} n$: Problems, Planes, and Lines of Flight

§119

After one of the annual-meeting dinners, guests are sitting at the usual small, round banquet tables, and catered food is sitting on the typical hot trays. A traditional barbershop quartet is in one corner of the room, providing entertainment by singing barbershop quartet classics (one offering: "It's Only a Paper Moon" is repurposed as a commentary that life and relationships still matter, even if this universe is just a computer simulation). As the evening goes on, plates are cleared off of the tables, but the conversation continues. A founding member of the MTA looks around, taking in the people who constitute the association, and he says, "You know, I'm pretty sure that at least half the people in this room are closet atheists." Everyone at the table laughs. After waiting a beat, he adds, "and about half the atheists in the room are closet theists."

What goes unsaid is that if the MTA's vision of the future is actualized, the categories of theism and atheism would no longer be antithetical.

§120

In this book, we have traced out the conjunctions and contradictions, the resonances and disharmonies, between two distinctly American movements— secular transhumanism and the Mormon faith—and how the interference patterns between them have allowed for the inauguration of a third movement, Mormon Transhumanism, which is an expression of the first two while not being limited to or entirely derivative of them. We followed the contours of these three objects, sometimes tracing the arc of a single development, sometimes leaping between scenes to show parallels and contrasts. And as we have done so, with each series of arcs, parallels, and contrasts, we have explored a greater narrative and found larger, imbricated structures.

First, Mormons. As first laid down by Joseph Smith, The Church of Jesus Christ of Latter-day Saints stood out in sharp contrast to the forms of protestant-inflected revival Christianity crisscrossing the frontier. In Joseph's Church, salvation was collective, not individual; it multiplied the number of sacred texts; it instituted a chain of prophets; it was at once low-church in the meeting-house and high-church in the mason-inflected-temple. And, like other utopian communities dotting the American landscape in the nineteenth century, it openly experimented with the family form, and particularly with marriage. Finally, after multiple instances of being denied protection—and sometimes even being actively attacked—by both states and the federal government, it rejected American secularism, and it collapsed religion and state into a single patriarchal, priestly form of governance.

In the end, this helped open up the frontier, where settlers displaced the indigenous inhabitants to create a culture and society peculiar to the inter-Mountain West. But it also opened up speculative frontiers, as first Joseph Smith, and then later his followers and successors, worked out what was implied by the system he had established. Joseph's formal innovation was a willingness to iterate operations that were singular in other forms of Christianity. It facilitated the creation of repetitions that were marked by differences in both their internal, generative logic and also on the face of their realized expressions. This working out of possibilities was a speculative exercise that opened up the path to infinities, accelerating the potential speed of thought as these intellectual lines of flight were carried to their fullest.

But these speculative exercises, and the social practices and theological and cosmological claims that they gave rise to, made the Church of Jesus Christ of Latter-day Saints toxic to an already hostile, outside culture. This gave rise to internal disciplinary apparatuses that helped the Church govern itself when it was a desert theocracy and maintain coherence when it was a worldwide religion. Outside hostility also encouraged the Church to take up the external mask of a certain kind of politically conservative, patriotic American pseudo-Protestantism. This was a form of cultural camouflage, allowing it to not stand out. But in time, the mask stuck to the face. And the conservative tendency to turn inward would always be at odds with the outward, cosmic expansiveness of the early days of the Church. Threats had to be anticipated and dealt with; after all, these were the latter days before the apocalypse and the millennium. Disciplining Saints, teaching Saints to discipline themselves, and fixing a particular image of thought through the installation of a correlated belief became the order of the day. But this narrowing meant a purposeful blindness towards aspects of the past. Later, thanks to new information technology, that blindness became readily visible to members; this new technology also made visible the extent of symbolic and real violence carried out by the early Church to both actualize these expansive dreams and to govern itself and fend off others. Allowed to speak frankly in these computer-virtual spaces, simmering problems of sexuality and gender built communities, but also anger and resentment, which came to the fore. All this pushed many believers into dangerous faith crises that threatened their belonging, their kinship networks, and sometimes their livelihoods and their marriages. The Mormon internet, and the Blogger-nacle in particular, became an engine of apostasy.

§121

Now Transhumanism. The fruit of a secular settlement that carved out a more constricted social space for religion, transhumanism allowed other claimants to speak on ultimate matters. When science stepped up to that position in earlier late-modern moments, it opened up the doors to infinity—cosmological near-infinite space, geological near-infinite time, infinitesimal scales of quantum physics. But it did so in a way that eviscerated the possibility of meaning instead of creating it. We still had eschatologies, but we had lost salvation.

The possibility of using technology to overcome species-limits, to change us such that we would no longer be human, reopened the possibility of giving the universe a meaningful telos. We would overcome death, become something near gods, and make the universe intelligent. This allowed for the enchainment of near-term technological innovation and long-term science fiction vaporware; next year's computer chip was just another step on the way to some other year's general AI device, which was itself merely a way station to superhuman intelligences. This new authority to speak on ultimate matters also allowed for a lateral enchainment between disciplines: cryonicists, nanotechnologists, life-extension researchers, and workers in artificial intelligence now had cause to speak to one another as they jointly imagined a post-human world. But this post-human world was not a given. It was extrapolated from the present moment, but the direction these lines of flight could take were not evident in advance. Near-term transhumanist technology might be real to research projects, but transhumanist thought was ultimately speculative thought.

Standing as religion's rival when it came to ultimate matters made trans-humanist thought an anxious thought. Initially libertarian-leaning in its politics and contrarian and non-conformist in its attitude, it was suspicious of authorities and dismissive of received wisdom. The aspects of transhuman work that looked like fringe science opened up the project to professional ridicule and state intervention, although time (and Silicon Valley) eased this opprobrium's sting. And transhumanism also feared zealots standing in their way as they tried to reach up towards immortality. In time, transhumanism ran in two different directions—those who saw the religious as potential partners in making a co-alition of the impossible, and those who saw the religious as mortal enemies, crazed by an aestheticized celebration of death. While there were transhumanist varietals all over the globe, American transhumanism dreamed of ending scarcity and creating an inclusive utopia—but it also envisioned a near-future moment when, under attack, it would have to become a closed, and possibly hostile, society, mirroring how they imagined their enemies imagined them.

§122

Mormonism is not transhumanism, nor is transhumanism Mormonism—or at least not as far as we would understand it. But there are other opinions on the matter: a common statement in the MTA is that not all transhumanists are Mormons, but all Mormons are transhumanists. This saying points out that both Mormonism and transhumanism have a radical overcoming in the human as a common goal, or, to put it in less determinate language, as a common problem. But neither transhumanism nor Mormonism is shaped by this one problem alone. Transhumanism is also informed by the question of how far science can advance itself, the problem of meaning in secular society, the problem of whether intelligence is fated, aleatory, or earned, and the challenge of real scarcity and potential abundance. Mormonism inherited many of the general problems that come with Christianity: Is the believer justified by works or faith? What is the nature of God and Jesus, and what is their relation? Does Christianity mark a break with Judaism, or is it at its core a continuation? What are the nature, authority, and identity of scripture? This list is not exhaustive, but it is at least representative. Sometimes, the existence of some problems that are particular to Joseph Smith's restoration tradition also work to constrain the range of possible answers associated with Christian forms—as is the case with any instantiation of Christianity. And some of the problems that the Church deals with, such as its relation to the state and what kind of temporal authority should be enjoyed by the Church, are problems that, in the present moment, come constrained by the partial solutions, in the guise of the limits created by various available secularisms that have already been decided on by others—even if Mormonism kicked against these limits in the nineteenth century.

Therefore, at least from an anthropological perspective, insisting on an identity between these two formations is going too far—but it can safely be said that Mormonism and transhumanism rhyme. Rhyming involves a simultaneous identity and a difference between rhyming elements. How do we think, identify, and differ at once? Like three points that demarcate a plane, we could say that the two sets of constituting questions that make Mormonism and Transhumanism each form a metaphorical plane, a possibility space that consists of all the ways that these enchained sets of questions have been, or could be, answered. The use of the term plane is metaphorical, because both Mormonism and transhumanism obviously have more than two dimensions; the number of dimensions is most likely the same as the number of core questions that form these Mormon and Transhumanist possibility spaces. The qualitative, combinatory potentia that make up these possibility spaces suggest that there are multiple Mormonisms, multiple transhumanisms, none of which can be paradigmatic; anything capable of being situated on those planes is equally Mormon or equally transhuman.[1]

These two planes do not run parallel to each other. They could be said to intersect at the spot where they share particular questions, where Mormon and transhumanist problems overlap, as does a possibility of an infinite but still limitable set of shared potentialities. These shared elements do not destroy the difference between the planes. In this overlapping space of potentialities, formed by overlapping questions and intersecting planes, it is the linkages of particular answers to separate autonomous questions located elsewhere on each formation's plane that allow for some expressions to be immediately recognizable as being Mormon or being transhumanist, even as they overlap. All this is to say that the qualitative range of answers to the question of theosis found in any particular instance or expression of Mormonism or transhumanism is, of course, constrained by the other problems they are enchained with, and then by the particular answers given to those problems, which is why these approaches only can rhyme. And this is why the different ways that Mormonism and transhumanism actualize these problems seem like funhouse inversions of each other at times; this is why, for instance, Mormon and transhumanist methods of caring for the soon-to-be resurrected dead are at once similar and yet opposed to one another. In these foldings, inversions, and rotations of concepts and practices, in these strange chiralities, a form of qualitative, topological structuralism can be imagined that would make the differences and similarities readable.[2] Multiple axes in this n dimensional overlapping space could be mapped, each one consisting of continuums running between antithetical concepts: the mastery of religion by technology/the mastering of technology by religion, universal effectivity/individual effectivity, collective ethical responsibility/individuated ethical responsibility, authoritarianism/egalitarianism, divinely fashioned/human-fashioned, computer-virtual/material, infinite/

singular, continuous/discontinuous, fixed gendered and sexual belonging/ post-gendered and sexual indifference or distinction. There is even, in Mormon's attachment to sacred locales and transhumanism's occasional cosmism, as well as in Mormonism's concern for resurrection using the original matter versus a transhumanist indifference to what their resurrected bodies are comprised of, (recall "neuropatients" and the possibility of virtual resurrection) a generic/particularist opposition. Positions on these various continuums are not totalized; various secular transhumanisms, and separate expressions of Mormonism, are often situated at different spots on these continuums. Sometimes there are different actualizations of Mormonisms and transhumanisms that lean toward opposite poles from others' actualizations of their respective set of expressions. And then there are churn and drift along these axes caused by the buffeting effects from forces exterior to this immanent structure, meaning that it is actually probably best to think of this as an n-1 dimensional array, since there are always the shocks and distentions caused by the great outside, questions asked by others, or not asked at all.[3]

Given this structuralism, it would seem that Mormon Transhumanism should be readable, too.

§123

In classical structuralism, there is the possibility of something called a mediating term, some object that stands midway between the two objects, allowing it to function as a sort of conceptual bridge. This in-betweenness might be marked by being some halfway point on a continuum between the two opposing binaries or paradoxically bearing the traits of both of the antithetical extremes. It could be imagined that Mormon Transhumanism could be such a mediating figure.

In a sense, Mormon Transhumanism does have a new approach to one of the oppositions—rather than religion mastering technology or technology mastering religion, the two categories seem to simultaneously envelope each other, collapsing the distinction. But reading this as suggesting that Mormon Transhumanism just works as a mediating term would be wrong. Mormon Transhumanists mapped familiar religious imaginings onto scientific speculation, and they also mapped scientific speculation onto a specific expression of religion. But these mappings were always partial and incomplete. Remember that the breed of structuralism we have here is a qualitative, immanent structuralism, where the structure is formed by two different metaphorical problem-planes intersecting with one another. Made of overlapping potentialities that are overdetermined by the way other questions outside this intersection are answered, it is not too much of a surprise that there are moments where there is no mediating position, as in the failure to have a fully transhumanist and fully Mormon funeral. There was a reason why Larry and Shawn King couldn't agree (although, to be fair, that animosity was probably overdetermined, too).

But there is another reason why Mormon Transhumanism cannot simply be reduced to a mediating term. The act of governing thought—in such a way that limits solutions to the line of potentiality made by the overlap of these two planes—creates further problems, and hence a plane and potentialities that may be historically subsequent to the other planes but not logically subsequent. Situated at the space where Mormonism and transhumanism intersect, as a third plane that shoots out from that overlap, it fills its own space. This plane is formed by some (but not all) of the animating problems found in Mormonism and some of the animating problems found in transhumanism. Theosis? Yes. But Mormon problems of governance and discipline are absent, or at least latent, as the MTA aspires to operate more democratically and transparently. And Mormon Transhumanism also has other, additional problems specific to it. Some of these are social; consider the way that being a believing transhumanist allows for some MTA members to better navigate the Wasatch Front total social fact of Mormon marriage, kinship, belonging, and business. Or consider how it allows for rationalist bridges that open up a way of being recognizable and readable to atheist or agnostic fellow workers; in short, how to dilate two different closed societies at once. There is also the problem of creating a common space for those dedicated to Mormonism and those who have rejected it—or have been rejected by it. And then there are conceptual questions that can only be asked by Mormon transhumanism, such as the issue of whether virtual reality is material in the same way that spirit is material. These potentialities and problems that are unique to Mormon Transhumanism are why one member of the association, a genial and well-published Italian transhumanist, can say that while he is not a Mormon, he is a Mormon Transhumanist.

Mormon Transhumanism is thus an imaginative project that answers the problem of religion and technology by aspiring to collapse that distinction; thus, it breaks the grammar of Mormonism and Transhumanism, being entirely new while at the same time being wholly Mormon and Transhumanism. Part of that aspiration means finding, in the potentialities particular to it, candidate, concrete social forms that, as of now, do not exist. This means that one of their constituting challenges is how to wrestle with both practicing pure speculation and threading that speculation back into the social. Speculation can travel at the speed of thought, dreaming up a posthuman child with her heart turned towards resurrection. Or speculation can imagine nested ancestor simulations, making operational, contingent universes containing operational, contingent universes. But in time this speculation must fall into the social/actual, and it must try to reconfigure the particular places that it is threaded back into. Millennial speculation becomes an ecologically passive house, or ancestor simulation is haltingly actualized in embryo in Ancestory.com backroom projects. This falling into the social is not guaranteed. Dreams of universe creation cannot be bent back from speculation and deflected back down to quotidian social forms and practices. Yet they still remain as virtual potentialities, and perhaps in some epoch they will find ways to be actualized. Regardless of whether these musings ever find a way to be expressed concretely, these universe-making potentialities still exist in their virtual form.

Just as Mormon Transhumanism occupies the overlap of Mormonism and Transhumanism without being limited to it, there is the speculation that escapes the virtual space laid out in the Mormon transhumanist plane, ideas that may intersect with some of the potentialities found within Mormon Transhumanism. As Mormon Transhumanism is not limited to the imaginative spaces formed by the overlaps of Mormonism or Transhumanism, places where Mormon Transhumanism intersects with other planes can give rise to yet newer sets of problems, and thus unique planes. These are the moments when speculation is not just exploring the inherent possibilities in an already established concept, or in combinatory speculation, but rather speculation works yet again by changing the grammar of the concept itself. Such is queer polygamy. Born of technical imaginings regarding the transformability of gender and queer Mormon visions of—at once escaping from and hewing even closer to—a readable orthodoxy, this is an idea that escapes Mormon Transhumanism, landing and organizing a different space organized by different problems.

There is no limit to temporally prior or subsequent iterations, of course, nor is there any clearly grasped final horizon to these intersecting planes. Transhumanism is bisected by technology-sector capitalism, several scientific disciplines, and the military-industrial complex. Mormonism, too, is cut across by numerous conservative-leaning American political movements (and a few progressive-leaning movements as well), by other forms of Christianity, and by different governmental formations. Again, this list is partial, but hopefully the point is made. It would be possible, if this were a different book, to leap from plane to plane and to follow all the different expressions with each plane and the structural permutations between planes: a new *Mythologiques*, not for the original indigenous peoples but for the settler colonies that rapaciously dispossessed them.

None of these larval social formations discussed in this book are secure, especially those that foreground speculative thought. There is always the danger that these virtualities may eventually be abandoned and exist as unthought—thus effectively not existing at all. But then that is the nature of experimentation. It is the production of a multiplicity of speculative trajectories and contingent social practices that speculation is supposed to give birth to, and is not something that from the first instance has the solidity that comes with discipline and sure belief. True, it is a way of preserving Mormonism and mastering transhumanism. But this work is only done by speculating how to preserve and control these things by making it something new. Speculation and the production of the new are the science of Mormon Transhumanism, the religion of Mormon Transhumanism. And that work is the production and proliferation of virtualities that, if actualized, would give birth to novel ways to accelerate, decelerate, and redirect the social. Mormon Transhumanists are thinking about the future. But the work and effects of this future are in the here-and-now, in the forging of new myths, of better machines for making Gods.

Acknowledgments

First, I would like to thank both the board of directors and the members of the Mormon Transhumanist Association. Ever since I first contacted them, as an institution and as a set of individuals, they have been sharing almost to a fault, exhibiting great patience and good humor as I haltingly navigated the steep learning curves of both Mormonism and transhumanism. As is sometimes said on transhumanist birthdays: may you live forever!

The research for this book first began when I was teaching at the University of Edinburgh, where I was the happiest (intellectually) I have ever been. Had I not had to choose between being at that august institution and being with my wife and daughter, I would still be there today. Among my colleagues there, I was nurtured by conversations and fellowship with Richard Baxstrom, Stefan Ecks, John Harries, Luke Heslop, Laura Jeffrey, Tobias Kelly, Casey High, Lucy Lowe, Diego Malara, Adam Marshall, Rebecca Marsland, Maya Mayblin, Alice Nagel, Jonathan Spencer, and Dimitri Tsintijilonis. In particular, Tom Boylston, Naomi Haynes Zhu, Magnus Course, and Maya Mayblin shared an outsized burden when it came to fellowship with me (Magnus in fact may have put big chunks of this mess into motion when he slyly encouraged me one day to revisit Lévi-Strauss to see if the French anthropologist had "gotten any better" since I last read him). After I left Edinburgh, a kind invitation from Emma Teale to contribute an essay for a writing project allowed me to again think through Mormonism against the backdrop of Auld Reekie, albeit it that time at quite a distance.

Coming from outside the discipline, and writing in an anthropological paradigm, meant that this work could never properly be included in the long-running and nuanced conversation that constitutes Mormon studies.

Nonetheless, many scholars were kind enough to take the time to talk to me. I would like to thank Kathleen Flake, John Gustav-Wrathall, Ben Hertzberg, Patrick Mason, and Melissa Wei-Tsing Inouye. Stephan Betts showed a lot of greatly appreciated enthusiasm for this book when it was in its very final stages. While we have not met in person, I have learned a great deal through my online-mediated friendship with Taylor Petrey.

Transhumanism is a new topic in the social sciences; compared to my prior projects, it seemed that only a few were paying it any heed. But those few who were attending to it made for good colleagues: Anya Bernstein, Jacob Boss, Jeremy Cohen, Annelin Eriksen, John Evans, and Ian Lowrie. While I have never been fortunate enough to meet him in the flesh, I have learned a good deal from Abou Farman's work, as well as from our email communications. I also have benefited greatly from conversations with Debbora Battaglia, Richard Irvine, Deana Weibel, and Matthew Wolf-Meyer, whose work dovetails nicely with transhumanism.

By contrast, anthropologists of religion are thick on the ground; but that doesn't mean that some of them aren't excellent. Be it for conversation, invitations, or inspiration, I would like to thank Andreas Bandak, John Barker, James Bielo, Joshua Brahinksy, Suzanne Brenner, Fanella Cannell, Timothy Caroll, Liana Chua, Simon Coleman, Thomas Csordas, Girish Daswani, John Dulin, Omri Elisha, Matthew Engelke, Ayala Fader, Amy Flynn-Curran, Ilana Gershon, Courtney Handman, Jessica Hardin, Jordan Haug, Jacob Hickman, Eric Hoenes del Pinal, Ingie Hovland, Brian Howell, Nofit Itzhak, Jessica Johnson, Hillary Kael, Webb Keane, Arsalan Khan, David Knowlton, Brad Kramer, Ashley Lebner, Derrick Lemons, Lauren Leve, Mèadhbh McIvor, Patrick McKearney, Charles Nuckolls, Bruno Reinhardt, China Scherz, Bambi Schieffelin, Rupert Stasch, Brendan Jamal Thornton, Matt Tomlinson, Kathryn Woolard, Joseph Webster, and Jarrett Zigon. There are certainly far more than just enumerated. While not anthropologists of religion per se, Jason Danley, Nicole Peterson, and Gregory Thompson have also been good conversation partners. Charles Matthewes's constant barrage of good humor has buoyed my mood more than once. The novelist Dominique Nuñez-Barnet has spent more than one late night on the phone talking to me about the craft of writing. Tom Lay at Fordham University Press has been an unwavering advocate for this book and is the best editor I have ever had. I am surely forgetting other people who are owed some gratitude for this book's existence; if you are one of them, please forgive me for my lapse.

There are a few people to whom this project owes a special debt. Adam Reed was kind enough to read a draft of this manuscript; given that he is one of the best (possibly *the* best) living ethnographers, that was a true boon. Courtney Handman and E. Marshall Brooks both blind-reviewed this for Fordham

University Press, and then later shared this fact with me; this book is better
(and shorter!) thanks to their insight and energy. Roy Brooks at the University
of San Diego School of Law got me into this mess by convincing me to go into
academia. Tanya Luhrmann helped with some invaluable advice at a critical
juncture in this book, and Austin Choi-Fitzpatrick downed many a pint as I
bent his ear about writing it. Joel Robbins, first as a mentor, then as a friend,
has helped both the book and me more than he probably understands. After
leaving Edinburgh to be with my family, I felt as if I was in exile from the
academy; his frequent lunches and phone calls helped stave off some particularly dark moods. Finally, I also want to acknowledge Marcel Hènaff, who
passed as this book was being written. He taught me both Deleuze and Lévi-
Strauss, two thinkers who dominate both my thinking and this book. He was
an inspired teacher and an insightful philosopher, and he is missed.

Material related to this book was presented in talks at the University of
Edinburgh, The London School of Economics, UNC Charlotte, Claremont
Graduate University, and The University of North Carolina at Chapel Hill. I
thank them for their hospitality and their thoughts. *Platypus: The CASTAC
Blog* also served as a public venue for me to present some of my earliest thoughts
related to the issues covered here. Portions of series three was published in
Ethnos as "Kolob Runs on Domo: Mormon Secrets and Transhumanist Code"
(DOI: 10.1080/00141844.2020.1770311); two paragraphs from series seven will
appear as part of an essay, "The Mormon Dead," in *The Dynamic Cosmos*,
edited by Matan Shapiro and Diana Espirito Santo, in press with Bloomsbury
Press. I thank the editors of the journal and the presses for permission to republish that material here. I would like to thank Columbia University Press
and Bloomsbury Press for permission to use the quote by Gilles Deleuze as
this book's epigraph.

I should mention that I also know people who aren't academics. Some of
them are even family. I would like to thank my sister, Mari, and her two nieces
Faye and Kallie, as well as my parents, Herman and Dolores. Two of my three
cats should be thanked: the two littermates, Iphigenia and Antigone, or Iphy and
Tiggy as they are more frequently referred to; Iphy deserves special credit for
furiously sleeping next to me whenever I was writing at my desk. I guess Galindo,
the miscreant third cat, deserves praise as well, if you really want to press the
issue. I also want—need—to thank both my wife and my daughter. My daughter
Clio is whip-smart, creative, unbelievably kind, and has a sense of humor better
than any adult I know. I may miss being at Edinburgh, Clio, but I never for a
moment regret choosing to be with you. And then there is my wife, Judith. Any
list of her virtues and talents that I could write would seem like mad exaggeration,
and yet would still fall short of all she does, and all she is. She is far more than
human; she is also the most human person I know. This book is for you, Judy.

Notes

Preface

1. Evans 2018.

2. Hodder 2010. A second edited volume was more open to the idea of there being religion at Çatalhöyük; see Hodder 2014.

3. Bialecki 2017b.

4. Brautigan 1967.

5. Doctorow and Stross 2012.

6. Mormon Transhumanist Association 2007.

7. Antosca 2018; Chan 2016; Chumley 2018; Sathian 2016.

8. See Bernstein 2014, 2015, 2016, 2019a, 2019b; Geraci 2018.

9. For a similar skepticism about ethics, viewing subjectification as a process that limits as it subjectifies, see Strhan 2015.

10. Viveiros de Castro 2017: 204.

11. Lévi-Strauss 2001; Sahlins 1976.

12. Deleuze 1994, xx.

13. See similarly, Wolf-Meyer 2019.

14. "Fiction." 2021. In *Oxford English Dictionary*.

15. Brooks 2018.

16. Fader 2020.

17. Coviello 2019.

18. Bernstein 2019a.

A Note on Names and Terms

1. Signs and tokens are a series of gestures and phrases, described as "sacred, but not secret," that are revealed to Mormons during the endowment ordinance (ritual)

at the temple; the signs and tokens are to be performed at the gates of heaven, allowing for passage to the other side.

Series Zero: "Children of God would try to play God"

1. Warchol 2010.

2. In reality barely 50 percent of Salt Lake City's population actually belongs to the Church of Jesus Christ of Latter-day Saints (by comparison, Provo, the university town an hour's drive to the south, is roughly 90 percent Mormon).

3. In Mormon eschatological doctrine, the Telestial is the lowest of the three 'degrees of glory,' ranked beneath the terrestrial and the celestial degrees.

4. *Ordinance* is a common Mormon term for a particular set of rituals performed by members of the Mormon priesthood.

5. Sacrament meeting being the term for the regular Sunday services in the Church of Jesus Christ of Latter-day Saints.

6. Bostrom 2005; Mayor 2018.

7. On Silicon Valley and the changing status of transhumanist imaginings, see Farman 2016.

8. In linguistic anthropology, the way that these two languages index two different sorts of endeavors, sets of knowledge, and types of people would be the metapragmatic content of these languages. See Silverstein 1976.

9. See Warner 2002.

10. The best example of this is Teresem, a movement founded by the entrepreneur Martine Rothblatt; while its organizational structure is complex and it is engaged in many different projects, for believers, it is centered around the creation of mind-files and hopes to eventually create God, if such a being does not already exist.

11. Bostrom 2005.

12. See. e.g., Marcus 1995.

13. Specifically, I am thinking of the space which Deleuze and Guattari would refer to as the socious. See Deleuze and Guattari 1985; Lowrie 2017.

14. For a review of the production of anthropological methods regarding 'virtual' spaces, see Stewar 2017.

15. See, e.g., Bernstein 2014, 2015, 2016, 2019; Farman 2012, 2016, 2019; Huberman 2020.

16. See Harding 1991; Robbins 2003.

17. Bialecki, Haynes, and Robbins 2008.

18. See, e.g., Norget, Napolitano, and Mayblin 2017; Hann and Goltz 2010.

19. Cannell 2005, 339.

20. Cannell 2005.

21. Bennion 1998, 2004, 2011; Cannell 2005, 2010, 2013a, 2013b, 2019; Davies 2000, 2003, 2010; Knowlton 2007, 2012; see also Elisha 2002. Brooks (2018) contains an important and insightful depiction of ex-Mormons, which also sheds clear light

on believing Mormons in the Salt Lake–Provo corridor. For a full review of anthropological material on Mormonism, and on anthropological theory relevant to Mormonism, see Cannell 2017.

22. Kramer 2014; Smith 2007.

23. For a discussion of twentieth-century Mormon sociology, see Mauss 1984; for a more recent sociological picture of Mormonism, see Riess 2019.

24. See. e.g., Kent 2000, Mouw 2016.

25. See Garriott and O'Neil 2008.

26. Bialecki 2012.

27. See Deleuze 1994.

28. See e.g., Haeri 2017, 2020; Keane 2007; Robbins 2002.

29. See Bialecki 2009, 2010; Engelke 2010b; Meyer 1998; Robbins 2007; Robbins and Engelke 2010.

30. See Bielo 2016; Tomlinson 2014.

31. See Harding 2000, 110.

32. Cf. Engelke 2014.

33. See, not entirely dissimilarly, Bandak and Coleman 2018.

34. Latour 2012.

35. Taylor 2007; Keane 2007. See also Lebner 2015, for another reminder that the idea of the secular and secularity as a principle necessarily precedes secularism as a political arraignment.

36. Asad 2009, 46.

37. Asad 2009.

38. Coviello 2019.

39. Harrison 2005.

40. Noble and Segal 1997.

41. Farman 2012, 1079.

42. Particularly good work on this topic has been done by Deana Weibel. See Weibel 2007, 2015, 2017, 2019a, 2019b, 2019c.

43. Khun 2012; Latour and Woolger 2013.

44. See e.g., Kardashev 1996.

45. Farman 2012.

46. On Russia, see Bernstein 2019; Young 2012. On science fiction, see Geraci 2011.

47. See, e.g., Tirosh-Samueson 2012. For a full discussion of the theological classification of transhumanism as religion, see Bialecki 2018.

48. See, e.g., Geraci 2012; Singler 2017.

49. For a rearticulation of what is considered to be the local classical (Bainbridge 1982) of this argument, see Bainbridge 2009.

50. On technoscience spiritualities, see Battaglia 2005.

51. This is not to argue that religion is an entirely nominalist category, but merely that what is counted as 'religion' as currently defined in American culture does in no way exhaust religion as an analytic category.

52. I am thinking here particularly of the claims of a theological school that is self-titled as 'Radical Orthodoxy.' For the *locus classicus* of this argument that secularism is a form of religion, see Milbank 2013.

53. See Engelke 2014.

54. Crapanzano 2004, 180.

55. See Farman 2012.

56. The account of the different modes of speculative thought presented here is heavily influenced by Boden's (2004) discussion of human and machine creativity. Also, while crafted independently of this work, these categories bear a striking similarity to the speculative categories of extrapolation, intensification, and mutation presented in Wolf-Meyer 2019.

57. On this last point, see Mittermaier 2011.

58. For a detailed example of how a specific strain of religio-scientific speculation is shaped and informed by the media and genres through which it is expressed, see Bielo 2018.

59. As such, they are a special case of the phenomenon discussed in Bialecki 2017, 205–17.

60. See e.g., Miller, Job, and Vassilev 2000.

61. See Foucault 1989.

62. See Crapanzano 2004; Mittermaier 2011; Robbins 2010; Sneath, Holbraad, and Pedersen 2009; Strauss 2006.

63. Deleuze 1984; Lacan and Fink 2006, 75–81.

64. Anderson 2006.

65. Castoriadis 1997.

66. Taylor 2006.

67. See e.g., Ivy 1995.

68. See Sneath 2009; Strauss 2006; on relexification, see Brightman 1995.

69. See Deleuze and Guattari 1987 for a general critique of seeing the imagination as a separate faculty.

70. Humphreys 2009; Leach, Nafus, and Krieger 2009; Sneath 2009.

71. Stafford 2009.

72. It is the shape and powers of the tool or skill that controls not only how it works in the real world but also in the virtual realm of thought. Ingold 1993.

73. Sneath, Holbraad, and Pedersen 2009.

74. Robbins 2010, 312.

75. Robbins 2010, 312.

76. Lévi-Strauss of course worked on more than myth; he also produced structuralist accounts of totemism, ritual, and sacrifice. Ritual, which works to take discrete segments and make them continuous, would seem apropos to the synthetic energies of much of the speculation discussed herein (Lévi-Strauss 1990a). But ritual is a form of action, making it poorly suited to the purely discursive material that speculation is composed of. (Furthermore, Levi-Strauss's understanding of ritual seems poorly suited to more contemporary accounts that stress ritual as a form of

autopoiesis than as cosmological intervention—see e.g., Hirschkind 2006; Mahmood 2011.) Similarly, sacrifice is about collapsing oppositions (Lévi-Strauss 1966), but speculation often augments tensions between terms. Totemism also seems inapplicable, as it is about metaphor, the mapping of one domain to another domain (Lévi-Strauss 1991). But absent some hermeneutics of suspicion, there is no other domain for speculation to map onto; speculation, like myth, is about itself.

77. Stasch 2006.

78. Ferdinand de Saussure being, of course, the nineteenth-century scholar who tried to stabilize the slippery understanding of what constitutes phonemes by seeing each phoneme being negatively reciprocally constituted by some relationship to a phonetic element that it is not. See de Saussure 2011. Technically, Lévi-Strauss's reading of de Saussure was heavily influenced by his near lifelong friendship with the linguist Roman Jackobson; see Loyer 2018.

79. Lévi-Strauss 1973, 90 (note 12).

80. Thompson 1968a, 1968b.

81. Lévi-Strauss, Eribon, and Wissing 1991.

82. For a more complete discussion by Lévi-Strauss of the importance of Thompson and transformations, see Lévi-Strauss 1990a, 675–78.

83. Viveiros de Castro 2017, 198.

84. Viveiros de Castro 2017, 204.

85. While this argument focuses on the sort of speculative thought addressed here bordering on the mythic, it should be noted that some have freely used it for other purposes, such as ritual (Mosko 1985, 1991) and (especially relevant to our discussion here) cultural invention (Schwimmer 2001).

86. Lévi-Strauss 1995, xii.

87. Lévi-Strauss 1995, xviii.

88. Also see Danowski and Viveiros de Castro 2016, which classifies science fiction, transhumanism, accelerationism, and Amerindian cosmology as all being "myths."

89. Lévi-Strauss 1990a, 693–95.

90. *Les deux sources de la morale et de la religion*, Henri Bergson.

91. Bergson 1935, 275.

92. Bergson 1935.

93. Bergson 1911, 1914.

94. Bergson 1911, 1921, 1929.

95. Bergson 1946, 73–86

96. For a review of some of the literature associated with this line of thought, see Clark 2012.

97. Lévi-Strauss 1991, 97, 98.

98. Godelier 2018, 360.

99. Lévi-Strauss 1990a, 621–24.

100. For this definition of religion, Tyler 1874, 424. On Tyler and religion, see Larsen 2014.

101. For "superhuman beings," Spiro 1966, 89–90. For "relations with the more than human," Bialecki 2017.

102. Givens 2015, 256–315.

First Series: Mormonisms

1. The title given is Uchtdorf's as of 2015. In 2018, when then-president of the Church Thomas S. Monson died, and the mantle of presidency fell to Russell M. Nelson, Dieter F. Uchtdorf left the position of Second Counselor to the Church's First Presidency, regained the title of Elder Uchtdorf, and returned to his status as one of the Quorum of the Twelve Apostles, which is still one of the most senior positions in the Church.

2. See Paul 2018.

3. For a discussion of a prior attempt by Former First President Gordon B. Hinckley to reject the use of the term 'Mormon,' see Shipps 2001.

4. Starting in October 2021, the Church discontinued the practice of having separate all-male-attendee priesthood-holder sessions and women's sessions in General Conference.

5. Stack 2014.

6. Peters 2012.

7. Howell 2015.

8. See e.g., Engelke 2010a, 2010b; Keane 1997, 2007; Meyers 2006, 2009; de Vries, Hent, and Samuel Webster 2001. For a general review of this literature, see Hoviland 2018.

9. The website had its name changed to ChurchofJesusChrist.org, as part of the push to move away from the term *Mormon*.

10. Boorstein 2011.

11. See LDS Search Optimization Standards. https://www.lds.org/bc/content/CP%20Training/PDFs/LDS%20Search%20Optimization%20Standards_April%202013.pdf?lang=eng.

12. As we will see later on, the person who coined the term is closely associated with the Mormon Transhumanist Association.

13. Uchtdorf 2015.

14. On gap between author, animator, and principle, see Goffman 1981.

15. Uchtdorf 2013.

16. Uchtdorf 2013.

17. Hardy 2016.

18. This last point has been acknowledged by the Church in an unsigned essay, released as part of a series addressing what is often seen as problematic aspects of Church history; see https://www.lds.org/topics/plural-marriage-in-kirtland-and-nauvoo?lang=eng&old=true.

19. The phrase "a spiritual light would shine" was taken from an 1879 account of the techniques used by Joseph Smith to translate the golden plates—The True

Latter-day Saints' Herald 26/22 (November 15th, 1879), by way of https://www.fair
mormon.org/evidences/Category:Book_of_Mormon/Translation/Method/Seer
_stone#cite_ref-22

20. Reeve 2015.

21. Flake 2005.

22. Mouw 2016.

23. Jeremiah Films 2005; on Cameron, see episode eight of season three of "The
Way of the Master."

24. Goodstein 2014.

25. Goodstein 2015.

26. Van Valkenburg 2016.

27. Jackson 2018.

28. See Cannell 2019. Technically, one member of the six, Mormon feminist
Lynne Kanavel Whiteside, was only disfellowshipped, which means that while she
is ostracized from the Church, she still technically is a member of it.

29. Goodstein 2016.

30. Mayblin et al. 2017.

31. Bialecki 2014.

32. There was a third priesthood, the "Patriarchal" priesthood, but that is now
understood as actually being a part of the more extensive Melchizedek priesthood.
Along with Joseph Smith, the Melchizedek and Aaronic priesthoods were also
granted to the early Mormon leader Oliver Cowdery by the three apostles.

33. Though it should be noted that according to Roman Catholic logic, it is
possible to be a proper part of the chain of apostolic succession while also being
excommunicated. See Mayblin 2019.

34. Cannell 2013a, 2013b; Kramer 2014.

35. There are separate temple recommends for Saints too young for an
endowment (the "limited-use" recommend, intended to allow youth to participate
in proxy baptisms for the dead) as well as for those who are attending for the
explicit purpose of receiving an endowment for the first time (the "recommend for
living ordinances").

36. In Spring of 2019, they waived the one-year wait requirement for couples who
had been married in a civil ceremony to have a temple sealing; this allowed for there
to be a public, secular wedding open to all, immediately preceding or following a
temple sealing. This has to a degree lessened the pain of non-Mormon or non-
temple-recommend-holding family members not being present at the wedding, but it
has not necessarily removed it entirely.

37. See the FAQs section of LDS.org; https://www.lds.org/church/temples/
frequently-asked-questions?lang=eng

38. See The Economist 2012.

39. Monroe 2017.

40. See Mauss 2002.

41. Doctrines and Convents 76:103.

42. Uchtdorf 2015
43. See "Discourse, April 7 1844, as Reported by Wilford Woodruff."

Second Series: Transhumanisms

1. Bostrom 2005, 3.
2. Astor 1894; Asimov and Côté 1986; Bellamy 1996; MacCulloch 1892.
3. Wolf-Meyers 2009; Sirius & Cornell 2015, 65–66; Overbye 2011; Long 2011.
4. Young 2012; Bernstein 2019.
5. See Anonymous 2018; Pellissier 2012.
6. Strictly speaking, intelligence and speed are different axes, and there is no formal reason why they should be coupled if a singularity were to occur. But these two continuums are still often conflated when the singularity is discussed.
7. Vinge 1993.
8. For one of the most widely circulated and elaborated discussions of the singularity as risk, see Bostrom 2014.
9. Good 1966; Solomonoff 1985.
10. Farman 2012, 2019.
11. Huxley 1957; on Julian Huxley's transhumanism, see Bashford 2013.
12. Harrison & Wolyniak 2015.
13. On theological readings of transhumanism, see Bialecki 2018.
14. Boyer 2009.
15. Fukuyama 2003.
16. See infra, § nine.
17. https://hpluspedia.org/wiki/Transhumanist_political_organisations; https://sites.google.com/site/transhumanistpartyglobal/tp-national-level-groups.
18. Istvan 2016a; McGinness, 2016.
19. Roussi 2016.
20. Istvan 2015a.
21. Istvan 2015a.
22. Reason 2013.
23. Reason 2013.
24. Istvan 2017a.
25. Istvan 2015a.
26. See Evans 2010, 2014.
27. Istvan 2015b; Mack 2015.
28. Istvan 2015c.
29. Messerly 2015.
30. Pellissier 2012.
31. Pellissier 2015a.
32. Pellessier 2015b.
33. Pellissier 2015b.
34. Kelly 2015.

35. Twyman 2015.

36. See iPetitions n.d.

37. More 1990, 4.

38. See Yudkowski 2002, 2009a, 2009b.

39. Saletan 2006. While not identified as such in the article, the "trans-bemanist" described as looking like a "shaved Willie Nelson" is Martine Bothblatt, a lawyer, CEO, head of the AI-supporting Terasen institute, and founder of one of the earliest and most successful satellite-to-car-radio networks; at one point, Ms. Rothblatt was the highest-paid female CEO in the United States. See Miller 2014.

40. Farman 2016.

41. While this is the sort of claim made in conversation, chatrooms, Facebook posts, and blogs, the most authoritative instantiations of this argument can be seen in Barrow & Tippler 1996 and Kurzweil 2016; See also Farman 2012, 2019.

42. Farman 2012, 2019.

Third Series: Mormon Transhumanism

1. Canham 2007; Pluralism Project n.d.

2. ARDA 2010.

3. Brigham Young University in Provo has to be distinguished from two other identically named, Church-owned Universities: Brigham Young University–Hawaii, known for its outreach to students from Oceania, and Brigham Young University–Idaho, a more conservative school that is sometimes referred to by its own students as the Taliban of the modern BYU system.

4. Actually, various Adam-God–centered posts came up multiple times on the Mormon boards of Belief.net, but they all tended to have nearly identical titles.

5. Pew Research Center 2009.

6. See Larson 1978.

7. Alma 42:13.

8. Lewis 2001, 5.

9. Widtsoe 1925, 521–22.

10. D&C 101:31.

11. See Paul 1992, chapter 4.

12. Smith 1948, 245; Book of Abraham 3:3.

13. Smith 1938, 10–11; Moses 1:29.

14. Huntington 1892, 263. Unlike Young's statements regarding solar life, Joseph Smith's purported discussion of lunar inhabitants is a secondhand report made after his death, so some skepticism might be warranted.

15. Evans & Grimshaw n.d., 271.

16. Robert 1986.

17. For more on this, see Bialecki 2020.

18. Jacobsen 1992, 273.

19. See https://deseretbook.com/p/marble-christus-statue-deseret-book-company -41038?variant_id=146932-9-inch (accessed October 25, 2019).

20. See Campbell 2017a, 2017b.

21. May 1997.

22. Foster 1997.

23. See Paul 1994.

24. Bergera 2002; Allen 1993.

25. See Bushman 2012.

26. Collett 2015a.

27. Collett 2015b.

28. See Farman 2019.

29. As we will see, the ties between these other topics and Mormon Transhumanism are stronger than first appears.

30. See, for example, Smith 2018.

Fourth Series: Kolob runs on Domo

1. See Ruiz 2007; on Utah, see Nielson 2017.

2. It should be noted that this absence has not kept some public intellectuals associated with the MTA from speculating on the contents of these missing pages; see Bradley 2019.

3. On the tensions between publication and secrecy, see Peters 2015, 409.

4. See, for example, Taussig 1999.

5. See Fader 2017.

6. On general sociality and identity in this virtual world, see Boellstorff 2015.

7. 3 Nephi 9:2–3; Scott 2011, 86. The nonvirtual Adam-ondi-Ahman is located in Daviess County, Missouri.

8. See De Groote 2008a, 2008b.

9. Richman, 2007.

10. Scott 2011, 95.

11. Asad 2009; Masuzawa 2005; Saler 1987.

12. On what religion does, see Lambek 2013; on religion as expression, see Bialecki 2016a, 2016b, 2012.

13. On markedness and absence, see Keane 1997a, 1997b, 2010; On secrecy and revelation, see, for example, Barth 1975; Robbins 2001.

14. Smith 2007.

15. Hicks 1986.

16. Hicks 1986.

17. Smith 2007, 208–9, 222–23.

18. Van Wagoner 2002.

19. Smith 2007.

20. Keane 2007.

21. Davies 2010; Quinn 1988; Hormer 2014. It should be noted that the alternate, more faithful Mormon reading is not that of temple practices borrowed from Freemasonry, but rather that Freemasonry and temple practices mirror one another because they both have a genealogy that runs back to Solomon's Temple.

22. This quote is taken from the congressional Reed Smoot hearings; see Probationary State 2010.

23. Kramer 2014.

24. Specifically, one must sustain the "President of the Church of Jesus Christ of Latter-day Saints as the Prophet, Seer, and Revelator," as well as the "members of the First Presidency and the Quorum of the Twelve Apostles," the "General Authorities," and local Church authorities.

25. See Kramer 2014, 72–73.

26. Kramer 2014, 72–73.

27. Kramer 2014, 67.

28. Kramer 2014, 71.

29. Compare McDannell 1998, with Mahmood 2011, and Brenner 1996.

30. See. e.g., Church of Jesus Christ of Latter-day Saints, "Temple Garments" n.d.

31. The racial uniformity of the Quorum of the Twelve has changed with the ascent of Garrett W. Gong.

32. This language is from the leaked internal Church document: see https://www .hrc.org/blog/leaked-mormon-policy-takes-aim-at-children-of-same-sex-couples -marriage-equ.

33. The bishop, which is technically a lay position that people rotate in and out of as they receive a *calling* from higher up in the Church hierarchy, is the rough equivalent of a protestant pastor; the stake president, also a lay position that people are *called* for, oversees the bishops in a particular geographical region.

34. Kramer 2014, 42–44.

35. While there is not space to fully flesh the idea out here, many Mormon trans-humanists show an interest in the nineteenth-century concept of there also being a "mother God," or alternately posit that God might be a collective rather than a singular identity. See also Bialecki 2020.

36. See https://new-god-argument.com (n.d.) for a full presentation of the argument; see also Cannon 2015.

37. Deleuze & Guattari 1987, 286–90.

Fifth Series: Discipline, Belief, and Speculative Religion

1. Pratt 2012.

2. Mormon Newsroom 2015.

3. Cannell 2019.

4. Brooks 2014.

5. Lunchwithandy 2016.

6. This is not to say that no members of the MTA have been excommunicated, of course. These excommunications are about a general rejection of the Church, and MTA membership was not a factor, or at least was not mentioned as one.

7. In the 2018 April general conference, it was announced that home visits would be replaced by *ministering*, a form of support intended to encourage flexibility and to allow women to be part of the home visiting teams.

8. Asad 1993.

9. Mahmood 2011; Hirschkind 2006; Reinhardt 2014.

10. Bielo 2017.

11. This was changed to two hours a week in 2018, though the educational component continues.

12. These are technically supposed to be boys who hold the Aaronic priesthood, but are still awaiting the Melchizedek priesthood.

13. Keane 2011.

14. Smith 2007.

15. White 2007, xi.

16. See Cannell 2005.

17. Coviello 2019.

18. See Inouye 2018.

19. See Geertz 1973.

20. On Pentecostal and Charismatic subjectification, see Bialecki 2017; Brahnisky 2012; Reinhardt 2014; Lurhmann 2012.

21. James 1912.

22. Farman 2012.

23. Bushman 2012.

24. Fader 2009; Lester 2005; Mahmood 2012; Robbins 2007. See similarly Hirschkind 2006.

25. Deleuze 1997.

26. Baxstrom 2008.

27. For this reason, discipline as an ethical exercise is often seen as being predicated on freedom. See Faubion 2001, 2011; Keane 2017; Laidlaw 2013, 2002; Robbins 2007b.

28. Compare with Luhmann, 2013. Note that religion, as such, would have no natural borders though, being an open set of possible assemblages.

29. Goldschmidt 1933; Bialecki 2014, S197–S198.

30. Petrey 2011.

31. Hoskins 2015.

Sixth Series: Freezing, Burying, Burning

1. Godlier 2018, 373–79.

2. Alcor 2016.

3. The SuperDs, true to their name, are larger, and can contain nine to eleven bodies and up to ninety heads. See Weinstein 2017.

4. There is some irony to this. When justifying the reason for the move in 1990, ALCOR promotional material stated that in addition to a need for new space due to growth, "A secondary, but important impetus for change is esthetic: the industrial warehouse look, in an area largely occupied by manufacturers and auto services, does not inspire confidence in the long-term strength and reliability of our organization. Many new and prospective members have made this clear to us with varying degrees of diplomacy. The new building we have in mind will be beautifully high-tech and have the look of institutional stability." See Mondragon 1990.

5. This phrase appeared on the ALCOR webpage for Scottsdale as of October 2017. The site now states that Scottsdale has "favorable weather year-round." See http://www.alcor.org/about

6. The dewars, it should be noted, do not need power to operate.

7. ALCOR membership flyer, on file with author.

8. My use of the word *archive* for the collection of cryopreserved patients is not a flight of fancy. Some transhumanists, both secular and religious, believe that it is unlikely that these bodies will be resurrected; rather, they hold that the preserved information, and particularly the preserved neural material, will be used to either build an entirely new body and brain imprinted with the traits, skills, and memories of the preserved, or to create a digital incarnation of the preserved individual that would exist in a virtual domain.

9. All quotes taken from video; see Galactic Public Archives 2016.

10. Tiffany 2010, 208.

11. Tiffany 2010, 202.

12. Tiffany 2010, 208.

13. Tiffany 2010, 207.

14. Alcor 2015. Using lower case for ALCOR was in the original document; capitalization is not consistent in the various ALCOR created or associated documents.

15. All quotes taken from video; see Galactic Public Archives 2016.

16. Shoffstall 2016, 3.

17. Walker 1979, 11.

18. Shoffstall 2016, 17.

19. Shoffstall 2016, 17.

20. Perry 1992c.

21. Shoffstall (2016), for instance, argues that a variety of factors, including a modern desire to sequester death, and magical thinking on the part of those either volunteering for cryonic preservation, or those who support those who volunteer for such preservation.

22. Suspension Failures: Lessons from the Early Years By R. Michael Perry https://alcor.org/Library/html/suspensionfailures.html. Accessed 25 October 2017.

23. "'Suspension Failures: Lessons from the Early Years' By R. Michael Perry" https://alcor.org/Library/html/suspensionfailures.html. Accessed 25 October 2017.

24. Mondragon 1994, 13.

25. Perry 1992a, 5.

26. Perry 1992a, 5.

27. Perry 1992a, 6.

28. Perry 1992a, 5.

29. Perry 1992a, 7.

30. Perry 1992b, 5.

31. Sahagun and Arax 1988.

32. Regis 1990, 143.

33. Johnson and Baldyga 2009; Alcor n.d., "Response."

34. Farman 2016.

35. Drexler 1996. The book, which first made the case for nanotechnology as a compliment to cryonics.

36. See Cryonics Newsletter #8 (March 1981): 1. Available at https://www.alcor.org /docs/cryonics-magazine-1981-03.txt

37. Darwin 1990, 11.

38. Kent 1985, 26.

39. Regis 1990.

40. Donaldson v. Lungren 2 Cal.App.4th 1614; see also Donaldson 1992.

41. Donaldson 1998, 40.

42. Donaldson 1998, 47.

43. Donaldson 1989, 26. The third sphere is a reference to Ptolemaic celestial theory, though resonances with the third Celestial degree of glory in Mormon eschatology are striking in the context of this book.

44. Donaldson 1989, 25, 32.

45. O'Connor 1989, 20.

46. Donaldson 1989, 31.

47. Lempert 1985, 3.

48. Laprade 1998, 37.

49. de Wolf 2010, 8, 9.

50. Plus 1998, 7.

51. Laprade 1998, 37.

52. Bostrom 2005, 11.

53. See Alcor's webpage for Alcor UK. http://alcor.org/AlcorUK.html

54. Bostrom 2005, 15.

55. In *Extropy* 4. Unfortunately, there is no original copy of this issue of *Extropy*— here, I'm relying on a republished version put out by the Libertarian Alliance (it was common for periodicals that were a part of this circuit at that time to republish essays that appeared in other similar journals). More, 1991, 1–2.

56. Petersen 2012.

57. Young 1997, 51.

58. More 1990b.

59. More 1990, 6.

60. Jaynes 2000.

61. All quotes in this paragraph from More 1990, 7–8.

62. More 1990, 8.

63. See Series 2, §28.

64. More 1990, 8–9.

65. More 1990, 9.

66. More 1990, 10.

67. Alcor 2010.

68. See Cerco Il Tuo Volto 2009.

69. More 2009.

70. Lucas 1992, 2–3.

71. All citations to the podcast from a file held by the author; unsurprisingly, this episode, along with the rest of the podcast, has been pulled from the internet.

72. Leibovich 2015.

73. Rosen 2015.

74. France 1991.

75. The specific attack Cannon was responding to was Istvan 2014; the Canadian law at issue was Bill 3—2004, Cremation, Internment and Funeral Services Act, passed by the Legislative Assembly of British Columbia.

76. See Cannon 2014a; Cannon 2014b. It is not unheard of for a blog-essay to be reposted to many different transhumanist venues, but it is rare, and it marks Cannon as being a part of a certain echelon of transhumanist public intellectuals.

77. I have been unable to determine if that visit evert took place, or what effect it had if so. I suspect that it did not, since inquiries about it with multiple group members did not trigger any memory.

78. See Church of Jesus Christ of Latter-day Saints, "The Purpose of Seminary."

79. Cannon 2018.

80. Cannon 2018.

81. Cannon 2018.

82. Cannon 2018.

83. There are additional regulations regarding death in 7.10.2, 9.10.3, 21.3.1–2, and 21.3.7 of the handbook. It is also referenced in section 3.4.9 of the first handbook. The first handbook is only distributed to Saints who are in leadership positions, so this cannot be independently confirmed without going against the explicit will of the Church.

84. See Church of Jesus Christ of Latter-day Saints *General Handbook* 18.6.1.

85. *General Handbook* 18.6.1

86. *General Handbook* 18.6.1

87. *General Handbook* 18.6.2.

88. *General Handbook* 18.6.4.

89. *General Handbook* 18.6.4.

90. *General Handbook* 18.6.4

91. *General Handbook* 18.6.4

92. *General Handbook* 18.6.6.

93. *General Handbook* 9.10.3.

94. Toomer-Cook 2008.

95. Brown, Murphy, and Malony 1997.

96. On the origins of this belief in Joseph Smith and early Mormonism, see Brown 2011.

97. Widtsoe 1925, 570–71.

98. On the difficulty of inferring actual practice or understandings based merely on official pronouncements, see Jenkins 2012.

99. It is perhaps no accident that Lévi-Strauss granted that there was an "implicit mythology" to ritual, even as he saw the main work of ritual not as exploring and proliferating vying potentialities but instead collapsing the tension between them. Lévi Strauss 1990, 670–74.

Seventh Series: "as if awakening from a night's sleep"

1. See, for example, Faubion 2014.

2. See Church of Jesus Christ of Latter-day Saints, "Provident Living" n.d.

3. Brown 2011.

4. Nelson 2016.

5. On the origin of apocalyptic material, and tensions between leadership and lay apocalyptic imaginings, see Blythe 2020.

6. Prisco 2018.

7. All quotes from Lincoln Cannon, *The Consolation* (n.d.), document on file with the author.

8. The hymn Cannon is quoting is "Should I hie to Kolob."

9. Kurzweil 2001; Cannon n.d., 3–4.

10. Spiro 1966.

11. Smith 1938, 181, 326.

12. Sahlins 1996.

13. Tipler 1994, xiii.

14. Tipler 1994, 56–57.

15. Tipler 1994, 33

16. Tipler 1994, 57.

17. Biron 2015; Sean 2001. It should be noted that continual expansion is a more fatal scenario than one where the universe is flat; there are multiple topological models that would allow for a flat but still finite universe. See Lachieze-Ray and Luminet 1995.

18. Associated Press 2005.

19. Kurzweil 2016.

20. Kurzweil and Grossman 2005, iv, 4.

21. Solman 2012.

22. Miles 2015.

23. Berman 2011.

24. Solman 2012.

25. Singularity Group n.d.

26. "Redeeming the Dead Redeemed Me" 2014

27. Google search conducted December 11, 2018.

28. Pugmire 2013. It should be noted that this is not the same as having billions of digitized documents: a record consists of a reference to a single individual, or in some cases, such as newspapers, an estimation of how many references to individuals can be found per page. See Ancestry 2009.

29. Marks 2009, 243–44.

30. McCracken and Grossman 2013.

31. Ancestry 2015.

32. Ruby et. al. 2018.

33. Brodwin 2018.

34. Cannon 2013.

35. (*Vaporware* being a computer-industry term for purely notional projects announced years ahead of their production, digital commodities that often never gets to even the hesitant, larval moments of development and distribution.)

36. From a structuralist standpoint, resurrection concerns are the phenomenon closest to the double twist we have in this book, if one includes in the sequence Mormon accounts of the resurrection and assumes a sequence that runs: Mormon resurrection → Tippet's cosmic transhumanist resurrection → Cannon's vision of the resurrection → Kurzweil's resurrection. In such a sequence, you have an inversion from a Divine autonomous, sure mechanism (in orthodox Mormon resurrection) as a marked category to (in Kurzweil's) a human resurrection not as a marked but sure category but as the conditioning background assumption; hence a reversal and a conversion from being a variable and a state (in the first instance in the series) to a function (in the last), as resurrection becomes less a fact and more something performed. There is also a corresponding shift from collective labor and agency to individuated labor and individuated agency, with the leap from Mormon Transhumanism to Kurzweil's resurrection marking the moment when these values are inverted. It should be noted that it is only the last two in series—Kurzweil's vision and the Mormon Transhumanist vision—that have begun to be actualized in any meaningful way.

Eighth Series: Worlds without End

1. See *infra*, Series 7, §96.

2. This piece eventually found a home on Lincoln Cannon's personal blog. See Cannon, 2013b.

3. In the post, he joshingly references the software used to explore that world as "FamilyHistory.com version 42"; the number forty-two being something of a recurrent inside joke among programmers, not unlike what the Wilhelm scream is for motion picture and television sound designers.

4. Bostrom 2003. An earlier version of this paper was circulated in 2001, which explains the otherwise science-fiction-like fact that responses to this essay were published before the essay itself.

5. Hanson 2001.

6. Mormon Transhumanist Association 2007, 25.

7. Mormon Transhumanist Association 2007, 26.

8. Mormon Transhumanist Association 2007, 26.

9. Mormon Transhumanist Association 2007, 36.

10. Mormon Transhumanist Association 2007, 31.

11. It should be noted that the idea of a community of Gods is part of one common Mormon interpretation of the temple endowment ordinance. This interpretation is based on the portion of the endowment that acts out the creation narrative; parts of the ordinance's script depict Elohim, Jehovah, and Michael working in conjunction during the seven-day creation period described in Genesis 1:1-2:3.

12. World Scientific, "Modern Physics Letters A, Aims & Scope," n.d.

13. Search "Nobel Prize" at https://www.worldscientific.com/worldscinet/mpla

14. Michigan State University, "Steven Hsu," n.d. https://www.pa.msu.edu/profile /hsu/

15. Wikipedia, "Anthony Zee," n.d.

16. Hsu and Zee 2006.

17. Hsu and Zee 2006, 1495.

18. Hsu and Zee 2006, 1495. For a more extensive discussion of this, see Merali 2017.

19. Hsu and Zee 2006, 1496.

20. Hsu and Zee 2006, 1497.

21. Hsu and Zee 2006, 1499.

22. Jones 2017.

23. See Levi-Strauss 1976, 253; Godelier 2018, 370.

24. The level of connection in the one physical Mormon transhumanist universe we see is less certain, but there seems to be a connection between universes; in a nod to the Mormon doctrine of premortal existence, humanity is presented as having already participated in the creation of this universe and may in time may have access to the memories of how we helped accomplish this.

Ninth Series: Queer Polygamy

1. On the plurality of gods, see Givens 2015: 102–105.

2. Paulsen & Pulido, 2011. See also Givens, 2015, pp. 106–116.

3. Erickson 1998, 190.

4. *Doctrines and Covenants*, Official Declaration 1.

5. Erickson 1998, 204–205.

6. van Wagoner 2002, 155.

7. van Wagoner 2002.

8. Deseret News 2005.

9. Mormon Newsroom 2006.

10. It should be noted that Russell M. Nelson is not the only Apostle sealed to multiple women; Dallin H. Oaks, first counselor in the LDS Church's governing First Presidency, is also sealed to his current wife, as well as his deceased spouse.

11. See *infra*, Series Four, § 44.

12. Mansunaga 2015.

13. Kane 2015.

14. Mansunaga 2015.

15. Levin 2015.

16. Mormon Newsroom 2015.

17. Stack 2016.

18. Iturriaga and Saguy 2017.

19. Park 2014.

20. Dehlin 2015.

21. Park 2018.

22. In 2019, Blaire Ostler would stop blogging with *Rational Faiths* in order to facilitate other projects, which include becoming one of the hosts of the Sunstone Education Foundation's *Sunstone Firesides Podcast*. See Ostler 2019a.

23. It should be noted that this maps onto the default anthropological categories for different sets of marriage.

24. This practice, referred to as the Law of Adoption, is understood by most of mainline Mormonism as the creation a father-son tie through a sort of spiritual adoption. It was placed in abeyance in 1984, just a handful of years after the First Manifesto.

25. Though contrast Coviello 2019, which reads 19th century Mormon polygamy through the lens of queer theory. Then, there is a difference between being read as queer posthumously and identifying one's contemporary project as queer.

26. As with many other things, use of the term *singleton* can be traced back to Nick Bostrom; specifically, to his repurposing the set theory concept as instead the existence of a "single decision-making agency at the highest level." See Bostrom 2006, 48.

27. Also see Kneese and Peters 2019.

28. Blair 2018

29. All quotations in this section from Ostler 2015a.

30. The text of this presentation can be found at Ostler 2019b.

31. KJV 1 Corinthians 15:55–57.

32. Ostler 2015b.

33. The Utah territories were the second in the nation to extend the vote to women, doing so in 1870, only a month later than the Wyoming territories granted women suffrage in late 1869. In 1887, the United States congress would disenfranchise all women in the territory, regardless of their marital status, as part of law designed to combat polygamy. See Alexander 1970.

34. Ostler 2017.

35. Ostler 2019c, 34.

36. Ostler 2019c, 42.

37. That does not mean that there is no ethnography—though it is notable that what little ethnography there is tends to lean into the observer rather than the participant mode, and even then, there is a limit on what was observed. See, e.g., Kimberly 2016.

38. See the anthropologist E. Marshall Brooks, who writes quite authoritatively on this topic: Brooks 2018.

39. 1 Corinthians 6:19–20.

40. See Aran 2015; Dark 2015; Lunchwithandy 2015.

41. See Brooks 2018.

42. See Foucault 1978.

43. See similarly Erzen 2006.

44. 1 Corinthians 15:52.

Series. $\lim_{n \to \infty} n$: Problems, Planes, and Lines of Flight

1. While it may not seem so, the construction presented here is similar to the thought of Claude Lévi-Strauss. He saw myth as a form of speculative thought that extrapolated far from the empirical work to grapple with insoluble problems that, while capable of being temporarily resolved in particular narratives, still insistently suggested other potentialities. He also situated myths as "spread out over a *hyperspace*, also occupied by other myths," (Lévi-Strauss 1990b, 105; emphasis added), much like this multiaxes virtual cognitive space discussed here.

2. See §10 *infra*.

3. Cf. Deleuze and Guattari 1987, 6.

Bibliography

Alcor. n.d. "Response to Larry Johnson Accusations." https://www.alcor.org/press
    /response-to-larry-johnson-allegations/
Alcor. 2010. Press release, December 23, 2010. Available at http://www.alcor.org/blog
    /alcor-life-extension-foundation-names-max-more-phd-as-chief-executive-officer/
Alcor. 2015. "Two-Year Old Thai Girl Becomes Alcor's 134th Patient." *Alcor.com*
    (blog), March 20, 2015. https://www.alcor.org/2015/03/two-year-old-thai-girl
    -becomes-alcors-134rd-patient/
Alcor. 2016. "SuperD Dewar Arrived Today." *Alcor.com* (blog), September 30, 2016.
    http://www.alcor.org/AtWork/p2facility.html; https://www.alcor.org/blog/superd
    -dewar-arrived-today/
Alexander, Thomas. 1970. "An Experiment In Progressive Legislation: The Granting
    of Woman Suffrage In Utah In 1870." *Utah Historical Quarterly* 36, no. 1: 20–30.
Allen, James B. 1993. "The Story of The Truth, The Way, The Life." *BYU Studies*
    33, no. 4: 1–36.
Anderson, Benedict. 2006. *Imagined Communities: Reflections on the Origin and
    Spread of Nationalism*. Brooklyn: Verso Books.
The Pluralism Project. America's New Religious Landscape: Salt Lake City, UT.
    The Pluralism Project. http://worldmap.harvard.edu/maps/pluralism-salt_lake
Ancestry. 2009. "How Many Billions of Records are on Ancestry.com?" *Ancestry.com*
    (blog), June 30, 2009. https://www.ancestry.com/corporate/blog/how-many
    -billions-of-records-are-on-ancestrycom/
Ancestry. 2015. "AncestryDNA and Calico to Research the Genetics of Human
    Lifespan." *Ancestry.com* (blog), July 21, 2015. https://www.ancestry.com/corporate
    /newsroom/press-releases/ancestrydna-and-calico-research-genetics-human
    -lifespan
Antosca, Albert. 2018. "Thinking Outside the Old Religious Box: Transhumanism
    is Complicating the Sometimes Antagonistic Relationship between Faith and

Science." *Slate*, March 21, 2018. https://slate.com/technology/2018/03/trans humanism-is-complicating-the-relationship-between-faith-and-science.html

Aran, Isha. 2015. "Inside the World of Mormon Porn." *Fusion*. Archived at https://web.archive.org/web/20160118063126/http://fusion.net/story/166513/mormon-lds-porn-sex/

ARDA: Association of Religion Data Archive. 2010. *Metro-Area Membership Report: Provo-Orem, UT, Metropolitan Statistical Area*. Data collected by Association of Statisticians of American Religious Bodies (ASARB). https://www.thearda.com/rcms2010/rcms2010a.asp?U=39340&T=metro&S=name&Y=2010

Asad, Talal. 2003. *Formations of the Secular: Christianity, Islam, Modernity*. Palo Alto: Stanford University Press.

Asad, Talal. 2009. *Genealogies of Religion: Discipline and Reasons of Power in Christianity and Islam*. Baltimore, MD: JHU Press.

Asimov, Isaac, and Jean Marc Côté. 1986. *Futuredays: A Nineteenth-Century Vision of the Year 2000*. New York: Henry Holt & Company.

Associated Press. 2005. "Never Say Die: Live Forever." *Wired*, February 2, 2015. https://www.wired.com/2005/02/never-say-die-live-forever/

Astor, John Jacob. 1894. *A Journey in Other Worlds: A Romance of the Future*. New York: D. Appleton and Company.

Bainbridge, William. 1982. "Religion for a Galactic Civilization." In *Science Fiction and Space Futures*, edited by Eugene M. Emme, 187-201. San Diego: American Astronautical Society.

Bainbridge, William. 2009 "Religion for a Galactic Civilization 2.0." *Institute for Ethics and Emerging Technologies*. https://ieet.org/index.php/IEET2/more/bainbridge20090820

Bandak, Andreas, and Tom Boylston. 2014. "The 'Orthodoxy' of Orthodoxy: On Moral Imperfection, Correctness, and Deferral in Religious Worlds." *Religion and Society* 5, no. 1: 25–46.

Bandak, Andreas, and Simon Coleman. 2018. "Different Repetitions: Anthropological Engagements with Figures of Return, Recurrence and Redundancy." *History and Anthropology* 30, no. 2: 119–32.

Barnhart, Duane, dir. 2008. *The Way of the Master*. Season 3, episode 8, "Mormonism." Aired October 8, 2008, Pyro Pictures, television.

Barth, Fredrik. 1975. *Ritual and Knowledge among the Baktaman of New Guinea*. New Haven: Yale University Press.

Bashford, Alison. 2013. "Julian Huxley's Transhumanism." In *Crafting Humans: From Genesis to Eugenics and Beyond*, edited by Marius Turda, 153–67. Taiwan: National Taiwan University Press.

Battaglia, Deborah. 2005. "For Those Who Are Not Afraid of the Future: Raelian Clonehood in the Public Sphere." In *E.T. Culture: Anthropology in Outer Spaces*, edited by D. Battaglia, 149–79. Durham, NC: Duke University Press.

Baxstrom, Richard. 2008. *Houses in Motion: The Experience of Place and the Problem of Belief in Urban Malaysia*. Stanford, CA: Stanford University Press.

Bellamy, Edward. 1996. *Looking Backwards, 2000–1887*. Mineola, NY: Dover.

Bennion, Janet. 1998. *Women of Principle: Female Networking in Contemporary Mormon Polygamy*. New York: Oxford University Press.

Bennion, Janet. 2004. *Desert Patriarchy: Mormon and Mennonite Communities in the Chihuahua Valley*. Tucson: University of Arizona Press.

Bennion, Janet. 2011. *Polygamy in Primetime: Media, Gender and Politics in Mormon Fundamentalism*. Lebanon, NH: Brandeis University Press.

Bergson, Henri. 1911. *Creative Evolution*. New York: Henry Holt & Company.

Bergson, Henri. 1914. *Laughter: An Essay on the Meaning of the Comic*. New York: Macmillan.

Bergson, Henri. 1921. *Time and Free Will: An Essay on the Immediate Data of Consciousness*. London: George Allen & Unwin.

Bergson, Henri. 1929. *Matter and Memory*. London: George Allen & Unwin.

Bergson, Henri. 1935. *The Two Sources of Morality and Religion*. London: Macmillan.

Bergson, Henri. 1946. *The Creative Mind: An Introduction to Metaphysics*. New York: The Philosophical Library.

Bergera, G. J. 2002. *Conflict in the Quorum: Orson Pratt, Brigham Young, Joseph Smith*. Salt Lake City: Signature Books.

Berman, John. 2011. "Futurist Ray Kurzweil Says He Can Bring His Dead Father Back to Life Through a Computer Avatar," *ABC News*, August 9, 2011. https://abcnews.go.com/Technology/futurist-ray-kurzweil-bring-dead-father-back-life/story?id=14267712

Bernstein, Anya. 2014. "Cyborgs, Weak Cosmists, and a Russian Planet." *NYU Jordan Center for the Advanced Study of Russia*. http://jordanrussiacenter.org/news/cyborgs-weak-cosmists-russian-planet/#.WKoVUBiZOV4

Bernstein, Anya. 2015. "Freeze, Die, Come to Life: The Many Paths to Immortality in Post–Soviet Russia." *American Ethnologist* 42, no. 4: 766–81.

Bernstein, Anya. 2016. "Love and Resurrection: Remaking Life and Death in Contemporary Russia." *American Anthropologist* 118, no. 1: 12–23.

Bernstein, Anya. 2019. *The Future of Immortality: Remaking Life and Death*. Princeton, NJ: Princeton University Press.

Bernstein, Anya. 2019b. "Life, Unlimited: Russian Archives of the Digital and the Human." *JRAI: Journal of the Royal Anthropological Institute* 25, no. 4: 676–97.

Bialecki, Jon. 2009. "Disjuncture, Continental Philosophy's New 'Political Paul,' and the Question of Progressive Christianity in a Southern California Third Wave Church." *American Ethnologist* 36, no. 1: 110–23.

Bialecki, Jon. 2010. "Angels and Grass: Church, Revival, and the Neo-Pauline Turn." *South Atlantic Quarterly* 109, no. 4: 695–717.

Bialecki, Jon. 2012. "Virtual Christianity in an Age of Nominalist Anthropology." *Anthropological Theory* 12, no. 3: 295–319.

Bialecki, Jon. 2014. "After the Denominozoic: Evolution, Differentiation, Denominationalism." *Current Anthropology* 55, no. S10: S193–S204.

Bialecki, Jon. 2014. "Does God Exist in Methodological Atheism? On Tanya Lurhmann's When God Talks Back and Bruno Latour." *Anthropology of Consciousness* 25, no. 1: 32–52.

Bialecki, Jon. 2016a. "Protestant Language, Christian Problems, and Religious
    Realism." *Suomen Antropologi: Journal of the Finnish Anthropological Society* 40,
    no. 4: 37–42.
Bialecki, Jon. 2016b. "Ethnography at Its Edges: The Insistence of Religion."
    American Ethnologist 43, no. 2: 369–73.
Bialecki, Jon. 2017a. A *Diagram for Fire: Miracles and Variation in an American
    Charismatic Movement*. Oakland: University of California Press.
Bialecki, Jon. 2017b. "'Religion' after Religion, 'Ritual' after Ritual." In *The Routledge
    Companion to Contemporary Anthropology*, edited by Simon Coleman, Susan
    Hyatt, and Ann Kingsolver, 183–200. London: Routledge.
Bialecki, Jon. 2018. "Anthropology, Theology, and the Challenge of Immanence."
    In *Theologically Engaged Anthropology*, edited by J. Derrick Lemons, 156–88.
    Oxford: Oxford University Press.
Bialecki, Jon. 2020. "Future-Day Saints: Abrahamic Astronomy, Anthropological
    Futures, and Speculative Religion." *Religions* 11, no. 11: 612.
Bialecki, Jon, Naomi Haynes, and Joel Robbins. 2008. "The Anthropology of
    Christianity." *Religion Compass* 2, no. 6:1139–58.
Bielo, James. 2016. "Replication as Religious Practice, Temporality as Religious
    Problem." *History and Anthropology* 28, no. 2:131–48.
Bielo, James. 2018. *Arc Encounter: The Making of a Creationist Theme Park*. New
    York: New York University Press.
Biron, Lauren. 2015. "Our Flat Universe." *Symmetry*. https://www.symmetrymaga
    zine.org/article/april-2015/our-flat-universe?email_issue=725
Blaire, Gabrielle. 2018. "Men Cause 100% of Unwanted Pregnancies." *Human
    Parts*, September 24, 2018. https://humanparts.medium.com/men-cause-100-of
    -unwanted-pregnancies-eboe8288a7e5
Blythe, Christopher James. 2020. *Terrible Revolution: Latter-day Saints and the
    American Apocalypse*. Oxford: Oxford University Press.
Boden, Margaret. 2004. *The Creative Mind: Myths and Mechanisms*. London: Routledge.
Boellstorff, Tom. 2015. *Coming of Age in Second Life: An Anthropologist Explores the
    Virtually Human*. Princeton: Princeton University Press.
Boorstein, Michelle. 2011. "Mormons Using the Web to Control Their Own Image."
    Washington Post, August 17, 2011. https://www.washingtonpost.com/local/
    mormons-using-the-web-to-control-their-own-image/2011/08/11/gIQA1J6BMJ
    _story.html
Bostrom, Nick. 2003. "Are You Living in a Computer Simulation?" *Philosophical
    Quarterly* 53, no. 211: 243–55.
Bostrom, Nick. 2005. "A History of Transhumanist Thought." *Journal of Evolution
    and Technology* 14, no. 1. https://jetpress.org/volume14/bostrom.html
Bostrom, Nick. 2006. "What Is a Singleton?" *Linguistic and Philosophical Investi-
    gations* 5, no. 2: 48–54.
Bostrom, Nick. 2014. *Superintelligence: Paths, Dangers, Strategies*. Oxford: Oxford
    University Press.

Boyer, Paul. 2009. *When Time Shall Be No More: Prophecy Belief in Modern American Culture.* Cambridge, MA: Harvard University Press.

Bradley, Don. 2019. *The Lost 116 Pages: Reconstructing the Book of Mormon's Missing Stories.* Salt Lake City: Greg Kofford Books.

Brahinsky, Joshua. 2012. "Pentecostal Body Logics: Cultivating a Modern Sensorium." *Cultural Anthropology* 27,no. 2: 215–38.

Brautigan, Richard. 1967. *All Watched Over by Machines of Loving Grace.* San Francisco: Communication Company.

Brenner, Suzanne. 1996. "Reconstructing Self and Society: Javanese Muslim Women and 'the Veil.'" *American Ethnologist* 23, no. 4:673–97.

Brightman, Robert. 1995. "Forget Culture: Replacement, Transcendence, Relexification." *Cultural Anthropology* 10, no. 4: 509–46.

Brodwin, Erin. 2018. "A Collaboration between Google's Secretive Life-Extension Spinoff and Popular Genetics Company Ancestry has Quietly Ended." *Business Insider* August 1, 2018. https://www.businessinsider.com/google-calico-ancestry -dna-genetics-aging-partnership-ended-2018-7

Brooks, Joanna. 2014. *The Book of Mormon Girl: A Memoir of an American Faith.* New York: Free Press.

Brooks, E. Marshall. 2018. *Disenchanted Lives: Apostasy and Ex-Mormonism among the Latter-day Saints.* New Brunswick, NJ: Rutgers University Press.

Brown, Samuel Morris. 2011. *In Heaven As It Is On Earth: Joseph Smith and the Early Mormon Conquest of Death.* Oxford: Oxford University Press.

Brown, W.S., N.C. Murphy, and H.N. Malony, eds., 1997. *Whatever Happened to the Soul? Scientific and Theological Portraits of Human Nature.* Minneapolis: Fortress Press.

Bushman, Richard Lyman. 2007. *Joseph Smith: Rough Stone Rolling.* New York: Vintage Books.

Bushman, Richard Lyman. 2012. Foreword to *Parallels and Convergences: Mormon Thought and Engineering Vision.* Edited by A. Scott Howe. Salt Lake City: Greg Kofford Books.

Campbell, Mary. 2017a. "Salt Lake City Spaceman: Christus-Creation at the LDS Visitors's Center." *Cosmologics Magazine.* https://cosmologicsmagazine.com /mary-campbell-salt-lake-city- spaceman/

Campbell, Mary. 2017b. "Saints in Space: Mormonism, Masculinity, and the Mercury 7." *Cosmologics Magazine.* https://cosmologicsmagazine.com/mary -campbell-saints-in-space/

Canham, Matt. 2007. "Utah Less Mormon than Ever." *The Salt Lake Tribune,* November 18, 2007. http://archive.sltrib.com/article.php?id=7496034&itype =NGPSID

Cannell, Fenella. 2005. "The Christianity of Anthropology." *JRAI: Journal of the Royal Anthropological Institute,* 11, no. 2: 335–56.

Cannell, Fenella. 2010. "The Anthropology of Secularism," *Annual Review Anthropology* 39: 85–100.

Cannell, Fenella. 2013a. "The Blood of Abraham: Mormon Redemptive Physicality and American Idioms of Kinship." *JRAI: Journal of the Royal Anthropological Institute* 19, no. s1: S77–S94.

Cannell, Fenella. 2013b. "The Re-Enchantment of Kinship." In *Vital Relations: Modernity and the Persistent Life of Kinship*, edited by Fenella Cannell and Susie Mckinnon, 228–51. Santa Fe, NM: SAR Press.

Cannell, Fenella. 2017. "Mormonism and Anthropology: On Ways of Knowing." *Mormon Studies Review* 4, no. 1: 1–15.

Cannell, Fenella. 2019. "Latter-Day Saints and the Problem of Theology." In *Theologically Engaged Anthropology*, edited by J. Derrick Lemons, 244–65. Oxford: Oxford University Press.

Cannon, Lincoln. 2013a. "Imagine Future Technology for Family History Simulations." *Ancestry.com* (blog), November 11, 2019. https://blogs.ancestry .com/ancestry/2013/11/19/imagine-future-technology-for-family-history-simu lations/

Cannon, Lincoln. 2013b. "Are We Living in a Family History Simulation?" *Lincoln Cannon* (blog), December 13, 2013. https://lincoln.metacannon.net/2013/12/are -we-living-in-family-history.html

Cannon, Lincoln. 2014a. "Resuscitation, by Cryonics or Otherwise, Is a Religious Mandate." *IEET: Institute for Ethics and Emerging Technologies*, August 11, 2014. https://ieet.org/index.php/IEET2/more/cannon20140811

Cannon, Lincoln. 2014b. "Resuscitation, by Cryonics or Otherwise, Is a Religious Mandate." *H+*, August 11, 2014. https://hplusmagazine.com/2014/08/11/resus citation-cryonics-otherwise-religious-mandate/

Cannon, Lincoln 2018. "My Funeral (God Forbid)." *Lincoln Cannon* (blog), September 6, 2013. https://lincoln.metacannon.net/2018/09/my-funeral-god -forbid.html

Cannon, Lincoln, n.d. "The Consolation." Document on file with the author.

Cannon, Donald Q. 1978. "The King Follett Discourse: Joseph Smith's Greatest Sermon in Historical Perspective." *BYU Studies* 18, no. 2: 1–14.

Carroll, Sean, 2001. "The Cosmological Constant." *Living Reviews in Relativity* 4, no. 1: 1.

Castoriadis, Cornelius. 1997. *World in Fragments: Writings on Politics, Society, Psychoanalysis, and the Imagination*. Palo Alto: Stanford University Press.

Cerco il Tuo Volto. 2009. "Diocesi di Pistoia : una settimana sull'immortalità terrena." *Cerco il tuo volto* (post). September 10, 2009. https://www.cercoiltuo volto.it/video/diocesi-di-pistoia-una-settimana-sullimmortalita-terrena/

Chan, Dawn. 2016. "The Immortality Upgrade." *The New Yorker*, April 20, 2016. https://www.newyorker.com/tech/annals-of-technology/mormon-transhumanism -and-the-immortality-upgrade?verso=true

Chumley, Cheryl. 2018. "Christians, Beware the Cult of Transhumanism." *The Washington Times*, September 7, 2018. https://www.washingtontimes.com/news /2018/sep/7/christians-beware-cult-transhumanism/

Church, Nate. 2016. "Transhumanist Presidential Candidate Zoltan Istvan: Tech Giants Will Make 'Billions and Billions' Off Machines Replacing Humans." *Breibart,* April 8, 2016. https://www.breitbart.com/tech/2016/04/08/transhumanist -presidential-candidate-zoltan-istvan-tech-giants-will-make-billions-and-billions -off-machines-replacing-humans/

Church of Jesus Christ of Latter-day Saints. 2021. *General Handbook: Serving the Church of Jesus Chrsit of Latter-day Saints.* Version 7/21. Salt Lake City: Church of Jesus Christ of Latter-day Saints. https://www.churchofjesuschrist.org/study /manual/general-handbook?lang=eng

Church of Jesus Christ of Latter-day Saints. "Provident Living." n.d. https://provident living.churchofjesuschrist.org/?lang=eng

Church of Jesus Christ of Latter-day Saints. "The Purpose of Seminary." n.d. https:// www.lds.org/si/seminary/about/purpose-of-seminary?lang=eng

Church of Jesus Christ of Latter-day Saints. "Temple Garments." n.d. https://news room.churchofjesuschrist.org/article/temple-garments

Clark, J. C. D. 2012. "Secularization and Modernization: The Failure of a 'Grand Narrative.'" *The Historical Journal* 55, no. 1: 161–94.

Cohen, Debra Nussbaum. 2005. "Faithful Track Questions, Answers and Minutiae on Blogs." *The New York Times,* March 5, 2005. https://www.nytimes.com/2005 /03/05/us/faithful-track-questions-answers-and-minutiae-on-blogs.html

Collett, Sarah. 2015a. "98: Lincoln Cannon on Mormon Transhumanism and Becoming One in the Body of Christ," June 9, 2015. A *Thoughtful Faith,* podcast. http://hwcdn.libsyn.com/p/f/d/b/fdb665cc2cc13b70/A_thoughtful_Faith-098 -LincolnCannonPt1.mp3?c_id=9172150&cs_id=9172150&expiration=1556743 729 &hwt=774115197adc6e6e5e7930f7eb188741

Collett, Sarah. 2015b. "99: Lincoln Cannon on Mormon Transhumanism and Becoming One in the Body of Christ." A *Thoughtful Faith,* podcast. http:// hwcdn.libsyn.com/p/4/9/3/493992f27b52328c/A_thoughtful_Faith-099-Lincoln CannonPt2.mp3?c_id=9172145&cs_id=9172145&expiration=1556743965&hwt =e4a76134b48b695fe573dccdadec92e8

Coviello, Peter. 2019. *Make Yourselves Gods: Mormons and the Unfinished Business of American Secularism.* Chicago: University of Chicago Press.

Crapanzano, Vincent. 2004. *Imaginative Horizons: An Essay in Literary-Philosophical Anthropology.* Chicago: University of Chicago Press.

Danowkski, Déboras, and Eduardo Viveiros De Castro. 2016. *The Ends of the World.* London: Polity.

Dark, Stephen. 2015. "Mormon-Themed Porn Unites Sex, Faith and Celluloid." *Salt Lake City Weekly,* July 20, 2015. https://www.cityweekly.net/BuzzBlog/archives /2015/07/20/mormon-themed-porn-unites-sex-faith-and-celluloid

Darwin, Mike. 1990. "Christian Cryonics: Support Group Needed." *Cryonics* 11, no. 3: 10–11.

Davies, Douglas. 2000. *The Mormon Culture of Salvation: Force, Grace, and Glory.* Farnham, UK: Ashgate.

Davies, Douglas. 2003. *An Introduction to Mormonism*. Cambridge: Cambridge University Press.

Davies, Douglas. 2010. *Joseph Smith, Jesus, and Satanic Opposition*. Surrey, UK: Ashgate.

De Groote, Michael. 2008a. "Virtual Mormons Get a 'Second Life,'—Part 1." *Deseret News*, July 1, 2008. https://www.deseretnews.com/article/705381362/ Virtual-Mormons-get-a-Second-Life--Part-1.html

De Groote, Michael. 2008b. "Virtual Mormons Get a 'Second Life,'—Part 2", *Deseret News*, July 3, 2008. https://www.deseretnews.com/article/705381363/ Virtual-Mormons-get-a-Second-Life--Part-2.html

Dehlin, John. 2015. "Mormon Stories #551: Lindsay Hansen Park's Year of Polygamy." *Mormon Stories*, produced by Open Stories Foundation, July 21, 2015, video. https://www.youtube.com/watch?v=JtMQLKjJh3E

Deleuze, Gilles. 1984. *Kant's Critical Philosophy: The Doctrine of the Faculties*. London: Athlone Press.

Deleuze, Gilles. 1994. *Difference and Repetition*. New York: Columbia University Press.

Deleuze, Gilles. 1997. *Cinema 2*. Minneapolis: University of Minnesota Press.

Deleuze, Gilles, and Felix Guattari. 1983. *Capitalism and Schizophrenia*. Minneapolis: University of Minnesota Press.

Deleuze, Gilles, and Felix Guattari. 1987. *A Thousand Plateaus*. Minneapolis: University of Minnesota.

Deleuze, Gilles, and Felix Guattari. 1994. *What Is Philosophy?* New York: Columbia University Press.

Deseret News. 2005. "Elder Nelson's Wife, Dantzel, Dies at Age 78." *Deseret News*. https://www.deseretnews.com/article/600111826/Elder-Nelsons-wife-Dantzel-dies -at-age-78.html

"Discourse, April 7 1844, as Reported by Wilford Woodruff." The Joseph Smith Papers. Found at http://www.josephsmithpapers.org/paper-summary/discourse -7-april-1844-as-reported-by-wilford-woodruff/3.

Doctorow, Cory, and Charles Stross. 2012. *The Rapture of the Nerds: A Tale of the Singularity, Posthumanity, and Awkward Social Situations*. New York: Tor Books.

Donaldson, Thomas. 1989. "The Apocalypse Has Been Called Off." *Cryonics* 10, no. 6: 24–31.

Donaldson, Thomas. 1992. "Thomas Donaldson et al. v. Van De Kamp." *Cryonics* 13, no. 3: 13–16.

Donaldson, Thomas. 1998. "Nanotechnology as Science and as Religion." *Cryonics* 19, no. 2: 40–47.

Drexler, Eric. 1996. *Engines of Creation*. London: Fourth Estate.

Elisha, Omri. 2002. "Sustaining Charisma Mormon Sectarian Culture and the Struggle for Plural Marriage, 1852–1890." *Nova Religio: The Journal of Alternative and Emergent Religions* 6, no. 1: 45–63.

Engelke, Matthew. 2010a. "Number and the Imagination of Global Christianity; Or, Mediation and Immediacy in the Work of Alain Badiou." *South Atlantic Quarterly* 109, no. 4: 811–29.

Engelke, Matthew. 2010b. "Past Pentecostalism: Notes on Rupture, Realignment, and Everyday Life in Pentecostal and African Independent Churches." *Africa* 80, no. 2: 177–99.

Engelke, Matthew. 2011. "Response to Charles Hirschkind: Religion and Trans- duction." *Social Anthropology* 19, no. 1: 97–102.

Engelke, Matthew. 2014. "Christianity and the Anthropology of Secular Humanism." *Current Anthropology* 55, no. S10: S292–S301.

Erickson, Dan. 1998. *"As a Thief in the Night": The Mormon Quest for Millennial Deliverance.* Salt Lake City: Signature Books.

Erzen, Tanya. 2006. *Straight to Jesus: Sexual and Christian Conversions in the Ex-Gay Movement.* Berkley, CA: University of California Press.

Evans, John. 2010. *Contested Reproduction: Genetic Technologies, Religion, and Public Debate.* Chicago: University of Chicago Press.

Evans, John. 2014. "Faith in Science in Global Perspective: Implications for Transhumanism." *Public Understanding of Science* 23, no. 7: 814–32.

Evans, John. 2018. *Morals Not Knowledge: Recasting the Contemporary U.S. Conflict Between Religion and Science.* Berkeley, CA: University of California Press.

Evans, D. W., and John Grimshaw. 1871. *Journal of Discourses by President Brigham Young, His Two Counsellors, and the Twelve Apostles* vol. 13. London: Horace S. Eldredge.

Fader, Ayala. 2009. *Mitzvah Girls: Bringing up the Next Generation of Hasidic Jews in Brooklyn.* Princeton University Press.

Fader, Ayala. 2017. "The Counterpublic of the J(ewish) Blogosphere: Gendered Language and the Mediation of Religious Doubt among Ultra-Orthodox Jews in New York." *JRAI: Journal of the Royal Anthropological Institute* 23, no. 4: 727–47.

Fader, Ayala. 2020. *Hidden Heretics: Jewish Doubt in the Digital Age.* Princeton: Princeton University Press.

Farman, Abou. 2012. "Re-Enchantment Cosmologies: Mastery and Obsolescence in an Intelligent Universe." *Anthropological Quarterly* 85, no. 4: 1069–88.

Farman, Abou. 2016. "Cryonics in the Cradle of Technocivilization." *Platypus: The Castac Blog,* December 9, 2016. http://blog.castac.org/2016/12/cryonics/

Farman, Abou. 2019. "Mind out of Place: Transhuman Spirituality." *Journal of the American Academy of Religion* 87, no. 1: 57–80.

Farman, Abou. 2020. *On Not Dying: Secular Immortality in the Age of Technoscience.* Minneapolis: Minnesota University Press.

Faubion, James. 2001. "Toward an Anthropology of Ethics: Foucault and the Pedagogies of Autopoiesis." *Representations* 74, no. 1: 83–104.

Faubion, James. 2011. *An Anthropology of Ethics.* Cambridge, UK: Cambridge University Press.

Flake, Kathleen. 2005. *The Politics of American Religious Identity: The Seating of Senator Reed Smoot, Mormon Apostle*. Chapel Hill: University of North Carolina Press.

The Economist. 2012. "Fleecing the Flock." January 28, 2012. http://www.economist .com/node/21543526

Foster, Lawrence. 1997. "Free Love and Community: John Humphrey Noyes and the Onedia Perfectionists." In *America's Communal Utopias*, edited by D. E. Pitzer and P. S. Boyer, 253–78. Chapel Hill: University of North Carolina Press.

Foucault, Michel. 1978. *The History of Sexuality Volume 1: An Introduction*. New York: Pantheon Books.

Foucault, Michel. 1989. *The Order of Things: An Archaeology of the Human Sciences*. London: Routledge.

France, Lisa Respers. 2019. "Larry King Files for Divorce after Almost 22 Years of Marriage." *CNN*, August 21, 2019. https://www.cnn.com/2019/08/21/entertain ment/larry-shawn-king-divorce/index.html

Fukuyama, Francis. 2003. *Our Posthuman Future: Consequences of the Biotechnology Revolution*. New York: Farrar, Strauss, and Giroux.

Galactic Public Archives. 2016. "Cryonics at a Crossroads: A Tour of ALCOR Life Extension Foundation," from Galactic Public Archives posted March 21, 2016. Video (14:02).

Garr, Arnold K. 2009. "Joseph Smith: Campaign for President of the United States." *Ensign*, February, 2009. https://www.churchofjesuschrist.org/study/ensign/2009 /02/joseph-smith-campaign-for-president-of-the-united-states?lang=eng

Garriott, William, and Kevin Lewis O'Neill. 2008. "Who Is a Christian? Toward a Dialogic Approach in the Anthropology of Christianity." *Anthropological Theory* 8, no. 4: 381–98.

Geertz, Clifford. 1973. "Religion as a Cultural System." In *The Interpretation of Cultures: Selected Essays*. New York: Basic Books.

Geraci, Robert. 2011. "There and Back Again: Transhumanist Evangelism in Science Fiction and Popular Science." *Implicit Religion* 14, no. 2: 141–72.

Geraci, Robert. 2012. *Apocalyptic AI: Visions of Heaven in Robotics, Artificial Intelligence, and Virtual Reality*. Oxford: Oxford University Press.

Geraci, Robert. 2018. *Temples of Modernity: Nationalism, Hinduism, and Trans-humanism in South Indian Science*. Lanham, MD: Rowman & Littlefield.

Givens, Terryl. 2015. *Wrestling the Angel: The Foundations of Mormon Thought: Cosmos, God, Humanity*. Oxford: Oxford University Press.

Godelier, Maurice. 2018. *Claude Lévi-Strauss: A Critical Study of His Thought*. London: Verso.

Goffman, Erving. 1981. "Footing." In *Forms of Talk*, edited by E. Goffman, 124–59. Philadelphia: University of Pennsylvania Press.

Goldschmidt, Richard. 1933. "Some Aspects of Evolution." *Science* 78 (December 15): 539–46.

Good, I. J. 1966. "Speculations Concerning the First Ultraintelligent Machine." *Advances in Computers* 6: 31–88.

Goodstein, Laurie. 2014. "Mormons Expel Founder of Group Seeking Priesthood for Women." *New York Times*, June 24, 2014. https://www.nytimes.com/2014/06 /24/us/Kate-Kelly-Mormon-Church-Excommunicates-Ordain-Women-Founder .html?searchResultPosition=3

Goodstein, Laurie. 2015. "Mormon Church Expels Outspoken Critic." *New York Times*, February 10, 2015. https://www.nytimes.com/2015/02/11/us/mormon-church -expels-critic-for-apostasy.html?searchResultPosition=3

Goodstein, Laurie. 2016. "Leaked Videos Pull Back Curtain on Mormon Leadership." *New York Times*, October 6, 2016. https://www.nytimes.com/2016/10/07/us /mormon-videos-leaked.html

Gow, Peter. 2014. "Lévi-Strauss's 'Double Twist' and Controlled Comparison Transformational Relations between Neighboring Societies." *Anthropology of this Century* 10 (May). http://aotcpress.com/articles/lvistrausss-double-twist -controlled-comparison-transformational-relations-neighbouring/

H+pedia, s.v. "Transhumanist Demographics," Last modified May 1, 2018, 18:43. https://hpluspedia.org/wiki/Transhumanist_demographics

Hann, Chris, and Hermann Goltz, eds. 2010. *Eastern Christians in Anthropological Perspective* Vol. 9. Berkeley: University of California Press.

Hanson, Robin. 2001. "How to Live in a Simulation." *Journal of Evolution and Technology* 7, no. 1.

Haeri, Niloofar. 2017. "The Sincere Subject: Mediation and Interiority among a Group of Muslim Women in Iran." *HAU: Journal of Ethnographic Theory* 7, no. 1: 139–61.

Harding, Susan. 1991. "Representing Fundamentalism: The Problem of the Repugnant Cultural Other." *Social Research* 58, no. 2: 373–93.

Harding, Susan. 2000. *The Book of Jerry Falwell: Fundamentalist Language and Politics*. Princeton: Princeton University Press.

Hardy, Grant. 2016. "More Effective Apologetics." Lecture presentation at the 18th Annual FAIR Conference, Provo, UT, August 2016. https://www.fairmormon.org /conference/august-2016/more-effective-apologetics

Harrison, Peter. 2015. *The Territories of Science and Religion*. Chicago: University of Chicago Press.

Harrison, Peter, and Joseph Wolyniak. "The History of Transhumanism." *Notes and Queries* 62, no. 3: 465–67.

Hicks, Michael. 1986. "Minding Business: A Note on the Mormon Creed." *BYU Studies Quarterly* 26, no. 4:125–32.

Hirschkind, Charles. 2006. *The Ethical Soundscape: Cassette Sermons and Islamic Counterpublics*. New York: Columbia University Press.

Hodder, Ian, ed. 2010. *Religion in the Emergence of Civilization: Çatalhöyük as a Case Study*. Cambridge, UK: Cambridge University Press.

Hodder, Ian, ed. 2014. *Religion at Work in a Neolithic Society: Vital Matters*. Cambridge, UK: Cambridge University Press.

Hormer, Michael. 2014. *Joseph's Temples: The Dynamic Relationship between Freemasonry and Mormonism*. Provo: University of Utah Press.

Hoskins, Janet. 2015. *The Divine Eye and the Diaspora: Vietnamese Syncretism Becomes Transpacific Caodaism*. Honolulu, Hawaii: University of Hawaii Press.

Hovland, Ingie. 2018. "Beyond Mediation: An Anthropological Understanding of the Relationship Between Humans, Materiality, and Transcendence in Protestant Christianity." *Journal of the American Academy of Religion* 86, no. 2: 425–53.

Howell, Sharon. 2015. "Social Media and Technology Are Extending the Reach of General Conference." http://tech.lds.org/blog/678-social-media-and-technology-are-extending-the-reach-of-general-conference Last accessed October 21, 2019.

Hsu, Steve, and Anthony Zee. 2006. "Message in the Sky" *Modern Physics Letters* A 21, no. 19: 1495–1500.

Huberman, Jennifer. 2020. *Ancestors and Avatars: Anthropological Approaches to Transhumanism*. Cambridge, UK: Cambridge University Press.

Humphrey, Caroline. 2009. "The Mask and the Face: Imagination and Social Life in Russian Chat Rooms and Beyond." *Ethnos* 74, no. 1: 31–50.

Huntington, O.B. 1892. "Inhabitants of the Moon." In *The Young Women's Journal: Published by the Young Ladies' Mutual Improvement Association of Zion: Volume III*, edited by Susan Young Gates, 263–64. Salt Lake City: George Q. Cannon and Sons.

Huxley, Julian. 1957. "Transhumanism." Chap. 1, in *New Bottles for New Wine*. London: Chatto & Windus, 13–17.

Ingold, Tim. 1993. "Technology, Language, Intelligence: A Reconsideration of Basic Concepts." In *Tools, Language and Cognition in Human Evolution*, edited by Kathleen Rita Gibson and Tim Ingold, 449–72. Cambridge: Cambridge University Press.

Inouye, Melissa, 2018. "A Tale of Three Primaries: The Gravity of Mormonism's Informal Institutions." In, *Decolonizing Mormonism: Approaching a Postcolonial Zion*, edited by Gina Colvin and Joanna Brooks. Salt Lake City, UT: University of Utah Press.

iPetitions, n.d., "Transhumanists Disavow Zoltan Istvan Candidacy for US President," *Ipetions* https://www.ipetitions.com/petition/transhumanists-disavow-zoltan-istvan-candidacy

Istvan, Zoltan. 2013. *The Transhumanist Wager*. Futurity Imagine Media LLC.

Istvan, Zoltan. 2014. "Civil Rights Clash: Transhumanists Prepare to Challenge an Anti-Cryonics Law in Canada." *HuffPost*, September 27, 2014. https://www.huffpost.com/entry/transhumanists-are-prepar_b_5615580

Istvan, Zoltan, 2015a. "Why a Presidential Candidate Is Driving a Giant Coffin Called the Immortality Bus Across America," *HuffPost*, August 8, 2015 http://www.huffingtonpost.com/zoltan-istvan/why-a-presidential-candidate-is-driving-a-giant-coffin-called-the-immortality-bus-_b_7928826.html

Istvan, Zoltan, 2015b. "I Just Got a Computer Chip Implanted in My Hand—and the Rest of the World Won't Be Far Behind," *Business Insider*, September 25, 2015. http://www.businessinsider.com/presidential-candidate-zoltan-istvan-gets-an-rfid-chip-implant-2015-9

Istvan, Zoltan. 2015c. "Immortality Bus Delivers Newly Created Transhumanist Bill of Rights to the US Capitol." *HuffPost*, December 21, 2015. http://www.huffington post.com/zoltan-istvan/immortality-bus-delivers-_b_8849450.html

Istvan, Zoltan. 2016. "Why I'm Not Taking Any Campaign Contributions for My Presidential Run" *Business Insider*, March 16, 2016. http://www.businessinsider .com/zoltan-istvan-not-taking-campaign-contributions-for-presidential-run-2016-3

Istvan, Zoltan. 2017. "To Ensure a Future of Transhumanism, Atheists Should Confront the Deathist Culture Religion Has Sown." *Huffpost*, December 6, 2017. http://www.huffingtonpost.com/zoltan-istvan/should-atheists-confront-_b_930 3396.html

Iturriaga, Nicole, and Abigail C. Saguy. 2017. "'I Would Never Want to be an Only Wife': The Role of Discursive Networks and Post-Feminist Discourse in Reframing Polygamy." *Social Problems* 64, no. 1: 333–50.

Ivy, Marilyn. 1995. *Discourses of the Vanishing: Modernity, Phantasm, Japan*. Chicago: University of Chicago Press.

Jackson, Laurie. 2018. "Why a Former Mormon Bishop Was Excommunicated for Criticizing Sexually Explicit Youth Interviews." *CNN*, September 24, 2018. https:// www.cnn.com/2018/09/24/us/mormon-young-excommunicated/index.html.

Jacobsen, Florence S. 1992. "Christus Statue." In *Encyclopedia of Mormonism*, edited by Daniel H. Ludlow, 273–74. New York: Macmillan Publishing.

James, William. 1912. *The Will to Believe*. New York: Longsman, Green and Co.

Jaynes, Julian., 2000. *The Origin of Consciousness in the Breakdown of the Bicameral Mind*. Boston, MA: Houghton Mifflin Harcourt.

Jenkins, Timothy. 2012. "The Anthropology of Christianity: Situation and Critique." *Ethnos* 77, no. 4: 459–76.

Jeremiah Films. 2005. *God Makers*. Hemet, CA: Jeremiah Films, video.

Johnson, Larry, and Scott Baldyga, 2009. *Frozen: A True Story My Journey into the World of Cryonics, Deception, and Death*. New York: Vanguard Books.

Jones, Caleb. 2015. "Mormonism and the Fractal Lineage of Gods." *The Transfigurist*. https://www.transfigurist.org/2015/04/mormonism-and-fractal-lineage-of-gods.html

Kane, Rich. 2015. "Hundreds Rally against New Mormon Policy; Many File Forms to Quit the Faith." *The Salt Lake Tribune*, December 5, 2015. https://archive.sltrib.com /article.php?id=3178894&itype=CMSID

Kardashev, N. S. 1997. "Cosmology and Civilizations." *Astrophysics and Space Science* 252, no. 1/2: 25–40.

Keane, Webb. 1997a. "Religious Language." *Annual Review of Anthropology* 26: 47–71.

Keane, Webb. 1997b. *Signs of Recognition: Powers and Hazards of Representation in an Indonesian Society*. Chicago: University of Chicago Press.

Keane, Webb. 2007. *Christian Moderns: Freedom and Fetish in the Mission Encounter*. Berkeley: University of California Press.

Keane, Webb. 2010. "Marked, Absent, Habitual." In *Religion in the Emergence of Civilization: Çatalhöyük as a Case Study*, edited by I. Hodder, 187–219. Cambridge, UK: Cambridge University Press.

Keane, Webb. 2017. *Ethical Life: Its Natural and Social Histories*. Princeton, NJ: Princeton University Press.

Kent, Saul. 1985. "Programming People for Immortality." *Cryonics* 6, no. 1: 26–33.

Kelly, David. 2015. "Transhumanity.net Endorses Zoltan's Campaign for US President." *Transhumanity.net* http://transhumanity.net/transhumanity-net-endorses-zoltans-campaign-for-us-president/

Knowlton, David. 2007. "Hands Raised Up: Power and Context in Bolivian Mormonism." *Dialogue* (Winter): 47–71.

Knowlton, David. 2012 "Mormonism in Latin America." In Carl Mosser and Richard Sherlock, eds., *The Mormon World*. London: Routledge, 2012.

Kimberly, Claire. 2016. "Permission to Cheat: Ethnography of a Swingers' Convention." *Sexuality and Culture*, 20, no. 1: 56-68.

Kneese, Tamara, and Benjamin Peters. 2019. "Mormon Mommies Will Never Die." *Logic* 8: 171–80.

Kramer, Bradley. 2014. "Keeping the Sacred: Structured Silence in the Enactment of Priesthood Authority, Gendered Worship, and Sacramental Kinship in Mormonism" (doctoral dissertation, Univeristy of Michigan).

Kuhn, T. S., 2012. *The Structure of Scientific Revolutions*. Chicago: University of Chicago Press.

Kurzweil, Ray. 2001. "The Law of Accelerating Returns," *Kurzweil Accelerating Intelligence* (blog). http://www.kurzweilai.net/thelawofacceleratingreturns

Kurzweil, Ray. 2016. *The Singularity Is Near: When Humans Transcend Biology*. London: Duckworth.

Kurzweil, Ray, and Terry Grossman. 2005. *Fantastic Voyage: Live Long Enough to Live Forever*. Emmaus, PA: Rodale.

Lacan, Jacque, and Bruce Fink. 2006. *Ecrits: The First Complete Edition in English*. New York: W.W. Norton & Company.

Lachieze-Ray, Marc, and Jean-Pierre Luminet. 1995. "Cosmic Topology." *Physics Reports* 254, no. 3: 135–214.

Laidlaw, James. 2002. "For an Anthropology of Ethics and Freedom." *JRAI: Journal of the Royal Anthropological Institute* 8, no. 2: 311–32.

Laidlaw, James. 2013. *The Subject of Virtue: An Anthropology of Ethics and Freedom*. Cambridge, UK: Cambridge University Press.

Lambek, Michael. 2013. Introduction to A *Companion to the Anthropology of Religion* Chichester, edited by Janice Boddy and Michael Lambek. Hoboken, NJ: Wiley-Blackwell.

Laprade, Michael. 1998. "Can a Christian Be a Cryonicist?" *Cryonics* 19, no. 2:37.

Larsen, Timothy. 2014. *The Slain God: Anthropologists and the Christian Faith*. Oxford: Oxford University Press.

Latour, Bruno. 2012. *We Have Never Been Modern*. Cambridge, MA: Harvard University Press.

Latour, Bruno, and Steve Wolgan. 2013. *Laboratory Life: The Construction of Scientific Facts*. Princeton, NJ: Princeton University Press.

Leach, James, Dawn Nafus, and Berhnhard Krieger. 2009. "Freedom Imagined: Morality and Aesthetics in Open Source Software Design." *Ethnos* 74, no. 1: 51–71.

Lebner, Ashley. 2015. "The Anthropology of Secularity beyond Secularism." *Religion and Society: Advances in Research* 6: 62–74.

Leibovich, Mark. 2015. "Larry King Is Preparing for the Final Cancellation." *New York Times*, August 30, 2015. https://www.nytimes.com/2015/08/30/magazine/larry -king-is-preparing-for-the-final-cancellation.html

Lempert, Michel. 1985. "Letter to the Editor." *Cryonics* 6, no. 3: 3.

Lester, Rebecca. 2005. *Jesus in Our Wombs: Embodying Modernity in a Mexican Convent*. Berkeley: University of California Press.

Levin, Sam. 2016. "'I'm not a Mormon': Fresh 'Mass Resignation' over Anti-LGBT Beliefs." *The Guardian*, August 15, 2016. https://www.theguardian.com/world /2016/aug/15/mormon-church-lgbt-mass-resignation-protest-utah

Lévi-Strauss, Claude. 1963. *Structural Anthropology*. New York: Basic Books.

Lévi-Strauss, Claude. 1966. *The Savage Mind*. London: Weidenfeld and Nicolson.

Lévi-Strauss, Claude. 1973. *From Honey to Ashes*. New York: Harper & Row.

Lévi-Strauss, Claude. 1976. *Structural Anthropology*, vol. 2. New York: Basic Books.

Lévi-Strauss, Claude. 1983. *The Raw and the Cooked*. Chicago: University of Chicago Press.

Lévi-Strauss, Claude. 1990a. *The Naked Man*. Chicago: University of Chicago Press.

Lévi-Strauss, Claude. 1990b. *The Origin of Table Manners*. Chicago: University of Chicago Press.

Lévi-Strauss, Claude. 1991. *Totemism*. London: Merlin Press.

Lévi-Strauss, Claude. 1995. *The Story of the Lynx*. Chicago: University of Chicago Press.

Lévi-Strauss, Claude. 1998. *The Jealous Potter*. Chicago: University of Chicago Press.

Lévi-Strauss, Claude. 2001. "Hourglass Configurations." In *The Double Twist: From Ethnography to Morphodynamics*, edited by Pierre Maranda, 5–32. Toronto: University of Toronto.

Lévi-Strauss, Claude, Didier Eribon, and Paula Wissing. 1991. *Conversations with Claude Levi-Strauss*. Chicago: University of Chicago Press.

Lewis, C. S. 2001. *Miracles: A Preliminary Study*. San Francisco: HarperOne.

Long, Kelvin. 2011. "Towards an Interstellar Institute." *Centaruri Dreams*. https:// www.centauri-dreams.org/2011/06/08/long-toward-an-interstellar-institute/

Lowrie, Ian. 2017. "What Sort of Thing Is the Social? Or, Durkheim and Deleuze on Organization and Infrastructure." In *The New Politics of Materialism: History, Philosophy, Science*, edited by S. Ellenzweig and J.H. Zammito, 154–77. London: Routledge.

Loyer, Emmanuelle. 2018. *Lévi-Strauss: A Biography*. Cambridge, UK: Polity Press.

Lucas, Arel. 1992. "Letter to the Editor." *Cryonics* 13, no. 6: 2–3.

Luhmann, Niklas. 2013. *A Systems Theory of Religion*. Palo Alto: Stanford University Press.

Luhrmann, T. M. 2012. *When God Talks Back: Understanding the American Evangelical Relationship with God.* New York: Vintage.

Lunchwithandy. 2015. "My bishop threatened to excommunicate me because of my blog. I don't have a blog" Reddit, May 28, 2015. https://www.reddit.com/r/exmormon /comments/37mfdt/my_bishop_threatened_to_excommunicate_me_because/

MacCulloch, J. 1892. *Golf in the Year 2000; Or, What We Are Coming To.* London: Unwin.

Mack, Eric. 2015. "Presidential candidate suggests microchips for Syrian refugees." *CNET,* November 20, 2015. https://www.cnet.com/news/presidential-candidate -suggests-microchips-for-syrian-refugees/

Mahmood, Saba. 2011. *Politics of Piety: The Islamic Revival and the Feminist Subject.* Princeton, NJ: Princeton University Press.

Mansunaga, Samantha. 2015. "LGBT Anti-Discrimination Bill Passes Utah State Senate." *Los Angeles Times,* March 6, 2015. https://www.latimes.com/nation/la -na-utah-lgbt-20150306-story.html

Marantz, Andrew. 2016. "The Transhumanists' Nominee for President," *The New Yorker* November 6, 2016. https://www.newyorker.com/magazine/2016/11/14/the -transhumanists-nominee-for-president

Marcus, George. 1995. "Ethnography in/of the World System: The Emergence of Multi-Sited Ethnography." *Annual Review of Anthropology* 24: 95–117.

Marcus, Solomon. "The Logical and Semiotic Status of the Canonic Formula of Myth." *Semiotica* 116, no. 2–4: 115–88.

Marks, Jonathan. 2009. *Why I Am Not a Scientist: Anthropology and Modern Knowledge.* Berkeley: University of California Press.

Marx, Karl. 1844. "Contribution to the Critique of Hegel's Philosophy of Right." *Deutsch-Französische Jahrbücher* 7, no. 10: 71–85.

Masuzawa, Tomoko. 2005. *The Invention of World Religions, or, How European Universalism Was Preserved in the Language of Pluralism.* Chicago: University of Chicago Press.

Maughan, Tim. 2015. "Meet Zoltan, the Presidential Candidate Who Drives a Coffin." *BBC News,* November 29, 2015. http://www.bbc.com/future/story/2015 1127-meet-zoltan-the-strangest-candidate-running-for-president

Mauss, Armand. 1984. "Sociological Perspectives on the Mormon Subculture." *Annual Review of Sociology* 10: 437–60.

Mauss, Marcel. 2002. *The Gift: The Form and Reason for Exchange in Archaic Societies.* London: Routledge.

May, Dean. 1997. "One Heart and Mind: Communal Life and Values Among the Mormons." In *America's Communal Utopias,* edited by D. E. Pitzer and P. S. Boyer, 135–38. Chapel-Hill: University of North Carolina Press.

Mayblin, Maya. 2019. "The Ultimate Return: Dissent, Apostolic Succession, and the Renewed Misinterpreted of Roman Catholic Women Priests." *History and Anthropology* 30, no. 2: 133–48.

Mayblin, Maya, K. Norget, and Valentina Napolitano. 2017. "Introduction: The Anthropology of Catholicism." In *The Anthropology of Catholicism: A Reader*, edited by Maya Mayblin, K. Norget, and Valentina Napolitano,1–29.

Mayor, Adrienne. 2018. *Gods and Robots: Myths, Machines, and Ancient Dreams of Technology*. Princeton, NJ: Princeton University Press.

McCracken, Harry, and Lev Grossman. 2013. "The Audacity of Google." *Times International* 182, no. 14: 22.

McDannell, Colleen. 1998. *Material Christianity: Religion and Popular Culture in America*. New Haven, CT: Yale University Press.

McGinnness, Brett. 2016. "Zoltan Istvan 2016: Let's Make Americans Immortal." *USA Today*, June 15, 2016. http://www.usatoday.com/story/news/politics/onpolitics/2016/06/15/zoltan-istvan-2016-lets-make-americans-immortal/85929946/

Merali, Zeeya. 2017. *A Big Bang in a Little Room: The Quest to Create New Universes*. New York: Basic Books.

Messerly, John. 2015. "Transhumanism and Religion." *Institute for Ethics and Emerging Technologies*. Archived at https://ieet.org/index.php/IEET2/more/messerly20150118

Meyer, Birgit, 1998. "'Make a Complete Break with the Past': Memory and Post-Colonial Modernity in Ghanaian Pentecostalist Discourse." *Journal of Religion in Africa/Religion en Afrique* 28, no. 3: 316–49.

Meyer, Birgit. 2006. "Religious Sensations: Why Media, Aesthetics and Power Matter in the Study of Contemporary Religion." Inaugural lecture presented at Vrije Universiteit, Amsterdam, October 6, 2006.

Meyer, Birgit. 2009. *Aesthetic Formations: Media, Religion, and the Sense*. New York: Palgrave.

Michigan State University. n.d. "Steven Hsu" (faculty profile). https://www.pa.msu.edu/profile/hsu/

Milbank, John. 2013. *Theology and Social Theory: Beyond Secular Reason*. Malden, MA: Blackwell.

Miles, Kathleen. 2015. "Ray Kurzweil: In the 2030s, Nanobots in Our Brains Will Make Us 'Godlike'" *Huffington Post*, October 10, 2015. https://www.huffingtonpost.com/entry/ray-kurzweil-nanobots-brain-godlike_us_560555a0e4b0af3706dbe1e2

Miller, Julian, Dominic Job, and Vesselin K. Vassilev. 2000. "Principles in the Evolutionary Design of Digital Circuits—Part I." *Genetic Programming and Evolvable Machines* 1, no. 1–2: 7–35.

Miller, Lisa. "The Trans-Everything CEO." *New York Magazine*. https://nymag.com/news/features/martine-rothblatt-transgender-ceo/#

Mittermaier, Amira. 2011. *Dreams That Matter: Egyptian Landscapes of the Imagination*. Berkeley: University of California Press.

Mondragon, Carlos. 1990. "Alcor Begins Planning a New Facility." *Cryonics* 11, no. 3: 5–7. Available at https://www.alcor.org/docs/cryonics-magazine-1990-03.txt

Mondragon, Carlos. 1994. "Paperwork Counts: Alcor Is Forced to Surrender a Body for Burial?" *Cryonics* 15, no. 3: 13.

Monroe, Rachel. 2017. "Something in the Air: How Essential Oils Became the Cure for Our Age of Anxiety." *The New Yorker,* October 9, 2017. https://www.newyorker .com/magazine/2017/10/09/how-essential-oils-became-the-cure-for-our-age-of -anxiety

More, Max. 1990a. "Changing Times." *Extropy* 6: 1.

More, Max. 1990b. "Transhumanism: Towards a Futurist Philosophy." *Extropy* 6: 6–12.

More, Max. 1991. "In Praise of the Devil." *Atheist Notes* 3: 1–2.

More, Max. 2009. "Why Catholics Should Support the Transhumanist Goal of Extended Life." *Max More's Strategic Philosophy* (blog), September 9, 2009. http:// strategicphilosophy.blogspot.com/2009/09/why-catholics-should-support.html

Mormon Newsroom. 2006. "Elder Russell M. Nelson Marries Wendy L. Watson." *Mormon Newsroom.* https://www.mormonnewsroom.org/article/elder-russell-m .-nelson-marries-wendy-l.-watson

Mormon Newsroom. 2015. "Church Provides Context on Handbook Changes Affecting Same-Sex Marriage." *Mormon Newsroom,* November 6, 2015. https:// newsroom.churchofjesuschrist.org/article/handbook-changes-same-sex-marriages -elder-christofferson

Mormon Transhumanist Association. 2007. "Transfiguration: Parallels and Complements Between Mormonism and Transhumanism." *Sunstone* 145, 25–39.

Mosko, Mark. 1985. *Quadripartite Structures: Categories, Relations and Homologies in Bush Mekeo Culture.* Cambridge: Cambridge University Press.

Mosko, Mark.1991. "The Canonical Formula of Myth and Nonmyth." *American Ethnologist* 18, no. 11: 126–51.

Mouw, Richard. 2016. "Mormons Approaching Orthodoxy." *First Things.* https:// www.firstthings.com/article/2016/05/mormons-approaching-orthodoxy

Nelson, Russell M. 2016. "Stand as True Millennials." *Ensign Magazine* 46, no. 10: 24–31.

Nielson, Liesl. 2017. "Utah No. 6 in US for Number of Plastic Surgeons Per Capita." *KSL.com,* March 15, 2017. https://www.ksl.com/article/43505715/utah-no-6-in-us -for-number-of-plastic-surgeons-per-capita

Noble, David F., and Howard Segal. 1997. *The Religion of Technology: The Divinity of Man and the Spirit of Invention.* New York: Alfred A. Knopf.

Nongbri, Brent. 2013. *Before Religion: A History of a Modern Concept.* New Haven, CT: Yale University Press.

Norget, Karen, Valentina Napolitano, and Maya Mayblin, eds. 2017. *The Anthropology of Catholicism: A Reader.* Berkeley: University of California Press.

O'Connell, Mark. 2017a. "600 Miles in a Coffin-Shaped Bus, Campaigning Against Death Itself: Zoltan Istvan Ran for President with a Modest Goal in Mind: Human Immortality." *The New York Times,* February 9, 2017. https://www.nytimes.com /2017/02/09/magazine/600-miles-in-a-coffin-shaped-bus-campaigning-against -death-itself.html

O'Connell, Mark. 2017b. *To Be a Machine: Adventures Among Cyborgs, Utopians, Hackers, and the Futurists Solving the Modest Problem of Death*. New York: Doubleday.

O'Connor, Max T. 1989. "The Apocalypse, Technology, and Human Problems. *Cryonics* 10, no. 8: 19–21, 20.

Olson, Adam. 2005 "Tabernacle Choir Getting to Know Unique Conference Center." *Liahona*. https://www.lds.org/liahona/2005/08/news-of-the-church /tabernacle-choir-getting-to-know-unique-conference-center?lang=eng&_r=1

Ostler, Blaire. 2015a. "How a Mother Became a Transhumanist." *The Transfigurist*, June 6 2015. http://www.transfigurist.org/2015/06/how-mother-became-trans humanist.html

Ostler, Blaire. 2015b. "Raising My Daughter in the Church." *Feminist Mormon Housewives*. https://www.feministmormonhousewives.org/2015/06/raising-my -daughter-in-the-church/

Ostler, Blaire. 2017 "A Feminist Defense of Polygamy." Blaire Ostler (blog), October 21. http://www.blaireostler.com/journal/2016/10/21/a-feminists-defense-of-polygamy

Ostler, Blaire. 2019a. "Farewell, Rational Faiths." *Rational Faiths: Keeping Mormonism Weird*. https://rationalfaiths.com/farewell-rational-faiths

Ostler, Blaire. 2019b. "O Death, Where Is Thy Sting?" Blaire Ostler (blog). http:// www.blaireostler.com/journal/2019/4/15/o-death-where-is-thy-sting

Ostler, Blaire. 2019c. "Queer Polygamy." *Dialogue: A Journal of Mormon Thought* 52, no. 1:33–43.

Overbye, Dennis. 2011. "Offering Funds, U.S. Agency Dreams of Sending Humans to Stars." *New York Times.*, August 18, 2011. https://www.nytimes.com/2011/08/18 /science/space/18starship.html?pagewanted=all

Park, Lindsay Hansen. 2014. "Episode 01: Fanny Alger." *My Year of Polygamy* (podcast). https://www.yearofpolygamy.com/year-of-polygamy/year-of-polygamy -fanny-alger-episode-01/

Park, Lindsay Hansen. 2018. "Episode 146: Queer Polygamy." *My Year of Polygamy* (podcast). https://www.yearofpolygamy.com/year-of-polygamy/episode-146-queer -polygamy/

Paul, Deanna. 2018. "Using the Term 'Mormon' Is a 'Victory for Satan,' says President of 'The Church of Jesus Christ.'" *Washington Post*, October 8, 2018. https://www.washingtonpost.com/religion/2018/10/08/using-mormon-is-victory -satan-says-president-church-jesus-christ/?utm_term=.f2fb08102942

Paul, Robert. 1986. "Joseph Smith and the Plurality of Worlds Idea." *Dialogue: A Journal of Mormon Thought* 19, no. 2: 13–36.

Paul, Robert. 1992. *Science, Religion, and Mormon Cosmology*. Urbana and Chicago: University of Illinois Press.

Paul, Robert. 1994. "B.H. Roberts on Mormonism and Science." In *The Truth, The Way, The Life an Elementary Treatise on Theology: The Masterwork of B. H. Roberts*, edited by Stan Larson. Salt Lake City: Signature Books.

Paulsen, David L., and Martin Pulido. 2011. "A Mother There: A Survey of Historical Teachings about Mother in Heaven." *BYU Studies* 50, no. 1: 70–97.

Peters, John Durham. 2012. *Speaking into the Air: A History of the Idea of Communication*. Chicago: University of Chicago Press.

Pellissier, Hank. 2012. "Transhumanists: Who Are They? What Do They Want, Believe, and Predict?" *IEET: Institute for Ethics and Emerging Technologies*. http://ieet.org/index.php/IEET/more/pellissier20120909

Pellissier, Hank. 2015a. "Transhumanism: There Are at Least Ten Different Philosophical Categories; Which Ones Are You?" *IEET: Institute for Ethics and Emerging Technologies*. https://ieet.org/index.php/IEET2/more/pellissier20151213

Pellissier, Hank. 2015b. "15 Question Zoltan Istvan is Avoiding—Why? What Are the Answers?" *Transhumanity.net* http://transhumanity.net/15-question-zoltan-istvan-is-avoiding-why-what-are-the-answers-opinion/

Petersen, Boyd. 2012. "'One Soul Shall Not Be Lost': The War in Heaven in Mormon Thought." *Journal of Mormon History* 38, no. 1: 1–50.

Petrey, Taylor. 2011. "Towards a Post-Heterosexual Mormon Theology." *Dialogue: A Journal of Mormon Thought* 44, no. 4: 106–41.

Pew Research Center. 2009. "A Portrait of Mormons in the U.S." http://www.pewforum.org/2009/07/24/a-portrait-of-mormons-in-the-us/#4

Perry, Mike. 1992a. "For the Record: Our Finest Hour, Part One." *Cryonics* 13, no. 9: 4–7.

Perry, Mike. 1992b. "For the Record: Our Finest Hour, Part Two." *Cryonics* 13, no. 10: 4–7.

Perry, Mike. 1992c. "Suspension Failures: Lessons from the Early Years."Alcor.com (post). Originally published in *Cryonics* 13, no. 2; last updated October 2014 at https://www.alcor.org/library/suspension-failures-lessons-from-the-early-years/

Plus, Mark. 1998. "Letters to the editor: Cryonics and Christianity." *Cryonics* 19, no. 4: 7–9.

Pratt, Timothy. 2012. "Mormon Women Set Out to Take a Stand, in Pants." *New York Times*, December 20, 2012. https://www.nytimes.com/2012/12/20/us/19mormon.html?mtrref=www.google.com&gwh=59E40D4BF58861209CC9845476924339&gwt=pay

Prisco, Giulio. 2018. "Transhumanist Spirituality, Again." Giulio Prisco (blog). http://giulioprisco.blogspot.com/2008/10/transhumanist-spirituality-again.html

Probationary State. 2010. "Reed Smoot Hearings – Day 11: B.H. Roberts." *Probationary State* (blog). October 11, 2010. http://probationarystate.blogspot.com/2010/10/reed-smoot-hearings-day-11-bh-roberts.html

Pugmire, Genelle. 2013. "Ancestry.com Joins Forces with LDS Owned FamilySearch." *Herald Extra*. https://www.heraldextra.com/news/local/central/provo/ancestry-com-joins-forces-with-lds-owned-familysearch/article_4ce0e3fc-067a-5546-8f00-a0d0e111033a.amp.html

Quinn, D. Michael. 1998. *Early Mormonism and the Magic World View*. Salt Lake City: Signature Books.

Racine, Luc. 2001. "Analogy and the Canonical Formula of Mythic Transformations." In *The Double Twist: From Ethnography to Morphodynamics*. edited by Pierre Maranda, 33–55. Toronto: University of Toronto Press.

Reason. 2013. "Is '"Deathism" a Useful Term?" *Fight Aging!* https://www.fightaging
.org/archives/2013/03/is-deathism-a-useful-term/

"Redeeming the Dead Redeemed Me." 2014. Interview with Stephen Jezek. From
Church of Jesus Christ of Latter-day Saints, posted July, 2014. Video, 4:55. https://
www.churchofjesuschrist.org/family-history/video/redeeming-dead-redeemed
-me?lang=eng

Reeve, Paul. 2015. *Religion of a Different Color: Race and the Mormon Struggle for
Whiteness.* Oxford, UK: Oxford University Press.

Regis, Ed. 1990. *Great Mambo Chicken and the Transhumanist Condition.* Boston:
Addison-Wesley.

Reinhardt, Bruno. 2014. "Soaking in Tapes: The Haptic Voice of Global Pentecostal
Pedagogy in Ghana." *JRAI: Journal of the Royal Anthropological Institute* 20, no.
2: 315–36.

Richman, Larry. 2007. "Mormons Active in Second Life." *LDS 365* (blog), May 11,
2007. https://lds365.com/2007/05/11/mormons-active-in-second-life/

Riess, Jana. 2019. *The Next Mormons: How Millennials Are Changing the LDS
Church.* Oxford: Oxford University Press.

Robbins, Joel. 2001. "God Is Nothing but Talk: Modernity, Language and Prayer in
a Papua New Guinea Society." *American Anthropologist* 103, no. 4:901–12.

Robbins, Joel. 2002. "My Wife Can't Break Off Part of Her Belief and Give It to Me:
Apocalyptic Interrogations of Christian Individualism among the Urapmin of
Papua New Guinea. *Paideuma* 48: 189–206.

Robbins, Joel. 2003. "What Is a Christian? Notes toward an Anthropology of
Christianity." *Religion* 33, no. 3: 191–99.

Robbins, Joel. 2007a. "Continuity Thinking and the Problem of Christian Culture:
Belief, Time, and the Anthropology of Christianity." *Current Anthropology* 48,
no. 1:5–38.

Robbins, Joel. 2007b. "Between Reproduction and Freedom: Morality, Value, and
Radical Cultural Change." *Ethnos* 72, no. 3: 293–314.

Robbins, Joel. 2010. "On Imagination and Creation: An Afterword." *Anthropological
Forum* 20, no. 3: 305–13.

Robbins, Joel, and Mathew Engelke. 2010. "Introduction to the Special Issue:
Global Christianity, Global Critique." *South Atlantic Quarterly* 109, no. 4:
623–31.

Romain, Tiffany. 2010. "Extreme Life Extension: Investing in Cryonics for the Long,
Long Term." *Medical Anthropology* 29, no. 2, 194–215.

Rosen, Judy. 2015. "King of All Media: Larry King's Internet Afterlife." *New York
Times,* July 1, 2015. https://www.nytimes.com/2015/07/01/t-magazine/larry-king
-internet-afterlife.html

Roussi, Antoneta. 2016. "Now This Is an 'Outsider Candidate': Zoltan Istvan, A
Transhumanist Running for President, Wants to Make You Immortal." *Salon,*
February 19, 2016. http://www.salon.com/2016/02/19/now_this_is_an_outsider
_candidate_zoltan_istvan_a_transhumanist_running_for_president_is_promoting
_the_facilitation_of_immortality/

Ruby, J., Kevin M. Wright, Kristin A. Rand, Amir Kermany, Keith Noto, Don Curtis, Neal Varner, Daniel Garrigan, Dmitri Slinkov, Ilya Dorfman, Julie M. Granka, Jake Byrnes, Natalie Myres, and Catherine Ball Graham. 2018. "Estimates of the Heritability of Human Longevity Are Substantially Inflated due to Assortative Mating." *Genetics* 210, no. 3: 1109–24.

Ruiz, Rebecca. 2007. "In Pictures: America's Vainest Cities." *Forbes*. November 29. https://www.forbes.com/2007/11/29/plastic-health-surgery-forbeslife-cx_rr_1129 health_slide.html#3bc086836b1b

Sahagun, Louis, and Marx Arax. 1988. "Coroner Says It Was Homicide by Drugs in Frozen Head Case." *Los Angeles Times*, February 24. http://articles.latimes.com /1988-02-24/news/mn-11787_1_saul-kent.

Sahlins, Marshall. 1976. "La pensée bourgeoise." Chap. 4 in *Culture and Practical Reason*, 166–204. Chicago: University of Chicago Press.

Sahlins, Marshall. 1996. "The Sadness of Sweetness: The Native Anthropology of Western Cosmology." *Current Anthropology* 37, no. 3: 395–415.

Saler, Benson. 1987. "*Religio* and the Definition of Religion." *Cultural Anthropology* 2 (3): 395–99.

Saletan, William. 2006. "Among the Transhumanists." *Slate* http://www.slate.com /articles/health_and_science/human_nature/2006/06/among_the_trans humanists.html

Sathian, Sanjena. 2016. "The Intersection of Tech and Church." *OZY* https://www .ozy.com/fast-forward/the-intersection-of-tech-and-church/65342/

de Saussure, Ferdinand. 2011. *Course in General Linguistics*. New York: Columbia University Press.

Schwimmer, Eric. 2001. "Is the Canonic Formula Useful in Cultural Description?" In *The Double Twist: From Ethnography to Morphodynamics*. edited by Pierre Maranda, 56–96. Toronto: University of Toronto Press.

Scott, David W. 2011. "The Discursive Construct of Virtual Angels, Temples, and Religious Worship: Mormon Theology and Culture in Second Life." *Dialogue: A Journal of Mormon Thought* 44, no. 1: 85–104.

Shipps, Jan. 2001. "Signifying Sainthood, 1830–2001." *Arrington Annual Lecture*. Paper 6. https://digitalcommons.usu.edu/arrington_lecture/6

Shoffstall, Grant. 2016. "Failed Futures, Broken Promises, and the Prospect of Cybernetic Immortality: Towards an Abundant Sociological History of Cryonics Suspension, 1962–1979" (doctoral dissertation, University of Illinois at Urbana-Champaign).

Silverstein, Michael. 1976. "Shifters, Linguistic Categories, and Cultural Description." In *Meaning in Anthropology*, edited by Keith Basso and Henry A. Selby, 11–55. Albuquerque: University of New Mexico Press.

Singler, Beth. 2017. "Why Is the Language of Transhumanism and Religion So Similar?" *Aeon*. https://aeon.co/essays/why-is-the-language-of-transhumanists -and-religion-so-similar

Singularity Group. n.d. "About." https://www.su.org/about

Sirius, R. U., and Jau Cornell. 2015. *Transcendence: The Disinformation Encyclopedia of Transhumanism and the Singularity*. Newburyport, MA: Red Wheel/Weiser.

Smith, Daymon. 2007. "The Last Shall Be First and the First Shall Be Last: Discourse and Mormon History" (doctoral dissertation, University of Pennsylvania).

Smith, Gregory. 2018. "What Is Mormon Transhumanism? And Is It Mormon?" *Interpreter: A Journal of the Latter-day Saint Faith and Scholarship* 29: 161–190.

Smith, Joseph. 1938. *Teachings of the Prophet Joseph Smith*. Salt Lake City: Deseret Press.

Smith, Joseph. 1948. *Documentary History of the Church*. Salt Lake City: Deseret Press.

Sneath, David. 2009. "Reading the Signs by Lenin's Light: Development, Divination and Metonymic Fields in Mongolia." *Ethnos* 74, no. 1: 72–90.

Sneath, David, Martin Holbraad, and Morten Axel Pedersen. 2009. "Technologies of the Imagination: An Introduction." *Ethnos* 74, no. 1: 5–30.

Solman, Paul. 2012. "Ray Kurzweil on Bringing Back the Dead." *PBS NewsHour* July 12, 2012 (video). https://www.youtube.com/watch?v=ZlhYY3z5Hv8

Solomonoff, R. J. 1985. "The Time Scale of Artificial Intelligence: Reflections on social effects." *Human Systems Management*, 5(2): 149–153.

Spiro, Melford. 1966. "Religion: Problems of Definition and Explanation," in *Anthropological Approaches to the Study of Religion*, edited by Michael Banton, 85–126. London: Tavistock.

Stack, Peggy Fletcher. 2014. "New Almanac Offers Look at the World of Mormon Membership." *Washington Post*. https://www.washingtonpost.com/national /religion/new-almanac-offers-look-at-the-world-of-mormon-membership/2014 /01/13/7beb7888-7c86-11e3-97d3-b9925ce2c57b_story.html

Stack, Peggy Fletcher. 2015. "Some Mormons Stocking Up amid Fears that Doomsday Could Come this Month." *The Salt Lake Tribune*, September 15. http://www.sltrib .com/home/2935776-155/some-mormons-stocking-up-amid-fears

Stack, Peggy Fletcher. 2016. "New Mormon Policy on Gay Couples 'Analogous' to Its Polygamy Stance? Yes and No." *The Salt Lake Tribune*, January 4, 2016. https:// www.sltrib.com/news/mormon/2016/01/04/new-mormon-policy-on-gay-couples -analogous-to-its-polygamy-stance-yes-and-no/

Stasch, Rupert. 2006. "Structuralism in Anthropology." In *Encyclopedia of Anthropology (Second Edition)*, 167–70. Oxford: Elsevier.

Stewar, Anna. 2017. "Anthropology and the Internet." In *The Routledge Companion to Contemporary Anthropology*, edited by Simon Coleman, Susan B, Hyatt, and Ann Kingsolver, 92–106. London: Routledge.

Strauss, Claudia. 2006. "The Imaginary." *Anthropological Theory* 6, no. 3: 322–44.

Strhan, Anna. 2015. *Aliens and Strangers?: The Struggle for Coherence in the Everyday Lives of Evangelicals*. Oxford: Oxford University Press.

Taussig, Michael. 1999. *Defacement: Public Secrecy and the Labor of the Negative*. Palo Alto: Stanford University Press.

Taylor, Charles. 2004. *Modern Social Imaginaries*. Durham, NC: Duke University Press.

Taylor, Charles. 2007. *A Secular Age*. Cambridge, MA: Harvard University Press.

Tirosh-Samuelson, Hava. 2012. "Transhumanism as a Secularist Faith." *Zygon* 47, no. 4: 710–34.

Tipler, Frank. 1994. *The Physics of Immortality: Modern Cosmology, God and the Resurrection of the Dead*. New York: Anchor Books.

The Pluralism Project Archive. 2001. "Mapping Religious Diversity in Utah." *The Pluralism Project Archive*. https://hwpi.harvard.edu/pluralismarchive/mapping -religious-diversity-utah-2001

Thompson, D'Arcy W. 1968a. *On Growth and Form*, vol. 1. Cambridge: Cambridge University Press.

Thompson, D'Arcy W. 1968b. *On Growth and Form*, vol. 2. Cambridge: Cambridge University Press.

Tomlinson, Mathew. 2014. "Bringing Kierkegaard into Anthropology: Repetition, Absurdity, and Curses in Fiji." *American Ethnologist* 41, no. 1: 163–75.

Toomer-Cook, Jennifer. 2008. "Ashes to Ashes: More People Choosing Cremation." *Deseret News*. https://www.deseretnews.com/article/705381860/Ashes-to-ashes -More-people-choosing-cremation.html

Twyman, Amon. 2015. "Zoltan Istvan Does Not Speak for the Transhumanist Party." *Transhumanist Party* (blog) https://transhumanistparty.wordpress.com/2015/10/12 /zoltan-istvan-does-not-speak-for-the-transhumanist-party/

Tyler, Edward Burnett. 1874. *Primitive Culture*, vol. 1. Boston: Estes and Lauriat.

Uchtdorf, Dieter F., 2013. "Come, Join with Us." Church of Jesus Christ of Latter-day Saints (blog). https://www.lds.org/general-conference/2013/10/come-join-with -us?lang=eng

Uchtdorf, Dieter F., 2015. "It Works Wonderfully!" Church of Jesus Christ of Latter-day Saints (blog). https://www.lds.org/general-conference/2015/10/it-works -wonderfully?lang=eng

Van Valkenburg, Nancy. 2016. "Author Jeremy Runnells Resigns from LDS Church at Excommunication Hearing in American Fork." *GephardtDaily*, April 18, 2016. https://gephardtdaily.com/top-stories/author-jeremy-runnells-resigns-from-lds -church-at-excommunication-hearing-in-american-fork/

Van Wagoner, Richard S. 2002. *Mormon Polygamy: A History*. Salt Lake City: Signature Books.

Vice staff. 2014. "Mormon-themed Porn Is Apparently a Booming Business." *Vice*, December 5, 2014. https://www.vice.com/en/article/kwp97y/mormon-themed -porn-is-apparently-a-booming-business

Vinge, Vernor. 1993. "The Coming Technological Singularity: How to Survive in the Post-Human Era" (article for VISION-21 Symposium, 1993). https://www-rohan .sdsu.edu/faculty/vinge/misc/singularity.html

Viveiros de Castro, Eduardo. 2017. *Cannibal Metaphysics: For a Post-Structural Anthropology*. Minneapolis: London University of Minnesota Press.

de Vries, Hent, and Samuel Weber, eds. 2001. *Religion and Media*. Palo Alto: Stanford University Press.

Walker, David. 1979. "Valley Cryonic Crypt Desecrated, Untended." *The Valley News*, June 10: 11.

Warner, Michael. 2002. "Publics and Counterpublics." *Public Culture* 14, no. 1: 49–90.

Weibel, Deana. 2007. "Magic, Science and Religion in Space?" *Anthropology News* 48, no. 2.

Weibel, Deana. 2015. "'Up in God's Great Cathedral': Evangelism, Astronauts, and the Seductiveness of Outer Space." In *The Seductions of Pilgrimage: Afar and Astray in the Western Religious Tradition*, edited by Michael Di Giovine and David Picard, 223–56. Farnham, UK: Ashgate Publishing.

Weibel, Deana. 2017. "Space Exploration as Religious Experience: Evangelical Astronauts and the Perception of God's Worldview." *The Space Review*, August 21, 2017. http://www.thespacereview.com/article/3310/1#idc-cover

Weibel, Deana. 2019. "The NASA-Vatican Relationship Models a Bridge between Science and Religion." *The Space Review*, July 16, 2019. http://www.thespace review.com/article/3757/1

Weibel, Deana. 2019a. "Destiny in Space." *Anthropology News* 60, no. 4: 26–27.

Weibel, Deana. 2019b "Astronauts vs. Mortals: Space Workers, Jain Ascetics, and NASA's Transcendent Few." *The Space Review*, April 18, 2019. http://www.the spacereview.com/article/3690/1

Weinstock, Nicole. 2017. "Member Profile: Steve Graber." *Cryonics* 38, no. 2: 11–14. Available at http://www.alcor.org/profiles/graber.html

White, Kendall, Jr. 1987. *Mormon Neo-Orthodoxy: A Crisis Theology.* Salt Lake City: Signature Books.

Widtsoe, John. 1925. *Discourses of Brigham Young.* Salt Lake City: Deseret Book Company.

Wikipedia. "Anthony Zee." Last updated August 4, 2021. https://en.wikipedia.org /wiki/Anthony_Zee

Wilson, William. 2004. "Austin and Alta Fife, Pioneering Folklorists." In *Folklore in Utah: A History and Guide to Resources*, edited by David Stanley, 41–48. Logan, Utah State University Press.

de Wolf, Chana. 2010. "Member Profile: Mark Plus." *Cryonics* 23, no. 2: 8–10.

Wolf-Meyer, Matthew. 2009. "Fantasies of Extremes: Sports, War and the Science of Sleep." *BioSocieties* 4, no. 2: 257–71.

Wolf-Meyer, Matthew. 2019. *Theory for the World to Come: Speculative Fiction and Apocalyptic Anthropology.* Minneapolis: University of Minnesota Press.

World Scientific. n.d. "Modern Physics Letters A, Aims & Scope," n.d. https://www .worldscientific.com/page/mpla/aims-scope

Woodruff, Wilford. 1993. *Waiting for the World's End: The Diaries of Wilford Woodruff.* edited by Susan Staker. Salt Lake City: Signature Books.

Young, Brigham. 1997. *Teachings of the Presidents of the Church: Brigham Young.* Salt Lake City: The Church of Jesus Christ of Latter-day Saints.

Young, George M. 2012. *The Russian Cosmists: The Esoteric Futurism of Nikolai Fedorov and His Followers.* Oxford: Oxford University Press.

Index

Jon Bialecki is Lecturer in Anthropology at the University of California, San Diego. He is the author of *A Diagram for Fire: Miracles and Variation in an American Charismatic Movement,* which won the Sharon Stephens Prize and was a finalist for the Clifford Geertz Prize.

www.ingramcontent.com/pod-product-compliance
Lightning Source LLC
Chambersburg PA
CBHW022132020426
42334CB00015B/863